Operator Theory: Advances and
Applications
Vol. 164

Pseudo-Differential Operators and Related Topics

Paolo Boggiatto
Luigi Rodino
Joachim Toft
M.W. Wong
Editors

Birkhäuser Verlag
Basel · Boston · Berlin

Editors:

Paolo Boggiatto
Dipartimento di Matematica
Università di Torino
Via Carlo Alberto, 10
10123 Torino
Italy
e-mail: boggiatto@dm.unito.it

Joachim Toft
School of Mathematics and Systems Engineering
Växjö University
SE-351 95 Växjö
Sweden
e-mail: Joachim.Toft@msi.vxu.se

Luigi Rodino
Dipartimento di Matematica
Università di Torino
Via Carlo Alberto, 10
10123 Torino
Italy
e-mail: rodino@dm.unito.it

M. W. Wong
Department of Mathematics and Statistics
York University
4700 Keele Street
Toronto, Ontario M3J 1P3
Canada
e-mail: mwwong@mathstat.yorku.ca

2000 Mathematics Subject Classification: Primary 35A17, 35L15, 35L40, 35L45, 35L60, 35L80, 35P05, 35S05, 35S30, 42C15, 43A80, 46F05, 47G30, 53D12, 58J40, 60G12, 60G15, 81R30; Secondary 17B45, 35B65, 35C20, 35J50, 35L30, 42C40, 44A05, 45C05, 46G10, 53D05, 53D50, 58B15, 58D30, 58J30, 60G57

A CIP catalogue record for this book is available from the Library of Congress, Washington D.C., USA

Bibliographic information published by Die Deutsche Bibliothek
Die Deutsche Bibliothek lists this publication in the Deutsche Nationalbibliografie; detailed bibliographic data is available in the Internet at <http://dnb.ddb.de>.

ISBN 3-7643-7513-2 Birkhäuser Verlag, Basel – Boston – Berlin

Member of the BertelsmannSpringer Publishing Group
Printed on acid-free paper produced from chlorine-free pulp. TCF∞
Cover design: Heinz Hiltbrunner, Basel
Printed in Germany
ISBN-10: 3-7643-7513-2
ISBN-13: 978-3-7643-7513-3
e-ISBN: 3-7643-7514-0

9 8 7 6 5 4 3 2 1

www.birkhauser.ch

Contents

Preface

As a satellite conference to the Fourth Congress of European Mathematics held at Stockholm University in 2004, the International Conference on Pseudo-Differential Operators and Related Topics was held at Växjö University in Sweden from June 22 to June 25, 2004. The conference was supported by Växjö University, the FIRB Research Group on Microlocal Analysis of Università di Torino, and the International Society for Analysis, its Applications and Computation (ISAAC). The conference was well attended by about 50 mathematicians from Bulgaria, Canada, Denmark, England, Finland, Germany, Italy, Japan, Mexico, Serbia and Montenegro, Russia and Sweden.

The conference covered a broad spectrum of topics related to pseudo-differential operators such as partial differential equations, quantization, Wigner transforms and Weyl transforms on Lie groups, mathematical physics, time-frequency analysis and stochastic processes. The speakers were enthusiastic about the prospect of publishing articles based on their presentations in a volume to be published in Professor Israel Gohberg's prestigious series entitled "Operator Theory: Advances and Applications".

All contributions from speakers have been carefully refereed and the articles collected in this volume give a representative flavour of the mathematics presented at the conference. This volume is a permanent record of the conference and a valuable complement to the volume "Advances in Pseudo-Differential Operators" published in the same series in 2004, which is devoted to the Special Session on Pseudo-Differential Operators at the Fourth ISAAC Congress held at York University in August 2003.

Operator Theory:
Advances and Applications, Vol. 164, 1–21

Strongly Elliptic Second Order Systems with Spectral Parameter in Transmission Conditions on a Nonclosed Surface

M.S. Agranovich

Abstract. We consider a class of second order strongly elliptic systems in $\mathbb{R}^n$, $n \geq 3$, outside a bounded nonclosed surface S with transmission conditions on S containing a spectral parameter. Assuming that S and its boundary γ are Lipschitz, we reduce the problems to spectral equations on S for operators of potential type. We prove the invertibility of these operators in suitable Sobolev type spaces and indicate spectral consequences. Simultaneously, we prove the unique solvability of the Dirichlet and Neumann problems with boundary data on S.

Mathematics Subject Classification (2000). Primary 35P05; Secondary 35J50, 45C05.

Keywords. Strong ellipticity, transmission condition, spectral equation, Lipschitz surface, surface potential, variational approach, Wiener–Hopf method.

1. Introduction

1.1. Statement of the Problems

We consider the second order system of partial differential equations

$$L_\omega(\partial)u(x) := L_0(\partial)u(x) + \omega^2 u(x) = 0 \tag{1.1}$$

in $\mathbb{R}^n \setminus S$, where S is an $(n-1)$-dimensional surface with $(n-2)$-dimensional boundary γ. More precisely, we assume that S is a part of a closed surface[1] Γ; the latter consists of two open parts $S = \Gamma_+$ and Γ_- without common points and of their common boundary γ. By ∂ we denote the "vector" of partial derivatives $\partial_j = \partial/\partial_j$, $j = 1, \ldots, n$. The operator L_0 is homogeneous with respect to differentiations and has constant coefficients. The numerical parameter $\omega = \omega_1 + i\omega_2$ belongs to

The work was supported by the grant of RFFI No. 04-01-00914.

[1] A closed surface is a compact surface without boundary.

the closed upper half-plane ($\omega_2 \geq 0$) and is nonnegative if it is real. At infinity, the solutions are subjected to natural radiation or decay conditions depending on ω. The assumptions are formulated more precisely below in Subsection 1.2. We assume that $n \geq 3$ to avoid the consideration of logarithmic potentials. The surface Γ divides the complement of itself into a bounded simply connected domain Ω^+ and an unbounded domain Ω^-. The superscripts $^+$ and $^-$ will also be used to denote the boundary values of functions on the inner and outer sides of Γ, respectively. By $\nu = \nu(x)$ we denote the unit outward normal at points $x \in \Gamma$.

Our main goal is to consider two spectral problems for (1.1) with transmission conditions on S containing a spectral parameter λ. The corresponding spaces will be specified later.

Problem I.

$$u^+ = u^-, \quad Tu^- - Tu^+ = \lambda u^\pm \quad \text{on } S. \tag{1.2}$$

Problem II.

$$Tu^+ = Tu^-, \quad Tu^\pm = \lambda[u^- - u^+] \quad \text{on } S. \tag{1.3}$$

Here Tu is the conormal derivative, see (1.8) below. In the simplest case of the Helmholtz equation, Tu is the normal derivative $\partial_\nu u$. We wish to describe some spectral properties of these problems. We will see shortly that they are closely connected with the non-spectral Dirichlet and Neumann problems for system (1.1):

The Dirichlet problem.

$$u^\pm = f \quad \text{on } S. \tag{1.4}$$

The Neumann problem.

$$Tu^\pm = g \quad \text{on } S. \tag{1.5}$$

The surfaces Γ and γ are assumed to be either C^∞ or Lipschitz; Γ is connected, while γ can consist of several components. The normal $\nu(x)$ is defined almost everywhere in the Lipschitz case.

In the case of a closed surface $S = \Gamma$, Problems I and II and some other problems for the Helmholtz equation were posed by the physicists Katsenelenbaum and his collaborators Sivov and Voitovich in the 60s. See the book [1] and its revised English edition [2]. In [1], a mathematical supplement [3] is contained, written by the author of the present paper, with the analysis of these and similar problems in acoustics and electrodynamics by tools of the theories of surface potentials and pseudo-differential operators. The surface was assumed to be smooth. The initial results were obtained by the author in collaboration with his graduate student Golubeva; see also her note [4]. Conditions (1.2) and (1.3) can be interpreted as related to a half-transparent screen.

Similar problems with boundary and transmission conditions on a closed Lipschitz surface for the Helmholtz equation were considered in [5] and for the Lamé system in elasticity theory (and $n = 3$) in [6]. The general case of systems (1.1) was considered in [7] and [8]. In [8], systems with variable coefficients were included into consideration. More precisely, in the last three papers the surface is assumed to be either smooth or Lipschitz. Of course, no theory of elliptic pseudodifferential operators exists in the case of a Lipschitz surface, but there is an extensive theory of classical surface potentials and non-spectral problems; see [7, 8] and numerous references therein.

In Section 2, we recall some definitions and technical tools from [7, 8] related to the case of a closed surface. We also add some supplementary material. In particular, we introduce the hypersingular operator and Calderón projections for general systems (1.1) following, e.g., the paper [9] on the Laplace and Helmholtz equations.

Non-spectral problems (Dirichlet, Neumann, and more general) with data on a nonclosed surface for the Helmholtz equation and the Lamé system were first considered by Stephan [10, 11] and then by Costabel and Stephan [12]. In elasticity theory, a non-closed surface has the meaning of a crack. The Lamé system models an isotropic medium. Cracks in anisotropic elastic media were considered by Duduchava, Natroshvili and Shargorodski [13] and by some other authors. Moreover, non-spectral problems in elasticity were considered with much more general conditions on S and in much more general spaces than in the present paper, see also, e.g., [14] and references therein.

These authors followed Vishik and Eskin (e.g., see [15], [16]) and Eskin [17] and used the Wiener–Hopf method assuming that Γ and γ are sufficiently smooth.

In Section 3, the main in the present paper, we will show that it is possible to consider Lipschitz surfaces Γ and γ using the simplest Sobolev type spaces $H^{\pm 1/2}(S)$ and $\widetilde{H}^{\pm 1/2}(S)$. (The spaces are defined in Subsection 2.5.) Instead of the Wiener–Hopf method, a modification of the classical variational approach is used (see our Propositions 3.2 and 3.4 for the case of pure imaginary ω). We prove the unique solvability of the Dirichlet and Neumann problems for general systems (1.1) and the invertibility of the potential type operators on S corresponding to these problems. It seems to us that these non-spectral results are of interest even for the Helmholtz and Lamé equations.

Exactly the same invertible operators occur in the spectral equations on S corresponding to our spectral problems I and II. Thus the simplest spectral results become available in the Lipschitz case; see Subsection 3.4.

In Section 4, we briefly mention some further results; they will be published elsewhere.

1.2. Exact Statement of the Assumptions

(See [7] for details.) The operator $L_0(\partial)$ is an $m \times m$ matrix. Replacing ∂ by $\xi = (\xi_1, \ldots, \xi_n)$, we obtain the principal symbol of the operator $-L_0(\partial)$:

$$L_0(\xi) = \sum \sum A_{jk}\xi_j\xi_k, \quad A_{jk} = (a_{jk}^{rs}). \tag{1.6}$$

Here the A_{jk} are real matrices satisfying the symmetry condition $a_{jk}^{rs} = a_{kj}^{sr}$. The matrix $L_0(\xi)$ is assumed to be positive definite for $\xi \neq 0$, which is the strong ellipticity condition for the operator $-L(\partial)$. Besides, as in [7, 8], we impose the additional condition

$$\sum a_{jk}^{rs}\xi_j^r\xi_k^s \geq C_1 \sum |\xi_j^r + \xi_r^j|^2 \tag{1.7}$$

for all real ξ_j^r; here and below, the C_j are positive constants, and summation over all indices is implied. If $m \neq n$, then the ξ_j^r with $j > n$ or $r > m$ are assumed to be zero. Condition (1.7) is satisfied, in particular, for the Helmholtz and Lamé equations.

The conormal derivative at a point x of the boundary is the matrix operator

$$T_x = \sum \nu_j(x) A_{jk}\partial_k. \tag{1.8}$$

If Γ is smooth, then it follows from (1.7) that the Neumann problems, interior and exterior, are elliptic. As to the interior and exterior Dirichlet problems, their ellipticity is the well-known consequence of the strong ellipticity. However, these problems can be considered in Lipschitz domains as well. Here it is essential that, under Condition (1.7), the expression

$$\left\{\int_{\Omega^\pm} \sum A_{jk}\partial_k u \cdot \partial_j \overline{u}\, dx + \|u\|_{0,\Omega^\pm}^2\right\}^{1/2}$$

is equivalent to the usual norm $\|u\|_{1,\Omega^\pm}$ in the Sobolev space $H^1(\Omega^\pm)$. This follows from the well-known Korn inequalities (e.g., see [18]).

The conditions at infinity are connected with the choice of a fundamental solution $\mathcal{E}_\omega(x)$. In all cases, this is a matrix function analytic outside the origin.

If $\omega = 0$, then the fundamental solution is defined by the formula

$$\mathcal{E}_0(x) = -F^{-1}L_0(\xi), \tag{1.9}$$

where F^{-1} is the inverse Fourier transform in the sense of distributions. This matrix function is positively homogeneous of degree $2-n$. Accordingly, the conditions at infinity have the form

$$u(x) = O(|x|^{-n+2}), \quad \partial_k u(x) = O(|x|^{-n+1}) \tag{1.10}$$

for all k.

If $\omega_2 > 0$, then the complete symbol $\omega^2 E - L_0(\xi)$ of system (1.1) is a nondegenerate matrix, and the fundamental solution is defined by the formula

$$\mathcal{E}_\omega(x) = F^{-1}[\omega^2 E - L_0(\xi)]. \tag{1.11}$$

It decays exponentially, and accordingly the conditions at infinity have the form

$$u(x) = O(\exp(-\delta|x|)), \quad \partial_k u(x) = O(\exp(-\delta|x|)) \tag{1.12}$$

for all k with some $\delta > 0$.

If $\omega > 0$, an additional condition is imposed on the matrix $L_0(\xi)$ for $\xi \neq 0$. Let $d_l(\xi)$ $(l = 1, \ldots, q)$ be all its pairwise distinct eigenvalues. They are positive.

Condition for $\omega > 0$. The eigenvalues $d_l(\xi)$ have constant multiplicities. The surfaces S_l in $\mathbb{R}^n$ defined by the equations $d_l(\xi) = \omega^2$ are, roughly speaking, convex.

More precisely, they are star-shaped with respect to the origin, the principal curvatures are positive at each point $x \in S_l$, and the radius vector drawn from the origin to $x \in S_l$ forms an acute angle with the outward normal to S_l at x.

Under this condition, the fundamental solution $\mathcal{E}_\omega(x)$ is defined as the limit of $\mathcal{E}_{\omega+i\varepsilon}(x)$ as $\varepsilon \downarrow 0$. Its analysis leads to the following radiation conditions:

$$u(x) = u_1(x) + \cdots + u_q(x), \tag{1.13}$$

where

$$u_l(x) = O(|x|^{-(n-1)/2}), \quad \partial_k u_l(x) - i\xi_k^l(\alpha)u_l(x) = O(|x|^{-(n+1)/2}). \tag{1.14}$$

Here $\alpha = x/|x|$ and $\xi^l = \xi^l(\alpha)$ is the radius vector drawn from the origin to the point on S_l, at which the unit exterior normal vector coincides with α.

In all three cases, $\mathcal{E}_\omega(x)$ satisfies the corresponding conditions at infinity. In the second and third cases, the difference $\mathcal{E}_\omega(x) - \mathcal{E}_0(x)$ has smaller order of singularity at the origin than $\mathcal{E}_0(x)$ and $\mathcal{E}_\omega(x)$, the difference of the orders being equal to 1.

In what follows, the solutions of (1.1) in Ω^- are subjected to conditions at infinity just indicated.

2. Preliminary Material

Here we recall some facts and technical tools related to the case of a closed surface $S = \Gamma$. The omitted details and proofs or references can be found mainly in [7] and [8]. Subsections 2.2–2.4 contain the supplementary material used later in the present paper. In Subsection 2.5, we recall the definitions and some properties of the spaces $H^t(S)$ and $\widetilde{H}^t(S)$ in the case of a nonclosed S, cf., e.g., [17], Section 4, and [19], Section 3.

2.1. Surface Potentials and Integral Formulas in the Case of a Smooth Closed Boundary

In the first three subsections, we assume that Γ and all functions are infinitely smooth. System (1.1) and boundary problems are considered in the classical setting.

First, we note that the exterior Dirichlet and Neumann problems are uniquely solvable. The corresponding classes of functions will be indicated in Subsection 2.3.

Using the fundamental solution described above, we introduce the single layer potential ($x \in \mathbb{R}^n$) and the double layer potential ($x \notin \Gamma$):

$$\mathcal{A}_\omega\varphi(x) = \int_\Gamma \mathcal{E}_\omega(x-y)\varphi(y)\,dS_y, \quad \mathcal{B}_\omega\psi(x) = \int_\Gamma [T_y\mathcal{E}_\omega(x-y)]'\psi(x)\,dS_y, \tag{2.1}$$

where $'$ stands for transposition. These functions are solutions of (1.1) in $\Omega^\pm$ and satisfy our conditions at infinity.

Denote by $A_\omega\varphi(x)$ the restriction of the single layer potential to Γ and by $B_\omega\psi(x)$ the direct value of the double layer potential on Γ. The first of these operators is an integral operator with a weak singularity; it is an elliptic pseudo-differential operator of order -1. The second operator is a pseudo-differential operator of order not greater than 0; it can be a singular integral operator.

For the boundary values of these potentials and their conormal derivatives, we have

$$\begin{gathered}(\mathcal{A}_\omega\varphi)^\pm = A_\omega\varphi, \quad (\mathcal{B}_\omega\psi)^\pm = (B_\omega \pm \tfrac{1}{2}I)\psi,\\ (T\mathcal{A}_\omega\varphi)^\pm = (B'_\omega \mp \tfrac{1}{2}I)\varphi, \quad (T\mathcal{B}_\omega\psi)^+ = (T\mathcal{B}_\omega\psi)^-.\end{gathered} \tag{2.2}$$

Here the operator B'_ω is the transpose of B_ω:

$$B'_\omega\psi = \int_\Gamma T_x E_\omega(x-y)\, dS_y. \tag{2.3}$$

It is again a pseudo-differential operator of order not greater than 0.

The following formulas for solutions of system (1.1) in $\Omega^\pm$ are true:

$$\begin{aligned}\mathcal{B}_\omega u^+ - \mathcal{A}_\omega(Tu^+) &= \begin{cases} u \text{ in } \Omega^+,\\ 0 \text{ in } \Omega^-,\end{cases}\\ \mathcal{A}_\omega(Tu^-) - \mathcal{B}_\omega u^- &= \begin{cases} u \text{ in } \Omega^-,\\ 0 \text{ in } \Omega^+.\end{cases}\end{aligned} \tag{2.4}$$

Passing to Γ in the upper formulas, we obtain the following relations between the Cauchy data:

$$(\tfrac{1}{2}I - B_\omega)u^+ = -A_\omega(Tu^+), \quad (\tfrac{1}{2}I + B_\omega)u^- = A_\omega(Tu^-). \tag{2.5}$$

Under condition (1.7), the zero order pseudo-differential operators $\frac{1}{2}I \pm B_\omega$ are elliptic.

2.2. The Hypersingular Operator and Calderón Projections

Definition 2.1. We introduce *the hypersingular operator*

$$D_\omega\psi = -(T\mathcal{B}_\omega\psi)^\pm \tag{2.6}$$

on Γ. It is a pseudo-differential operator of order 1.

Applying the operator T to both sides of the upper formulas in (2.4), passing to the boundary Γ, and using (2.2) and (2.5), we obtain

$$Tu^+ = -Du^+ + (\tfrac{1}{2}I - B'_\omega)Tu^+, \quad Tu^- = Du^- + (\tfrac{1}{2}I + B'_\omega)Tu^-. \tag{2.7}$$

The left-hand sides are replaced by zero if we interchange $^+$ and $^-$ in the right-hand sides.

Definition 2.2. We introduce the matrix operators

$$P^+ = \begin{pmatrix} \frac{1}{2}I + B_\omega & -A\\ -D & \frac{1}{2}I - B'_\omega\end{pmatrix}, \quad P^- = \begin{pmatrix} \frac{1}{2}I - B_\omega & A\\ D & \frac{1}{2}I + B'_\omega\end{pmatrix}. \tag{2.8}$$

They possess the following properties.

1. They are bounded operators in the space of column vectors $(\varphi, \psi)'$ with components $\varphi \in H^t(\Gamma)$, $\psi \in H^{t-1}(\Gamma)$ for all t. (Here we mean that φ and ψ are, in turn, column vectors of dimension m.) This space will be denoted by $\mathcal{H}^t(\Gamma)$.

2. If the vector $w^\pm = (u^\pm, Tu^\pm)'$ consists of the Cauchy data of system (1.1) in $\Omega^\pm$, then

$$P^+ w^+ = w^+, \quad P^+ w^- = 0, \quad P^- w^- = w^-, \quad P^- w^+ = 0. \tag{2.9}$$

This follows from (2.5) and (2.7). Here we can assume that $w^\pm \in \mathcal{H}^t(\Gamma)$, with $t \geq 3/2$ in the classical setting of the Dirichlet and Neumann problems and even with $t > 1$ if we additionally use the approach in [20]. The first and third equalities in (2.9) are not only necessary but also sufficient for vectors $w^\pm \in \mathcal{H}^t(\Gamma)$ to consist of Cauchy data for (1.1) in Ω^+ and Ω^-. Indeed, if these relations are satisfied, then the corresponding solutions are reconstructed by upper formulas in (2.4).

3. The relations

$$P^+ + P^- = I_2, \quad (P^+)^2 = P^+, \quad (P^-)^2 = P^-, \quad P^+P^- = P^-P^+ = 0 \tag{2.10}$$

are satisfied, where I_2 is the unit operator in the space of vector functions in question. These relations follow from (2.8) and (2.9).

Thus $P^\pm$ *are the Calderón projections for system* (1.1). (Concerning this notion for general elliptic equations, e.g., see [21].) They define the decomposition of the space $\mathcal{H}^t(\Gamma)$ into the direct sum of Cauchy data subspaces for system (1.1) in Ω^+ and Ω^-.

It follows, say, from the relation $(P^+)^2 = P^+$ that

$$(\tfrac{1}{2}I + B_\omega)(\tfrac{1}{2}I - B_\omega) = A_\omega D_\omega, \quad A_\omega B'_\omega = B_\omega A_\omega, \quad D_\omega B_\omega = B'_\omega D_\omega. \tag{2.11}$$

The second of these relations was obtained in [7] in a different way. From the first relation we see that *the operator* D_ω *is elliptic. Its principal symbol is expressed by the formula*

$$\sigma_{D_\omega} = \sigma_{A_\omega}^{-1} \sigma_{\frac{1}{2}I + B_\omega} \sigma_{\frac{1}{2}I - B_\omega}. \tag{2.12}$$

The right-hand side is known, see formulas for the principal symbols σ_{A_ω} and $\sigma_{\frac{1}{2}I \pm B'_\omega}$ in [7] or [8]. If A_ω is invertible (has the inverse of order 1), then

$$D_\omega = A_\omega^{-1}(\tfrac{1}{2}I + B_\omega)(\tfrac{1}{2}I - B_\omega) \tag{2.13}$$

(cf. (4.6) and (4.7) in [7]). However, in definition (2.6) we did not assume that A_ω is invertible.

2.3. The Dirichlet and Neumann Problems and Problems of the Form I and II in the Case of a Smooth Closed Boundary

If u is a solution of system (1.1) outside S with $u^+ = u^-$, then it follows from (2.4) that $u = \mathcal{A}_\omega(Tu^- - Tu^+)$. Therefore, *the Dirichlet problems* (1.1), (1.4) (interior and exterior simultaneously) *are reduced to the equation*

$$A_\omega \varphi = f \tag{2.14}$$

on Γ *by the substitution*

$$\varphi = [Tu] := Tu^- - Tu^+. \tag{2.15}$$

Here $f \in H^t(\Gamma)$, $\varphi \in H^{t-1}(\Gamma)$, $u \in H^{t+1/2}(\Omega^+)$, and $u \in H^{t+1/2}_{\mathrm{loc}}(\Omega^-)$,[2] $t > 1$. Besides, the solution u is analytic inside $\Omega^\pm$. Once equation (2.14) is solved, the corresponding solution of the Dirichlet problem is given by the formula

$$u = \mathcal{A}_\omega \varphi. \tag{2.16}$$

The interior Dirichlet problem is uniquely solvable for all ω except for some positive values ω_j tending to $+\infty$ (for which ω^2 is an eigenvalue of the operator $-L_0(\partial)$ in Ω^+ under the homogeneous Dirichlet condition $u^+ = 0$).

The operator A_ω has the inverse (a pseudo-differential operator of order 1) precisely for non-exceptional ω. This inverse can also be defined as follows. We find the solution u of the interior and exterior Dirichlet problems with $u^\pm = f$ and set $A_\omega^{-1} f = [Tu]$.

Problem I *for eigenfunctions is reduced to the equation*

$$\varphi = \lambda A_\omega \varphi \tag{2.17}$$

by the same substitution (2.15). In [7], the spectral properties of A_ω are described.

If u is the solution of (1.1) outside S with $Tu^+ = Tu^-$, then it follows from (2.4) that $u = -\mathcal{B}_\omega(u^- - u^+)$. Taking (2.6) into account, we see that *the Neumann problems* (1.1), (1.5) (interior and exterior) *are reduced to the equation*

$$D_\omega \psi = g \tag{2.18}$$

by the substitution

$$\psi = [u] := u^- - u^+. \tag{2.19}$$

Once this equation is solved, the corresponding solution of the Neumann problem is given by the formula

$$u = -\mathcal{B}_\omega \psi. \tag{2.20}$$

Here $g \in H^{t-1}(\Gamma)$, $\psi \in H^t(\Gamma)$, $u \in H^{t+1/2}(\Omega^+)$, and $u \in H^{t+1/2}_{\mathrm{loc}}(\Omega^-)$ for $t > 1$. Besides, the solution is analytic inside $\Omega^\pm$.

The interior Neumann problem is uniquely solvable for all ω except for some nonnegative values $\omega = \omega_j'$ tending to $+\infty$ (such that $(\omega_j')^2$ is an eigenvalue of $-L_0(\partial)$ in Ω^+ under the homogeneous Neumann condition $Tu^+ = 0$).

The unique solvability of the interior Dirichlet problem is equivalent to the invertibility of the operators A_ω, $\frac{1}{2}I + B_\omega$, and $\frac{1}{2}I + B'_\omega$, and the unique solvability of the interior Neumann problem is equivalent to the invertibility of the operators $\frac{1}{2}I - B_\omega$ and $\frac{1}{2}I - B'_\omega$. In this case, the following formulas are true for the Neumann-to-Dirichlet operators (see (2.5)):

$$u^+ = -(\tfrac{1}{2}I - B_\omega)^{-1} A_\omega T u^+, \quad u^- = (\tfrac{1}{2}I + B_\omega)^{-1} A_\omega T u^-. \tag{2.21}$$

[2]Here and below, the subscript $_{\mathrm{loc}}$ may be omitted if $\omega_2 > 0$. We will not repeat this remark.

However, as we already mentioned, the exterior Dirichlet and Neumann problems are always uniquely solvable. The second formula in (2.21) can be modified if ω is exceptional with respect to the interior Dirichlet problem. See [7].

The operator D_ω has the inverse (a pseudo-differential operator of order -1) for all ω except for $\omega = \omega_j'$. It can also be defined as follows. If u is the solution of the interior and exterior Neumann problems with $Tu^\pm = g$ on Γ, then $\psi = D_\omega^{-1} g = [u]$.

Problem II *for eigenfunctions is reduced to the equation*

$$D_\omega \psi = \lambda \psi \tag{2.22}$$

on Γ *by the substitution* (2.19). Essentially, the spectral properties of D_ω were investigated in [7]. More precisely, the operator D_ω^{-1} was there considered. Namely, assuming that the interior Dirichlet and Neumann problems are uniquely solvable for given ω and using (2.21), we considered the operator

$$u^- - u^+ = (\tfrac{1}{2}I + B_\omega)^{-1}(\tfrac{1}{2}I - B_\omega)^{-1} A_\omega T u^\pm. \tag{2.23}$$

If ω is exceptional only for the interior Dirichlet problem, then the corresponding generalization of the second formula in (2.21) can be used. However, it is easier to consider D_ω.

In conclusion, we note that A_ω and D_ω are analogs of the Neumann-to-Dirichlet and Dirichlet-to-Neumann operators for the problems in question with transmission conditions on Γ.

2.4. The Case of a Lipschitz Boundary

Here we again follow [7] and [8], see also references therein. We now assume that the surface Γ is Lipschitz. (In particular, it can have edges and conical points.) Then the Sobolev spaces $H^\tau(\Gamma)$ are defined in general only for $|\tau| \leq 1$. The Dirichlet and Neumann problems remain meaningful in a generalized (weak) setting. The solution belongs to $H^{t+1}(\Omega^+)$ and $H^{t+1}_{\text{loc}}(\Omega^-)$, $|t| < 1/2$ (but is analytic in $\Omega^\pm$). The boundary values $u^\pm$ obviously belong to $H^{t+1/2}(\Gamma)$, while the values of $Tu^\pm$ are defined by Green's formulas and belong to $H^{t-1/2}(\Gamma)$. Namely, the following Green formula in Ω^+ is well known (cf., e.g., [22]):

$$-\omega^2 \int_{\Omega^+} u \cdot \overline{v} \, dx = (Tu^+, v^+)_{0,\Gamma} - \int_{\Omega^+} E(u, \overline{v}) \, dx, \tag{2.24}$$

where

$$E(u, v) = \sum A_{jk} \partial_k u \cdot \partial_j v \tag{2.25}$$

and $(\cdot,\cdot)_{0,\Gamma}$ is the extension of the standard inner product in $H^0(\Gamma)$ to $H^{t-1/2}(\Gamma) \times H^{-t+1/2}(\Gamma)$ (the latter two spaces are dual with respect to this extension). Assuming that u is a solution to (1.1) in $H^{t+1}(\Omega)$ and v is an arbitrary function in $H^{-t+1}(\Omega^+)$, we define $Tu^+ \in H^{t-1/2}(\Gamma)$ as an anti-linear continuous functional on the space $H^{-t+1/2}(\Gamma)$ of functions v^+ on Γ. To define Tu^-, we write out a similar formula for $\Omega_R^- = \Omega^- \cap O_R$, where O_R is a large ball containing Γ and centered at the origin:

$$-\omega^2 \int_{\Omega_R^-} u \cdot \overline{v}\, dx = \int_{S_R} Tu \cdot \overline{v}\, dS - (Tu^-, v^-)_{0,\Gamma} - \int_{\Omega_R^-} E(u, \overline{v})\, dx. \tag{2.26}$$

If $\omega = 0$ or $\omega_2 > 0$, then we can pass to the limit as $R \to \infty$ and use the formula

$$-\omega^2 \int_{\Omega^-} u \cdot \overline{v}\, dx = -(Tu^-, v^-)_{0,\Gamma} - \int_{\Omega^-} E(u, \overline{v})\, dx. \tag{2.27}$$

Variational arguments are used for $t = 0$, and the unique solvability of the interior Dirichlet and Neumann problems is proved for all ω except for some positive values $\omega_j \to +\infty$ and nonnegative values $\omega_j' \to +\infty$, respectively. These results admit partial extensions to t with $|t| \leq 1/2$ and complete extensions to t with $|t| < 1/2$. Namely, in the case of the Dirichlet problem for all systems and in the case of the Neumann problem for some systems, the value $t = 1/2$ can be considered by means of the Rellich identities, which gives the results for $|t| \leq 1/2$; see, e.g., [19]. In the case of the Neumann problem for all systems, the extensions to t with $|t| < 1/2$ are obtained with the use of the Savaré theorem on the smoothness of solutions to variational problems in Lipschitz domains. See [7, 8] and references therein, starting from [23].

This approach should be compared with that based on the use of potentials (cf. [22]). Formulas (2.2) remain true for $\varphi \in H^{-1/2+t}(\Gamma)$ and $\psi \in H^{1/2+t}(\Gamma)$ with $|t| < 1/2$. For solutions in $H^{1+t}(\Omega^+)$ and $H^{1+t}_{\mathrm{loc}}(\Omega^-)$, $|t| < 1/2$, the integral representations (2.4) and relations (2.5) on Γ remain true. The formulas for the Calderón projections also remain true. Of course, we have no calculus of pseudo-differential operators. However, A_ω is a bounded operator from $H^{t-1/2}(\Gamma)$ to $H^{t+1/2}(\Gamma)$ and is invertible for all ω except for ω_j. Similarly, D_ω is a bounded operator from $H^{t+1/2}(\Gamma)$ to $H^{t-1/2}(\Gamma)$ and is invertible for all ω except for ω_j'. All this is always true for $|t| < 1/2$; the values $t = \pm 1/2$ will not be used in the present paper.

Now assume that $t = 0$ and $\omega = i\tau$, $\tau > 0$. For the solution of the Dirichlet problems, we have the formulas

$$u = \mathcal{A}_\omega A_\omega^{-1} u^\pm \tag{2.28}$$

and the two-sided estimates

$$\|u^\pm\|_{1/2,\Gamma} \leq C_1 \|u\|_{1,\Omega^\pm} \leq C_2 \|u^\pm\|_{1/2,\Gamma}. \tag{2.29}$$

Here the first inequality is well known for all $u \in H^1(\Omega^\pm)$ and the second follows from (2.28).

Similarly, for the solution of the Neumann problems we have the formulas

$$u = -\mathcal{B}_\omega D_\omega^{-1} Tu^\pm \tag{2.30}$$

and the two-sided estimates

$$\|Tu^\pm\|_{-1/2,\Gamma} \leq C_3 \|u\|_{1,\Omega^\pm} \leq C_4 \|Tu^\pm\|_{-1/2,\Gamma}. \tag{2.31}$$

Here the second inequality follows from (2.30). Let us explain the first inequality in (2.31). We have

$$\|Tu^\pm\|_{-1/2,\Gamma} \le C_5 \sup_{v:\|v^\pm\|_{1/2,\Gamma}=1} |(Tu^\pm, v^\pm)_{0,\Gamma}|.$$

Define v as the solution to the Dirichlet problems for system (1.1) with given $v^\pm$. Then the right-hand side is not greater than

$$C_6\|u\|_{1,\Omega^\pm} \sup \|v\|_{1,\Omega^\pm} \le C_7\|u\|_{1,\Omega^\pm}$$

(in view of (2.29) for v), which yields the desired inequality.

The operators A_ω and D_ω with $\omega_1 = 0$ are self-adjoint in the following sense:

$$(A_\omega\varphi_1, \varphi_2)_{0,\Gamma} = (\varphi_1, A_\omega\varphi_2)_{0,\Gamma} \quad (\varphi_j \in H^{-1/2}(\Gamma)), \tag{2.32}$$

$$(D_\omega\psi_1, \psi_2)_{0,\Gamma} = (\psi_1, D_\omega\psi_2)_{0,\Gamma} \quad (\psi_j \in H^{1/2}(\Gamma)). \tag{2.33}$$

Indeed, (2.32) is true for these ω and $\varphi_j \in H^0(\Gamma)$, since the matrix $\mathcal{E}_\omega(x)$ is real symmetric, and it is carried over $\varphi_j \in H^{-1/2}(\Gamma)$ by a passage to the limit. Formula (2.33) follows from the self-adjointness of the operators (2.21) for pure imaginary ω.

2.5. The Spaces $H^\tau(S)$ and $\widetilde{H}^\tau(S)$ on a Nonclosed Surface S

The *space* $H^\tau(S)$ consists, by definition, of the restrictions to S of functions[3] belonging to $H^\tau(\Gamma)$. Here S is considered as an open part of Γ, and for $\tau < 0$ the restriction is understood in the sense of distributions. The norm in $H^\tau(S)$ is defined by the formula

$$\|\psi\|_{\tau,S} = \inf\{\|\phi\|_{\tau,\Gamma} : \phi \in H^\tau(\Gamma), \ \phi|_S = \psi\}. \tag{2.34}$$

The *space* $\widetilde{H}^\tau(S)$ is defined as the subspace of $H^\tau(\Gamma)$ consisting of functions supported in $\overline{S}$. The norm in this space is defined by the formula

$$\|\psi\|_{\widetilde{H}^\tau(S)} = \|\psi\|_{\tau,\Gamma}. \tag{2.35}$$

The spaces $H^\tau(\Gamma_-)$ and $\widetilde{H}^\tau(\Gamma_-)$ are defined similarly.

The space $H^\tau(S)$ is the factor space $H^\tau(\Gamma)/\widetilde{H}^\tau(\Gamma_-)$.

If Γ and γ are Lipschitz, then these spaces are defined only for $|\tau| \le 1$. However, even in the case of smooth Γ and γ we need these spaces only for $|\tau| < 1$.

If $\tau_1 < \tau_2$, then $H^{\tau_1}(S) \supset H^{\tau_2}(S)$ and $\widetilde{H}^{\tau_1}(S) \supset \widetilde{H}^{\tau_2}(S)$, the corresponding embeddings are compact, and the spaces with index 2 are dense in the corresponding spaces with index 1.

Obviously, for $\tau \ge 0$ we may view the spaces $H^\tau(S)$ and $\widetilde{H}^\tau(S)$ as spaces of functions defined on S. For $\tau < 0$, the space $H^\tau(S)$ consists of distributions defined in S, and $\widetilde{H}^\tau(S)$ consists of distributions on Γ with supports lying in $\overline{S}$. We allow ourself to say that these distributions are functions defined on S.

For $|\tau| < 1/2$, the spaces $H^\tau(S)$ and $\widetilde{H}^\tau(S)$ can be identified. For $0 \le \tau < 1/2$, this follows from the fact that the continuation by zero on Γ_- is a

[3]Usual functions or generalized functions, i.e. distributions.

bounded operator from $H^\tau(S)$ to $H^\tau(\Gamma)$ giving a function in $\widetilde{H}^\tau(\Gamma)$ with norm equivalent to the norm of the original function in $H^\tau(S)$. For $-1/2 < \tau < 0$, the possibility of identifying these spaces follows from duality arguments (see below). In [17], it was shown that for $|\tau| < 1/2$ the space $H^\tau(\Gamma)$ is the direct sum of the subspaces $\widetilde{H}^\tau(\Gamma_+)$ and $\widetilde{H}^\tau(\Gamma_-)$. This result for the smooth case is carried over to the Lipschitz case by means of a Lipschitz diffeomorphism.

If the surfaces Γ and γ are smooth, then the linear manifolds $C_0^\infty(S)$ and $C^\infty(\overline{S})$ are dense in $\widetilde{H}^\tau(S)$ and $H^\tau(S)$, respectively. Besides, $C_0^\infty(S)$ is dense in $H^\tau(S)$ for $\tau < 1/2$. (Indeed, the space $H^{\tau_2}(S)$ is dense in $H^{\tau_1}(S)$ for $\tau_1 < \tau_2$, while $C_0^\infty(S)$ is dense in $H^{\tau_2}(S) = \widetilde{H}^{\tau_2}(S)$ for $\tau_2 \in (-1/2, 1/2)$.) In the case of Lipschitz Γ and γ, the first assertion is true for linear manifolds of functions satisfying the Lipschitz condition on S and $\overline{S}$, respectively, with supports in S in the first case.

The spaces $H^\tau(S)$ and $\widetilde{H}^{-\tau}(S)$ are dual to each other with respect to the continuation of the inner product $(\varphi, \psi)_{0,S}$ in $H^0(S)$ from functions belonging to the dense linear manifolds just indicated to the direct product of these spaces. See, e.g., [19].

The spaces $H^\tau(S)$ and $\widetilde{H}^\tau(S)$ form two interpolation scales with respect to the real and complex interpolation methods. This means that if we take two points τ_1 and τ_2, then in each of these scales the space X^τ with intermediate index τ is obtained from the spaces with indices τ_1 and τ_2 by the interpolation rules: for $0 \leq \theta \leq 1$,

$$[X^{\tau_1}, X^{\tau_2}]_\theta = [X^{\tau_1}, X^{\tau_2}]_{\theta,2} = X^{(1-\theta)\tau_1 + \theta\tau_2}. \tag{2.36}$$

See, e.g., [24].

However, there are two more interpolation scales obtained by pasting together: 1) The scale consisting of $\widetilde{H}^\tau(S)$ for $\tau \leq \theta$ and $H^\tau(S)$ for $\tau \geq \theta$; 2) the scale consisting of $H^\tau(S)$ for $t \leq \theta$ and $\widetilde{H}^\tau(S)$ for $\tau \geq \theta$. Here the point θ of pasting together is an arbitrary point in $(-1/2, 1/2)$. The possibility of pasting together follows from Wolff's theorem (see [25])[4] and the reiteration theorem [24].

Similar spaces are defined in the case of the space and the half-space instead of Γ and S, respectively.

3. Problems with Transmission Conditions on a Lipschitz Nonclosed Surface in the Simplest Spaces

3.1. The Contents of This Section

In this section, assuming that Γ and γ are Lipschitz, we reduce the Dirichlet and Neumann problems to equations on S and prove the invertibility of the corresponding operators $A_{\omega,S}$ and $D_{\omega,S}$. We use only the spaces $H^{\pm 1/2}(S)$ and $\widetilde{H}^{\pm 1/2}(S)$ and apply a version of the variational approach (cf., e.g., [19] and see references

[4] I obtained this information from V.I. Ovchinnikov and use here a possibility to express him my sincere gratitude.

therein). At the end of the section, we discuss the spectral properties of these operators. They are analogs of the Neumann-to-Dirichlet and Dirichlet-to-Neumann operators for problems in question.

Note that if system (1.1) is satisfied outside S, then

$$[u] = 0 \text{ and } [Tu] = 0 \text{ on } \Gamma_-. \tag{3.1}$$

We denote by p_+ the operator of restriction of functions defined on Γ to S and by E_0 the operator of continuation of functions defined on S to Γ by zero outside S.

3.2. The Dirichlet Problem and Problem I

We introduce the operator

$$A_{\omega,S} := p_+ A_\omega. \tag{3.2}$$

It will be applied to functions defined on Γ and equal to zero on Γ_-. We also can consider it as acting on functions defined on S and extended by zero on Γ_-. Then the right-hand side of (3.2) should be rewritten in the form $p_+ A_\omega E_0$.

If u is the solution to the Dirichlet problem (1.1), (1.4) in $H^1_{\mathrm{loc}}(\mathbb{R}^n \setminus S)$, then we set

$$\varphi = [Tu] \tag{3.3}$$

as in (2.15), but now this is a function in $\widetilde{H}^{-1/2}(S)$, for which we obtain the equation

$$A_{\omega,S}\varphi = f \quad \text{on } S, \tag{3.4}$$

similar to (2.14). The operator $A_{\omega,S}$ acts boundedly from $\widetilde{H}^{-1/2}(S)$ to $H^{1/2}(S)$. For u we have the formula

$$u = \mathcal{A}_\omega \varphi. \tag{3.5}$$

All this follows from the discussion in Subsection 2.4. Conversely, if $f \in H^{1/2}(S)$ and φ is the solution to equation (3.4) in $\widetilde{H}^{-1/2}(S)$, then (3.5) is the solution of the Dirichlet problem (1.1), (1.4) in $H^1_{\mathrm{loc}}(\mathbb{R}^n \setminus S)$. Thus, *we have equivalently reduced the Dirichlet problem* (1.1), (1.4) *to equation* (3.4) *on* S.

Passing to the spectral problem I, we see that if u is an eigenfunction of this problem with eigenvalue λ, then φ satisfies the equation

$$\varphi = \lambda A_{\omega,S}\varphi \quad \text{on } S, \tag{3.6}$$

similar to (2.17). Here we mean that φ in the left-hand side is restricted to S. Conversely, if φ is a solution in $\widetilde{H}^{-1/2}(S)$ to this equation, then (3.5) is a solution of Problem I. We add that taking our identifications into account, we have

$$H^{1/2}(S) \subset H^0(S) = \widetilde{H}^0(S) \subset \widetilde{H}^{-1/2}(S). \tag{3.7}$$

Thus we can treat $A_{\omega,S}$ as a bounded operator in $\widetilde{H}^{-1/2}(S)$ with range contained in $H^{1/2}(S)$ (we will see that actually the range coincides with this space) and hence lying in $\widetilde{H}^{-1/2}(S)$. Therefore, the spectral equation (3.6) make sense. (Another point of view is also possible: in (3.6), we can replace the operator $A_{\omega,S} = p_+ A_\omega$ by $\theta_+ A_\omega$, where $\theta_+(x)$ is the characteristic function of S on Γ. Then φ can be treated as equal to zero in Γ_- on both sides in (3.6).)

Like (2.17), equation (3.6) relates only to eigenfunctions; if $A_{\omega,D}$ has associated functions, their relation to the corresponding solutions of Problem I is somewhat more complicated. We do not discuss it and further will discuss the spectral properties of the operator $A_{\omega,S}$. Thus *Problem I for eigenfunctions is equivalently reduced to equation* (3.6) *on* Γ.

The following theorem is known in the case of smooth S and γ at least for the Helmholtz equation, the Lamé system, and the system of anisotropic elasticity for $n = 3$. See additional references in Section 4.

Theorem 3.1. *The Dirichlet problem* (1.1), (1.4) *with* $f \in H^{1/2}(S)$ *has a unique solution in* $H^1_{\mathrm{loc}}(\mathbb{R}^n \setminus S)$, *and the operator* $A_{\omega,S} : \widetilde{H}^{-1/2}(S) \mapsto H^{1/2}(S)$ *is invertible.*

Proof. First, we check the uniqueness in the Dirichlet problem. For $\omega = 0$ or $\operatorname{Im}\omega > 0$, we use the standard arguments. We set $v = u$ in (2.24) and (2.27) and assume that this function belongs to $H^1(\mathbb{R}^n \setminus S)$ and has zero values $u^{\pm}$ on S. Adding left- and right-hand sides and taking (3.1) into account, we obtain

$$-\omega^2 \int_{\mathbb{R}^n} |u|^2 dx = -\int_{\mathbb{R}^n} E(u, \overline{u})\, dx. \tag{3.8}$$

If $\operatorname{Im}\omega > 0$, then $\int |u|^2 dx = 0$ and hence $u = 0$. If $\omega = 0$, then $\int E(u, \overline{u})\, dx = 0$ and hence (cf. [18], Section 3) $u = \text{const} = 0$. The case $\omega > 0$ is somewhat more complicated. In this case, from (2.24) and (2.26) we obtain

$$-\omega^2 \int_{O_R} |u|^2 dx = \int_{S_R} Tu \cdot \overline{u}\, dS - \int_{O_R} E(u, \overline{u})\, dx,$$

and hence

$$\operatorname{Im} \int_{S_R} Tu \cdot \overline{u}\, dx = 0. \tag{3.9}$$

Now the arguments from [7] can be applied, and we find that $u = 0$ in Ω^-. Since u is an analytic function outside S, it vanishes identically.

It follows from the uniqueness in the Dirichlet problem that $A_{\omega,S}$ annihilates only the zero function.

The form

$$\langle \varphi_1, \varphi_2 \rangle_{-1/2,S} = -(A_{\omega,S}\varphi_1, \varphi_2)_{0,S} = -(A_\omega \varphi_1, \varphi_2)_{0,\Gamma} \tag{3.10}$$

is defined and bounded on the space $\widetilde{H}^{-1/2}(S)$. To check the latter equality, it suffices to approximate φ_2 by functions with supports lying inside S. We now need the following assertion (cf. Proposition 7.10 in [7]):

Proposition 3.2. *For* $\omega = i\tau$, $\tau > 0$, *the form* (3.10) *is an inner product in* $\widetilde{H}^{-1/2}(S)$, *the corresponding norm being equivalent to* $\|\varphi\|_{\widetilde{H}^{-1/2}(S)}$.

Proof. From (3.10) and (2.32), we have

$$\langle \varphi_1, \varphi_2 \rangle_{-1/2,S} = -(A_{i\tau}\varphi_1, \varphi_2)_{0,\Gamma} = -(\varphi_1, A_{i\tau}\varphi_2)_{0,\Gamma} = \overline{\langle \varphi_2, \varphi_1 \rangle}_{-1/2,S}$$

for $\varphi_j \in \widetilde{H}^{-1/2}(S)$.

It remains to estimate $\langle\varphi,\varphi\rangle_{-1/2,S}$ by $\|\varphi\|^2_{\widetilde{H}^{-1/2}(S)}$ from below. We define the function u by (3.5). This is a solution to (1.1) outside S belonging to $H^1(\mathbb{R}^n \setminus S)$, and we have

$$[Tu] = \varphi \text{ and } u^{\pm} = A_{i\tau}\varphi \quad \text{on } \Gamma. \tag{3.11}$$

From (2.24), (2.27) with $v = u$ and (3.10), we obtain

$$\langle\varphi,\varphi\rangle_{-1/2,S} = \int_{\mathbb{R}^n} [\tau^2|u|^2 + E(u,\overline{u})]\, dx \geq C_1\|u\|^2_{1,\mathbb{R}^n\setminus\Gamma}. \tag{3.12}$$

On the other hand, from the same formulas (2.24), (2.27) with $v^+ = v^-$ on Γ, we have

$$\tau^2 \int_{\mathbb{R}^n} u \cdot \overline{v}\, dx = -([Tu], v)_{0,\Gamma} - \int_{\mathbb{R}^n} E(u,\overline{v})\, dx. \tag{3.13}$$

Assuming that

$$\|v\|_{1,\mathbb{R}^n} \leq C_2\|v^{\pm}\|_{1/2,\Gamma} \leq C_2'\|v\|_{1,\mathbb{R}^n}$$

(we construct v as the solution of the Dirichlet problems in $\Omega^{\pm}$ with given $v^{\pm}$, see (2.29)) and using the left inequality in (2.31), we obtain

$$\|\varphi\|_{-1/2,\Gamma} = \sup_{v:\|v^{\pm}\|_{1/2,\Gamma}=1} |([Tu], v^{\pm})_{0,\Gamma}| \leq C_3\|u\|_{1,\mathbb{R}^n\setminus\Gamma} \sup \|v\|_{1,\mathbb{R}^n} \leq C_4\|u\|_{1,\mathbb{R}^n\setminus\Gamma}.$$

Combining this with (3.12), we obtain the desired estimate. □

Now we return to the proof of the theorem.

If we take f in $H^{1/2}(S)$, then the relation

$$\langle\varphi_1,\varphi_2\rangle_{-1/2,S} = -(f,\varphi_2)_{0,S} \quad (\varphi_2 \in \widetilde{H}^{-1/2}(S))$$

uniquely determines $\varphi_1 \in \widetilde{H}^{-1/2}(S)$ by the Riesz theorem. Thus we have shown that the equation $A_{i\tau,S}\varphi_1 = f$ has a solution in $\widetilde{H}^{-1/2}(S)$. We see that the operator $A_{\omega,S}$ is invertible for pure imaginary ω.

For other ω, this operator is a weak perturbation of the operator just investigated, since the difference of their kernels is by order one less singular and hence is the kernel of a compact operator from $\widetilde{H}^{-1/2}(S)$ to $H^{1/2}(S)$. (Moreover, its range lies even in $H^1(S)$.) Therefore, $A_{\omega,S}$ is a Fredholm operator for each ω and its index is zero. Since $\operatorname{Ker} A_{\omega,S}$ is trivial, we conclude that $A_{\omega,S}$ is invertible. Simultaneously, we have proved the unique solvability of the Dirichlet problem. □

In addition, we note that for $\omega = i\tau$, $\tau > 0$, if we make the substitution $\varphi_j = A^{-1}_{\omega,S}\psi_j$ $(j = 1, 2)$ in (3.10), then we obtain an inner product

$$\langle\psi_1,\psi_2\rangle_{1/2,S} = -(\psi_1, A^{-1}_{\omega,S}\psi_2)_{0,S} \tag{3.14}$$

in $H^{1/2}(S)$. The corresponding norm is equivalent to the usual norm in this space.

3.3. The Neumann Problem and Problem II

Here our considerations are similar to those in the previous subsection (but the results do not follow from the results obtained there).

We introduce the operator

$$D_{\omega,S}\psi = p_+ D_\omega \psi. \tag{3.15}$$

Here ψ is a function on Γ equal to zero on Γ_-. We also can apply this operator to functions defined on S and extended by zero to Γ_-. Then we rewrite the right-hand side in (3.15) in the form $p_+ D_\omega E_0 \psi$.

If u is a solution to the Neumann problem (1.1), (1.5) in $H^1_{\mathrm{loc}}(\mathbb{R}^n \setminus S)$, then we set

$$\psi = [u] \tag{3.16}$$

as in (2.19), but now this function belongs to $\widetilde{H}^{1/2}(S)$. By (2.4) and (2.7), we have

$$u = -\mathcal{B}_\omega \psi. \tag{3.17}$$

According to definitions (2.6) and (3.15), we obtain the equation

$$D_{\omega,S}\psi = g \quad \text{on } S \tag{3.18}$$

for ψ. The operator $D_{\omega,S}$ acts boundedly from $\widetilde{H}^{1/2}(S)$ to $H^{-1/2}(S)$.

Conversely, if $g \in H^{-1/2}(S)$ and ψ is a solution to equation (3.18) in $\widetilde{H}^{1/2}(S)$, then the function (3.17) is a solution to the Neumann problem (1.1), (1.5) in $H^1_{\mathrm{loc}}(\mathbb{R}^n \setminus S)$. Thus, *the Neumann problem* (1.1), (1.5) *is equivalently reduced to equation* (3.18) *on* S, similar to equation (2.18) in the case of a closed surface.

Problem II *for eigenfunctions is equivalently reduced to the equation*

$$D_{\omega,S}\psi = \lambda\psi \quad \text{on } S, \tag{3.19}$$

similar to equation (2.22) in the case of a closed surface. Here we mean that ψ in the right-hand side is restricted to S. The operator $D_{\omega,S}$ can be considered as acting in the space $H^{-1/2}(S)$ and having the domain $\widetilde{H}^{1/2}(S)$. It lies in $H^{-1/2}(S)$, since with our identifications we have

$$\widetilde{H}^{1/2}(S) \subset \widetilde{H}^0(S) = H^0(S) \subset H^{-1/2}(S). \tag{3.20}$$

Therefore, the spectral equation (3.19) make sense. (Another possible point of view is that we can replace the operator $D_{\omega,S} = p_+ D_\omega$ by $\theta_+ D_\omega$. Then ψ on both sides in (3.19) will be equal to zero in Γ_-.) The following theorem is also known in the case of smooth surfaces for the Helmholtz equation, the Lamé system and the system of anisotropic elasticity (for $n = 3$). See also Section 4 for additional references.

Theorem 3.3. *The Neumann problem* (1.1), (1.5) *with* $g \in H^{-1/2}(S)$ *has a unique solution in* $H^1_{\mathrm{loc}}(\mathbb{R}^n \setminus S)$, *and the operator* $D_{\omega,S} : \widetilde{H}^{1/2}(S) \mapsto H^{-1/2}(S)$ *is invertible.*

Proof. The uniqueness in the Neumann problem is checked literally in the same way as in the Dirichlet problem. It follows that the operator $D_{\omega,S}$ annihilates only the zero function.

The form

$$< \psi_1, \psi_2 >_{1/2,S} = -(D_{\omega,S}\psi_1, \psi_2)_{0,S} = -(D_\omega \psi_1, \psi_2)_{0,\Gamma} \tag{3.21}$$

is defined and bounded on the space $\widetilde{H}^{1/2}(S)$. The latter equality is true because of possibility to approximate ψ_2 by functions with supports lying inside S.

Proposition 3.4. *For $\omega = i\tau$, $\tau > 0$, the form (3.21) is an inner product in $\widetilde{H}^{1/2}(S)$. The corresponding norm is equivalent to the original norm in $\widetilde{H}^{1/2}(S)$.*

Proof. From (3.21) and (2.33), we have

$$< \psi_1, \psi_2 >_{1/2,S} = -(D_\omega \psi_1, \psi_2)_{0,\Gamma} = -(\psi_1, D_\omega \psi_2)_{0,\Gamma} = \overline{< \psi_2, \psi_1 >_{1/2,S}}.$$

Now we estimate $< \psi, \psi >_{1/2,S}$ from below by $\|\psi\|^2_{\widetilde{H}^{1/2}(S)}$. We define the function u by (3.17). Then, according to the second formula in (2.2) and definitions (2.6) and (3.15), we have

$$[u] = \psi \quad \text{and} \quad Tu^\pm = D_\omega \psi \quad \text{on } \Gamma. \tag{3.22}$$

From (2.24), (2.27) with $v = u$, and (3.21) we have

$$< \psi, \psi >_{1/2,S} = \int_{\mathbb{R}^n} \left[\tau^2 |u|^2 + E(u, \overline{u})\right] dx \geq C_5 \|u\|^2_{1,\mathbb{R}^n \setminus \Gamma}. \tag{3.23}$$

On the other hand, from (2.24) and (2.27), by interchanging the letters u and v, we obtain

$$\tau^2 \int_{\mathbb{R}^n} v \cdot \overline{u}\, dx = -(Tv^\pm, [u])_{0,S} - \int_{\mathbb{R}^n} E(v, \overline{u})\, dx. \tag{3.24}$$

Here we assume that v belongs to $H^1(\mathbb{R}^n \setminus S)$ and satisfies the conditions

$$Tv^+ = Tv^- \quad \text{and} \quad \|v\|_{1,\Omega^\pm} \leq C_6 \|Tv^\pm\|_{-1/2,\Gamma} \leq C_7 \|v\|_{1,\Omega^\pm}$$

on Γ (v is constructed as the solution to the Neumann problems in $\Omega^\pm$ with given $Tv^\pm$; see (2.31)). Hence

$$\begin{aligned} \|\psi\|_{1/2,\Gamma} &= \sup_{v:\|Tv^\pm\|_{-1/2,\Gamma}=1} |(Tv^\pm, [u])_{0,\Gamma}| \leq C_8 \|u\|_{1,\mathbb{R}^n \setminus \Gamma} \sup \|v\|_{1,\mathbb{R}^n \setminus \Gamma} \\ &\leq C_9 \|u\|_{1,\mathbb{R}^n \setminus \Gamma}, \end{aligned}$$

and, using (3.23), we obtain the desired estimate

$$< \psi, \psi >_{1/2,S} \geq C_{10} \|\psi\|^2_{\widetilde{H}^{1/2}(S)}.$$

□

Now we finish the proof of Theorem 3.3 in the same way as the proof of Theorem 3.1. For $\omega = i\tau$, $\tau > 0$, taking $g \in H^{-1/2}(S)$, we define a function $\psi_1 \in \widetilde{H}^{1/2}(S)$ by the relation

$$< \psi_1, \psi_2 >_{1/2,S} = -(g, \psi_2)_{0,S} \quad (\psi_2 \in \widetilde{H}^{1/2}(S)). \tag{3.25}$$

It is a solution of the equation $D_{\omega,S}\psi_1 = g$. Thus the operator $D_{\omega,S}$ is invertible for $\omega = i\tau$, $\tau > 0$. For other ω, we obtain the invertibility treating $D_{\omega,S}$ as a weak perturbation of $D_{i\tau,S}$. Simultaneously, we conclude that the Neumann problem is uniquely solvable. □

The substitution $\psi_j = D^{-1}_{\omega,S}\varphi_j$ transforms the form (3.21) into the inner product

$$< \varphi_1, \varphi_2 >_{-1/2,S} = -(\varphi_1, D^{-1}_{\omega,S}\varphi_2)_{0,S} \tag{3.26}$$

in $H^{-1/2}(S)$. The corresponding norm is equivalent to the original norm in this space.

3.4. The Spectral Properties of the Operators $A_{\omega,S}$ and $D_{\omega,S}$

We list and comment these properties without trying to formulate a cumbersome theorem. In 6 and 7, we mention some expected results.

1. For $\omega = i\tau$, $\tau > 0$, the operator $A_{\omega,S}$ is a self-adjoint compact operator in the space $\widetilde{H}^{-1/2}(S)$ with inner product (3.10). Hence there exists an orthonormal basis $\{e_j\}$ in this space consisting of eigenfunctions. Since this operator maps this space continuously onto $H^{1/2}(S)$ and has a bounded inverse, it follows that the functions e_j belong to $H^{1/2}(S)$ and form an orthogonal basis there with respect to the inner product (3.14). More precisely, if $A_{\omega,S}e_j = \lambda_j e_j$, then $\{\lambda_j e_j\}$ is a basis in $H^{1/2}(S)$ orthonormal with respect to (3.14).

The power $(-A_{\omega,S})^\theta$, $0 < \theta < 1$, defines a continuous invertible mapping of $\widetilde{H}^{-1/2}(S)$ onto $H^{\theta-1/2}(S) = \widetilde{H}^{\theta-1/2}(S)$. Therefore, the same eigenfunctions form a basis in these spaces too. It is orthogonal with respect to the inner product

$$\langle \varphi_1, \varphi_2 \rangle_{\theta-1/2,S} = ((-A_{\omega,S})^{1-2\theta}\varphi_1, \varphi_2)_{0,S}.$$

In particular, in $L_2(S)$ it is the usual inner product. Here $|\lambda_j|^{1/2}e_j$ form an orthonormal basis.

2. Passing to other ω, we can treat $A_{\omega,S}$ in $\widetilde{H}^{-1/2}(S)$ as a weak perturbation of the operator $A_{i\tau,S}$ just considered. Unfortunately, we cannot go beyond the spaces with indices from $-1/2$ to $1/2$ and therefore can only state that $A_{\omega,S}$ is a relatively compact perturbation of the operator $A_{i\tau,S}$. This property is inherited in all spaces $H^t(S)$, $-1/2 < t \leq 1/2$.

3. The s-numbers of these operators admit the estimate

$$s_j \leq C j^{-1/(n-1)} \tag{3.27}$$

(see, e.g., [26]).

4. It follows from Assertions 2 and 3 that it is possible to form a basis for the Abel–Lidskii method of summability with brackets of order $n-1+\varepsilon$ in $\widetilde{H}^{-1/2}(S)$ with arbitrarily small $\varepsilon > 0$ consisting of root functions (see formulations and references in [27]). This property is inherited by the same system of root functions in all spaces $H^t(S)$, $-1/2 < t \leq 1/2$.

5. It also follows from Assertion 2 that the characteristic numbers (i.e. the inverses of the eigenvalues) of the non-self-adjoint operator $A_{\omega,S}$ lie in the union of an arbitrarily narrow sectorial neighborhood of the ray $\mathbb{R}_-$ and a neighborhood

of the origin (depending on the choice of the sectorial neighborhood). See, e.g., [27].

6. In the case of an "almost smooth" surface Γ (smooth outside a closed subset of zero measure, this is a definition from [26]), for the eigenvalues of the operator $A_{\omega,S}$, numbered in non-decreasing order of real parts with multiplicities taken into account, the asymptotic formula

$$\lambda_j(A_{\omega,S}) = -cj^{-1/(n-1)} + o(j^{-1/(n-1)}) \tag{3.28}$$

is apparently true with positive c (calculated as in the case of a smooth surface). Cf. [26] and [28]. However, the proof is so far incomplete.

7. Almost the same results are true for the operator $D_{\omega,S}$. The difference consists in the fact that the "extreme" spaces are now $H^{-1/2}(S)$ and $\widetilde{H}^{1/2}(S)$. In the first of them, with inner product (3.26), we have a non-bounded operator $D_{\omega,S}$ with domain $\widetilde{H}^{1/2}(S)$, self-adjoint for pure imaginary ω. An estimate of the form (3.27) is true for the s-numbers of the inverse operator. The expected asymptotic formula for its eigenvalues is again of the form (3.28).

4. Concluding Remarks

In a separate paper, the results of Section 3 will be generalized, in particular, to the spaces $H^{t\pm1/2}(S)$ and $\widetilde{H}^{t\pm1/2}(S)$ with $|t| < 1/2$ under additional smoothness assumptions. We discuss the construction of a parametrix for our operators $A_{\omega,S}$ and $D_{\omega,S}$ based on the factorization of the principal symbols at points of γ and the Wiener–Hopf method. This approach was successfully used in [29], [30] and [31] in more general situations. However, in our context the construction of the parametrix is especially explicit and transparent. We also note that there exists an analogy between problems considered in the present paper and mixed problems investigated by many authors. See [12], [32] and the references therein.

References

[1] N.N. Voitovich, B.Z. Katsenelenbaum and N.N. Sivov, *Generalized Method of Eigenoscillations in Diffraction Theory*, Nauka, Moscow, 1977. In Russian.

[2] M.S. Agranovich, B.Z. Katsenelenbaum, A.N. Sivov and N.N. Voitovich, *Generalized Method of Eigenoscillations in Diffraction Theory*, Wiley-VCH, Berlin, 1999.

[3] M.S. Agranovich, Spectral properties of diffraction problems, *Supplement to* [1], 218–416. In Russian.

[4] Z.N. Golubeva, Some scalar diffraction problems and the corresponding nonselfadjoint operators, *Radiotechn. i Electron.* **21** (1976), No. 2, 219–227. English transl. in *Radio Eng. Electron. Physics* **21** (1976).

[5] M.S. Agranovich and R. Mennicken, Spectral properties for the Helmholtz equation with spectral parameter in boundary conditions on a nonsmooth surface, *Mat. Sb.* **190** (1999), No. 1, 29–68. English transl. in *Russian Acad. Sci. Sb. Math.* **190** (1999), 29–69.

[6] M.S. Agranovich, B.A. Amosov and M. Levitin, Spectral problems for the Lamé system with spectral parameter in boundary conditions on smooth or nonsmooth boundary, *Russian J. Math. Phys.* **6** (1999), 247–281.

[7] M.S. Agranovich, Spectral properties of potential type operators for a certain class of strongly elliptic systems on smooth and Lipschitz surfaces, *Trudy Mosk. Mat. Obsch.* **62** (2001), 5–55. English transl. in *Trans. Moscow Math. Soc.* **62** (2001), 1–47.

[8] M.S. Agranovich, Spectral problems for second-order strongly elliptic systems in smooth and non-smooth domains, *Uspekhi Mat. Nauk* **57** (2002), No. 5, 3–79. English transl. in *Russian Math. Surveys* **57** (2002), No. 5, 847–920.

[9] M. Costabel and E. Stephan, A direct boundary integral method for transmission problems, *J. Math. Anal. Appl.* **106** (1985), 367–413.

[10] E.P. Stephan, Boundary integral equations for screen problems in $\mathbb{R}^3$, *Integral Equations Operator Theory* **10** (1987), 236–257.

[11] E.P. Stephan, A boundary integral equation method for three-dimensional crack problems in elasticity, *Math. Meth. Appl. Sci.* **8** (1986), 609–623.

[12] M. Costabel and E.P. Stephan, An improved boundary element Galerkin method for three-dimensional crack problems, *Integral Equations Operator Theory* **10** (1967), 467–504.

[13] R. Duduchava, D. Natroshvili and E. Shargorodsky, Boundary value problems of the mathematical theory of cracks, *Proceedings of I.N. Vekua Institute of Applied Mathematics* **39** (1990), 68–90.

[14] R. Duduchava, Mixed crack type problem in anisotropic elasticity, *Math. Nachr.* **191** (1998), 83–107.

[15] M.I. Vishik and G.I. Eskin, Convolution equations in a bounded domain, *Uspekhi Mat. Nauk* **20** (1965), No. 3, 89–152. English transl. in *Russian Math. Surveys* **20** (1965), No. 3, 85–151.

[16] M.I. Vishik and G.I. Eskin, Elliptic convolution equations in a bounded domain, *Uspekhi Mat. Nauk* **22** (1967), No. 1, 15–76. English transl. in *Russian Math. Surveys* **22** (1967), No. 1, 13–75.

[17] G.I. Eskin, *Boundary Value Problems for Elliptic Pseudodifferential Operators*, Nauka, Moscow, 1973. English transl. in *Translation of Mathematical Monographs*, vol. 52, AMS, 1981.

[18] V.A. Kondratiev and O.A. Oleinik, Boundary value problems for the system of elasticity theory in unbounded domains, Korn's inequalitites, *Uspekhi Mat. Nauk* **43** (1988), No. 5, 55–98. English transl. in *Russian Math. Surveys* **43** (1988), No.5, 65–119.

[19] W. McLean, *Strongly Elliptic Systems and Boundary Integral Equations*, Cambridge University Press, 2000.

[20] J.-L. Lions and E. Magenes, *Problèmes aux limites non homogènes and applications*, Dunod, 1968.

[21] R.T. Seeley, Singular integrals and boundary value problems, *Amer. J. Math.* **88** (1966), 781–809.

[22] M. Costabel, Boundary integral operators on Lipschitz domains: elementary results, *SIAM J. Math. Anal.* **16** (1988), 614–626.

[23] G. Savaré, Regularity results for elliptic equations in Lipschitz domains, *J. Funct. Anal.* **152** (1998), 176–201.

[24] J. Berg and J. Löfström, *Interpolation Spaces*, Springer, 1976.

[25] S. Janson, P. Nilson and J. Peetre. Notes on Wolff's note on interpolation spaces, *Proc. London Math. Soc.* **48** (1984), 283–299.

[26] M.S. Agranovich and B. A. Amosov, Estimates for s-numbers and spectral asymptotics for integral operators of potential type on nonsmooth surfaces, *Funkt. Anal. i Prilozhen.* **30** (1996), No. 2, 1–18. English transl. in *Funct. Anal. Appl.* **30** (1996), No. 2, 75–89.

[27] M.S. Agranovich, Elliptic operators on closed manifolds, *Itogi Nauki i Tekhniki, Sovr. Problemy Mat., Fund. Napravlenyja*, Vol. **63**, VINITI, Moscow, 1990, 5–129. English transl. in Partial Differential Equations VI, *Encyclopaedia Math. Sci.* Vol. **63**, Springer, 1994, pp. 1–130.

[28] T.A. Suslina, Asymptotics of the spectrum of variational problems to an elliptic equation in a domain with a piecewise smooth boundary, *Zapiski Nauchn. Sem. LOMI* **147** (1985), 179–183. In Russian.

[29] G.I. Eskin, Boundary value problems and the parametrix for systems of elliptic pseudodifferential equations, *Trudy Moscow Mat. Obsch.* **28** (1973), 75–116. English translation in *Trans. Moscow Math. Soc.* **28** (1973), 74–115.

[30] R. Duduchava and F.-O. Speck, Pseudodifferential operators on compact manifolds with Lipschitz boundary, *Math. Nachr.* **160** (1993), 149–191.

[31] R. Duduchava and W.L. Wendland, The Wiener–Hopf method for systems of pseudodifferential equations with an application to crack problems, *Integral Equations Operator Theory* **23** (1991), 294–335.

[32] B. V. Pal'tsev, Mixed problems with non-homogeneous boundary conditions in Lipschitz domains for second-order elliptic equations with a parameter. *Mat. Sb.* **187** (1996), No. 4, 59–116. English Transl. in *Russian Acad. Sc. Sb. Math.* **187** (1996), No.4, 525–580.

M.S. Agranovich
Moscow Institute of Electronics and Mathematics (MIEM)
Moscow 109028
Russia
e-mail: magran@orc.ru

Operator Theory:
Advances and Applications, Vol. 164, 23–41

Well-Posedness of the Cauchy Problem for Some Degenerate Hyperbolic Operators

Alessia Ascanelli and Massimo Cicognani

Abstract. We use an uniform approach to different kinds of degenerate hyperbolic Cauchy problems to prove well-posedness in C^∞ and in Gevrey classes. We prove in particular that we can treat by the same method a weakly hyperbolic problem, satisfying an intermediate condition between effective hyperbolicity and the Levi condition, and a strictly hyperbolic problem with non-regular coefficients with respect to the time variable.

Mathematics Subject Classification (2000). Primary 35L80, 35L15.

Keywords. Degenerate hyperbolic equations, Cauchy problem.

1. Introduction

Let us consider the Cauchy problem in $[0,T]\times\mathbf{R}^n$

$$\begin{cases} P(t,x,D_t,D_x)u(t,x)=0, \quad (t,x)\in[0,T]\times\mathbf{R}^n\,, \\ u(0,x)=u_0(x), \\ \partial_t u(0,x)=u_1(x), \end{cases} \tag{1.1}$$

for a second order operator

$$\begin{cases} P=D_t^2-a(t,x,D_x)+b(t,x,D_x)+c(t,x), \\ a(t,x,\xi)=\displaystyle\sum_{i,j=1}^n a_{ij}(t,x)\xi_i\xi_j, \\ b(t,x,\xi)=\displaystyle\sum_{j=1}^n b_j(t,x)\xi_j, \end{cases} \tag{1.2}$$

$D = \frac{1}{\sqrt{-1}}\partial$, that satisfies the hyperbolicity condition

$$a(t,x,\xi) \geq 0, \quad t \in [0,T], x, \xi \in \mathbf{R}^n \tag{1.3}$$

with coefficients $a_{ij} \in C([0,T];\mathcal{B}^\infty)$, $b_j, c \in C^0([0,T];\mathcal{B}^\infty)$, $\mathcal{B}^\infty = \mathcal{B}^\infty(\mathbf{R}^n)$ the space of all functions defined in $\mathbf{R}^n$ which are bounded together with all their derivatives.

We say that the Cauchy problem (1.1) is well-posed in the space X of functions in $\mathbf{R}^n$ if for every $u_0, u_1 \in X$ there is a unique solution $u \in C^1([0,T];X)$.

It is well known that in the strictly hyperbolic case

$$a(t,x,\xi) \geq a_0|\xi|^2, \qquad a_0 > 0,\ t \in [0,T], x, \xi \in \mathbf{R}^n \tag{1.4}$$

if the coefficients a_{ij} are Lipschitz continuous in the variable t, then the problem (1.1) is well-posed in the Sobolev spaces $H^{-\infty}(\mathbf{R}^n) = \bigcup_s H^s(\mathbf{R}^n)$ and $H^\infty(\mathbf{R}^n) = \bigcap_s H^s(\mathbf{R}^n)$. The C^∞ well-posedness follows by the existence of domains of dependence.

This may fail to be true either for a weakly hyperbolic equation, that is when $a(t,x,\xi) = 0$ at some point $(t,x,\xi), \xi \neq 0$, even if the coefficients are smooth in all variables, or for a strictly hyperbolic equation with non-Lipschitz coefficients. In general, for such types of degenerate hyperbolic equations, one can find well-posedness in suitable Gevrey classes $\gamma^s = \gamma^s(\mathbf{R}^n)$ provided that the coefficients have the same Gevrey regularity with respect to the space variable x. Here γ^s denotes the space of all functions f in $\mathbf{R}^n$ such that

$$|\partial_x^\beta f(x)| \leq CA^{|\beta|}\beta!^s, \quad C, A > 0$$

for all β.

In the weakly hyperbolic case, the C^∞ well-posedness holds for an effectively hyperbolic operator and it is stable under any perturbation of the lower order terms $b(t,x,D_x), c(t,x)$. Otherwise, the first order term $b(t,x,\xi)$ has to satisfy Levi conditions. From [19], the condition

$$|\partial_x^\beta b(t,x,\xi)| \leq C_\beta\sqrt{a(t,x,\xi)}, \qquad t \in [0,T], x, \xi \in \mathbf{R}^n\,,\ \beta \in \mathbf{Z}_+^n \tag{1.5}$$

is sufficient in dimension of space $n = 1$ assuming that the coefficients are analytic functions of the two variables t, x. The same holds true for any $n \geq 1$ with analytic coefficients $a_{ij}(t), b_j(t), c(t)$ depending only on the variable t, see [10].

An intermediate condition between effective hyperbolicity and (1.5) has been introduced in [9]. There the C^∞ well-posedness is proved taking C^∞ functions $a_{ij}(t), b_j(t), c(t)$ of the variable t and assuming that there is an integer $k \geq 2$ such that the symbols $a(t,\xi), b(t,\xi)$ satisfy

$$\sum_{j=0}^{k} |\partial_t^j a(t,\xi)| \neq 0, \quad |b(t,\xi)| \leq Ca(t,\xi)^\gamma, \qquad t \in [0,T], |\xi| = 1, \tag{1.6}$$

with

$$\gamma + \frac{1}{k} \geq \frac{1}{2}. \tag{1.7}$$

Notice that for $a = a(t,\xi)$ independent of x, the effective hyperbolicity is equivalent to

$$a(t,\xi) = 0 \Rightarrow \partial_t^2 a(t,\xi) > 0 \qquad t \in [0,T], |\xi| = 1,$$

that can be expressed also as follows

$$\sum_{j=0}^{2} |\partial_t^j a(t,\xi)| \neq 0, \qquad t \in [0,T], |\xi| = 1.$$

This is in line with the fact that for $k = 2$ the condition (1.7) is satisfied with $\gamma = 0$ (no Levi condition). On the other hand for $\gamma = 1/2$ one can take $k = \infty$ that means that under the Levi condition it is not necessary to assume that $a(t,\xi)$ has only zeros of finite order. Furthermore (1.7) can not be improved since the Cauchy problem for

$$P = D_t^2 - t^{2\ell} D_x^2 + t^\nu D_x \tag{1.8}$$

is well-posed in C^∞ if and only if

$$\nu \geq \ell - 1,$$

see [15].

The dependence on the space variable $x \in \mathbf{R}^n$, $n \geq 1$, of the lower order terms $b(t,x,D_x) + c(t,x)$ is allowed in [12]. There the C^∞ well-posedness is proved under the assumption

$$\sum_{j=0}^{k} |\partial_t^j a(t,\xi)| \neq 0, \quad |\partial_x^\beta b(t,x,\xi)| \leq C_\beta a(t,\xi)^\gamma, \tag{1.9}$$

$$t \in [0,T], x \in \mathbf{R}^n, |\xi| = 1, \ \beta \in \mathbf{Z}_+^n,$$

this time with the larger, for $k > 2$, value of γ

$$\gamma \geq \frac{1}{2} - \frac{1}{2(k-1)}. \tag{1.10}$$

Coming back to the case of coefficients depending only on time, Gevrey well-posedness has been proved in [9] for a Gevrey index

$$s < s_0 := \frac{1-\gamma}{\frac{1}{2} - \left(\gamma + \frac{1}{k}\right)} \tag{1.11}$$

assuming that (1.6) is satisfied with

$$\gamma + \frac{1}{k} < \frac{1}{2}.$$

We have the minimum value $s_0 = 2$ for $\gamma + 1/k = 0$ that is without any further condition besides weak hyperbolicity. This is in line with the results by [3] and [16]. On the other hand, we have $s_0 = +\infty$ for $\gamma + 1/k = 1/2$ as in (1.7) so, in this case, the Cauchy problem is well- posed in C^∞ and in every Gevrey class.

Moreover, condition (1.11) is optimal because for $\nu < \ell - 1$ the Cauchy problem for the operator (1.8) is well-posed in γ^s if and only if

$$s \leq \frac{2\ell - \nu}{\ell - \nu - 1},$$

see [15].

Another kind of degeneracy for the problem (1.1) occurs when the operator P is strictly hyperbolic but the coefficients in the principal term $a(t, x, D_x)$ are not Lipschitz continuous. From the pioneering work [6] and from [11] (see also [1]) we know that the Log-Lipschitz regularity gives the optimal modulus of continuity for the C^∞ well-posedness.

A further way to weaken the Lipschitz regularity has been introduced in [7] starting from the case of coefficients depending only on the time variable, namely the singular behaviour

$$|\partial_t a(t,\xi)| \leq \frac{C}{|t-t_0|^q}, \qquad q \geq 1, \quad t \in [0,T], t \neq t_0, |\xi| = 1, \tag{1.12}$$

of the first derivative as t tends to a point t_0, say $t_0 = 0$.
The optimal exponent for the C^∞ well-posedness is $q = 1$, whereas for $q > 1$ the sharp index of Gevrey well-posedness is given by

$$s < \frac{q}{q-1}. \tag{1.13}$$

The dependence on space variables has been allowed in [4]. In the case $a \in C^2(]0,T]; \mathcal{B}^\infty)$, the sharp bound

$$|\partial_t^j a(t,x,\xi)| \leq C \left(\frac{|\log |t||}{|t|} \right)^j, \quad j = 1, 2, \ t \in [0,T], t \neq 0, x \in \mathbf{R}^n, |\xi| = 1,$$

for C^∞ well-posedness has been established in [8], [18].

Notice that when the coefficients are not smooth in all variables, a general second order operator

$$P = D_t^2 + Q_1(t,x,D_x)D_t + Q_2(t,x,D_x), \tag{1.14}$$

Q_j of order j, $j = 1, 2$, can not be reduced to the canonical form (1.2). A recent counterexample in [14] shows that the factor $|\log |t||$ is not allowed in the general case $Q_1 \neq 0$, in particular the results of [5] for higher order equations are optimal.

The aim of this paper is to consider from a unified point of view the above different degenerate hyperbolic problems, proving the C^∞ or the Gevrey well-posedness by means of an energy estimate. The C^∞ well-posedness has been considered in [2].

In the weakly hyperbolic case we are able to do this assuming that the principal term $a(t, x, D_x)$ in (1.2) is of the form

$$\begin{cases} a(t,x,\xi) = \alpha(t)Q(x,\xi), \quad \alpha \in C^\infty, \\ \alpha(t) \geq 0, \quad Q(x,\xi) = \displaystyle\sum_{i,j=1}^{n} q_{ij}(x)\xi_i\xi_j \geq q_0|\xi|^2, \quad q_0 > 0, \\ t \in [0,T], x, \xi \in \mathbf{R}^n, \end{cases} \tag{1.15}$$

and that there are $\gamma \geq 0$ and an integer $k \geq 2$ such that

$$\begin{gathered} \sum_{h=0}^{k} |\alpha^{(h)}(t)| \neq 0, \quad |\partial_x^\beta b_j(t,x)| \leq C_\beta \alpha(t)^\gamma, \\ j = 1, \dots, n, \quad t \in [0,T], x \in \mathbf{R}^n, \ \beta \in \mathbf{Z}_+^n. \end{gathered} \tag{1.16}$$

For this class of operators we prove the optimal results of well-posedness in C^∞ for $\gamma + 1/k \geq 1/2$ and in γ^s for $\gamma + 1/k < 1/2$, s as in (1.11). In particular, this improves the results of [9] and [12] in the case of dimension $n = 1$.

In the strictly hyperbolic case, we can consider the Cauchy problem for a general operator (1.14). We assume

$$|\partial_t Q_{j,p}(t,x,\xi)| \leq \frac{C}{t^q}, \qquad q \geq 1, \quad t \in]0,T], x \in \mathbf{R}^n, |\xi| = 1, \ j = 1, 2, \tag{1.17}$$

$Q_{j,p}$ the principal part of Q_j, and we obtain the optimal results of well-posedness in C^∞ for $q = 1$ and in γ^s for $q > 1$, $s < q/(q-1)$.

Both in weakly and strictly hyperbolic Cauchy problems, we use the same unified approach that consists of the following steps:

1) Factorization of the principal part of P by means of regularized characteristic roots.

2) Reduction of the equation $Pu = f$ to an equivalent 2×2 system $LU = F$ with

$$L = \partial_t - i\Lambda(t,x,D_x) + A(t,x,D_x) \qquad (i = \sqrt{-1}) \tag{1.18}$$

where $\Lambda(t,x,\xi)$ is a real diagonal matrix of symbols of order 1 and the matrix $A(t,x,\xi)$ satisfies either

$$\int_0^t |A(\tau,x,\xi)|d\tau \leq c_0 + \delta \log\langle\xi\rangle, \quad c_0, \delta > 0, \ t \in [0,T], \ \langle\xi\rangle = (1+|\xi|^2)^{1/2}, \tag{1.19}$$

in the C^∞ case or

$$\int_0^t |A(\tau,x,\xi)|d\tau \leq \delta\langle\xi\rangle^{1/s}, \qquad \delta > 0, \ t \in [0,T] \tag{1.20}$$

in the Gevrey case.

3) Energy estimate in Sobolev or in Gevrey-Sobolev spaces for the operator L. In this step, the bound of the integral of A allows us to use the sharp Gårding inequality after a change of variables that carries a loss of derivatives.

This paper is organized as follows: in Section 2 we prove the energy estimates for an operator L as in (1.18). In Section 3 we take a weakly hyperbolic operator P satisfying (1.15), (1.16) and we reduce the equation $Pu = f$ to a system of the form (1.18). We perform the same reduction for a strictly hyperbolic operator of the general form (1.14) in Section 4 assuming the condition (1.17).

Finally, we would like to mention that microlocal method are used in [13] and [20] to consider some weakly or non-Lipschitz hyperbolic problems from a unified point of view.

2. Energy Estimates

In this section we prove energy estimates for some first order systems. In the next sections we will show that these results can be applied both to weakly hyperbolic and strictly hyperbolic (with non-Lipschitz coefficients) scalar equations.

Let us first consider

$$L = \partial_t - i \begin{pmatrix} \tilde{\lambda}_1(t,x,D_x) & & 0 \\ & \ddots & \\ 0 & & \tilde{\lambda}_\nu(t,x,D_x) \end{pmatrix} + A(t,x,D_x) \tag{2.1}$$

for $t \in [0,T]$, $T > 0$, $x \in \mathbf{R}^n$, where

$$\tilde{\lambda}_j(t,x,\xi) \in \mathbf{R}, \ \tilde{\lambda}_j \in L^1([0,T];S^1), \ j = 1,\dots,\nu, \tag{2.2}$$

A is a $\nu \times \nu$ matrix such that

$$\begin{cases} A \in L^1([0,T];S^1), \\ |A(t,x,\xi)| \le \varphi(t,\xi), \quad \varphi \in L^1([0,T];S^1), \\ \displaystyle\int_0^T \left|\partial_\xi^\alpha \varphi(t,\xi)\right| dt \le \delta_\alpha \langle\xi\rangle^{-|\alpha|} \log\left(1 + \langle\xi\rangle\right), \qquad \alpha \ge 0. \end{cases} \tag{2.3}$$

Here, as usual, $S^m = S^m(\mathbf{R}^n \times \mathbf{R}^n)$ denotes the space of all symbols $a(x,\xi)$ satisfying

$$|\partial_\xi^\alpha \partial_x^\beta a(x,\xi)| \le C_{\alpha\beta} \langle\xi\rangle^{m-|\alpha|},$$

which is the limit space as $\ell \to \infty$ of the Banach spaces S_ℓ^m of all symbols such that

$$|a|_{m,\ell} := \sup_{x,\xi} \sup_{|\alpha|+|\beta| \le \ell} |\partial_\xi^\alpha \partial_x^\beta a(x,\xi)| \langle\xi\rangle^{-m+|\alpha|} < +\infty.$$

For the operator (2.1) we have the following *a priori* estimate:

Theorem 2.1. *Under the hypotheses* (2.2), (2.3) *there exists* $\delta > 0$ *such that the energy estimate*

$$\|U(t)\|_\mu^2 \le C_\mu \left(\|U(0)\|_{\mu+\delta}^2 + \int_0^t \|LU(\tau)\|_{\mu+\delta}^2 d\tau \right) \tag{2.4}$$

holds for all $U \in C^1([0,T]; H^{\mu+\delta}) \cap C([0,T]; H^{\mu+\delta+1})$.

Remark 2.2. This result implies the well-posedness, with a δ-loss of derivatives, in $H^{\pm\infty}$ of the Cauchy problem for the operator L. The positive constant δ is determined by the δ_α's in (2.3) with $|\alpha| \le \ell_0$, ℓ_0 depending only on the dimension n.

Proof. Let us define the operator $w_0(t, D_x)$ with symbol

$$w_0(t,\xi) = e^{-\int_0^t \varphi(\tau,\xi)d\tau} \tag{2.5}$$

and consider the operator

$$L_{w_0} = w_0 L w_0^{-1}. \tag{2.6}$$

We have

$$L_{w_0} = \partial_t - i \begin{pmatrix} \tilde{\lambda}_1 & & 0 \\ & \ddots & \\ 0 & & \tilde{\lambda}_\nu \end{pmatrix} + \varphi(t, D_x)I + A + B^{(1)}, \tag{2.7}$$

where $B^{(1)}$ satisfies

$$\begin{cases} |\partial_\xi^\alpha \partial_x^\beta B^{(1)}(t,x,\xi)| \le b_{\alpha,\beta}^{(1)}(t)\langle\xi\rangle^{-|\alpha|} \log(1+\langle\xi\rangle), \\ b_{\alpha,\beta}^{(1)} \in L^1([0,T]). \end{cases} \tag{2.8}$$

We will refer to the property (2.8) in the following by writing

$$B^{(1)} \in L^1([0,T]; S^0_{\log}),$$
$$|B^{(1)}(t)|_\ell = \sup_{|\alpha+\beta|\le\ell} b_{\alpha,\beta}^{(1)}(t).$$

Moreover, from the sharp Gårding inequality for systems (e.g. [17], Theorem 4.4, page 134) we know that there exist

$$P \in L^1([0,T]; S^1), \quad B^{(2)} \in L^1([0,T]; S^0_{\log})$$

such that

$$\varphi(t,D_x)I + A(t,x,D_x) = P(t,x,D_t,D_x) + B^{(2)}(t,x,D_x) \tag{2.9}$$

with P a positive operator, that is with

$$\langle PU(t), U(t)\rangle \ge 0, \ U \in C([0,T]; H^1) \tag{2.10}$$

where $\langle \cdot, \cdot \rangle$ denotes the scalar product in $H^0(\mathbf{R}^n) = L^2(\mathbf{R}^n)$.

If we take

$$B = B^{(1)} + B^{(2)} \in L^1([0,T]; S^0_{\log})$$

we have from (2.8)

$$|\langle BU(t), U(t)\rangle| \le b(t)\langle \log(1+\langle D_x\rangle)U(t), U(t)\rangle \tag{2.11}$$

with the function $b \in L^1([0,T])$, $b(t) = |B(t)|_{\ell_0}$ and we can write

$$L_{w_0} = \partial_t - i \begin{pmatrix} \tilde{\lambda}_1 & & 0 \\ & \ddots & \\ 0 & & \tilde{\lambda}_\nu \end{pmatrix} + P + B.$$

Now we introduce another change of variable defining

$$\begin{cases} w_1(t, D_x) = (1+\langle D_x\rangle)^{-\int_0^t b(\tau)d\tau}, \\ w = w_1 w_0, \quad L_w = wLw^{-1}. \end{cases} \tag{2.12}$$

A calculation of L_w gives

$$L_w = \partial_t - i \begin{pmatrix} \tilde{\lambda}_1 & & 0 \\ & \ddots & \\ 0 & & \tilde{\lambda}_\nu \end{pmatrix} + P + B + b(t)\log(1+\langle D_x\rangle)I + R, \tag{2.13}$$

$$R \in L^1([0,T]; S^0),$$

where now, from (2.11), also $B + b(t)\log(1+\langle D_x\rangle)I$ is a positive operator.

Let us consider for $\mu = 0$

$$\frac{d}{dt}\|U(t)\|_0^2 = 2\Re\langle U'(t), U(t)\rangle.$$

From (2.13), (2.10), (2.11) we have

$$\frac{d}{dt}\|U(t)\|_0^2 \le \beta(t)\|U(t)\|_0^2 + C\|L_w U(t)\|_0^2$$

with a function $\beta \in L^1([0,T])$. So by Gronwall's inequality we find

$$\|U(t)\|_0^2 \le C_0\left(\|U(0)\|_0^2 + \int_0^t \|L_w U(\tau)\|_0^2 d\tau\right). \tag{2.14}$$

In order to generalize (2.14) to the case $\mu > 0$, we only need to notice that for each μ

$$\langle D_x\rangle^\mu L_w \langle D_x\rangle^{-\mu} = L_w + R_\mu, \quad R_\mu \in L^1([0,T]; S^0).$$

So we have also

$$\|U(t)\|_\mu^2 \le C_\mu\left(\|U(0)\|_\mu^2 + \int_0^t \|L_w U(\tau)\|_\mu^2 d\tau\right) \tag{2.15}$$

which gives (2.4) since, from (2.3) and (2.12), there are positive δ, C such that

$$\|wV(t)\|_\mu^2 \ge C\|V(t)\|_{\mu-\delta}^2$$

for all $V \in C^0([0,T]; H^\mu(\mathbf{R})^n)$. □

Now we give an estimate in Sobolev-Gevrey spaces

$$H^{\mu,\lambda,s}(\mathbf{R}^n) = e^{-\lambda\langle D_x\rangle^{1/s}} H^{\mu}(\mathbf{R}^n),\ \lambda > 0,\ s > 1,$$

with norm

$$\|u\|_{\mu,\lambda,s} = \|e^{\lambda\langle D_x\rangle^{1/s}} u\|_{\mu},$$

for a system as in (2.1) but with symbols

$$\tilde{\lambda}_j, A \in L^1([0,T]; S^{1,s}), \tag{2.16}$$

where $S^{m,s} = S^{m,s}(\mathbf{R}^n \times \mathbf{R}^n)$ denotes the space of all symbols $a(x,\xi)$ satisfying

$$|\partial_\xi^\alpha \partial_x^\beta a(x,\xi)| \leq C_\alpha M^{|\beta|} \beta!^s \langle\xi\rangle^{m-|\alpha|},$$

which is the limit space

$$S^{m,s} := \lim_{\substack{\longleftarrow \\ \ell\to+\infty}} S_\ell^{m,s}, \quad S_\ell^{m,s} := \lim_{\substack{\longrightarrow \\ M\to+\infty}} S_{\ell,M}^{m,s}$$

of the Banach spaces $S_{M,\ell}^{m,s}$ of all symbols such that

$$|a|_{m,s,M,\ell} := \sup_{|\alpha|\leq\ell, \beta\in\mathbf{Z}_+^n} \sup_{x,\xi} |\partial_\xi^\alpha \partial_x^\beta a(x,\xi)| M^{-|\beta|} \beta!^{-s} \langle\xi\rangle^{-m+|\alpha|} < +\infty.$$

Condition (2.3) is replaced by

$$\begin{cases} |A(t,x,\xi)| \leq \varphi(t,\xi), \quad \varphi \in L^1([0,T]; S^1), \\ \displaystyle\int_0^t \varphi(\tau,\xi)d\tau \in C([0,T]; S^{1/s}). \end{cases} \tag{2.17}$$

Theorem 2.3. *Under the hypothesis* (2.16) *and* (2.17) *there exist* $\lambda_0 > 0$ *and* $b(t) \in L^1[0,T]$, $b(t) \geq 0$, *such that for*

$$w(t,\xi) = \exp\left(\int_0^t \left(\varphi(\tau,\xi) + b(\tau)\langle\xi\rangle^{1/s}\right) d\tau\right) \tag{2.18}$$

the energy estimate

$$\begin{gathered} \|w(t,D_x)U(t)\|_{\mu,\lambda,s}^2 \leq C_\mu \left(\|U(0)\|_{\mu,\lambda,s}^2 + \int_0^t \|w(\tau,D_x)LU(\tau)\|_{\mu,\lambda,s}^2 d\tau\right), \\ 0 \leq t \leq T^*,\ w(T^*,\xi) \leq e^{\lambda\langle\xi\rangle^{1/s}}, \end{gathered} \tag{2.19}$$

holds for all $U \in C^1([0,T]; H^{\mu,\lambda,s}) \cap C([0,T]; H^{\mu+1,\lambda,s})$, $0 < \lambda \leq \lambda_0$.

Proof. For $w(t,\xi)$ given by (2.18) with $b(t)$ to be chosen later on, let us consider the operator

$$L_w = e^{\lambda\langle D_x\rangle^{1/s}} w^{-1} L w e^{-\lambda\langle D_x\rangle^{1/s}}. \tag{2.20}$$

From [16], there is $\lambda_0 > 0$ such that for $0 \le \lambda \le \lambda_0$ we have

$$L_w = \partial_t - i \begin{pmatrix} \tilde{\lambda}_1 & & 0 \\ & \ddots & \\ 0 & & \tilde{\lambda}_\nu \end{pmatrix} + \varphi(t, D_x) + A + b(t)\langle D_x\rangle^{1/s} + B, \tag{2.21}$$

where $B \in L^1([0,T]; S^{1/s})$ has seminorms such that

$$|B(t)|_\ell \le b_\ell(t), \;\; b_\ell \in L^1([0,T]) \tag{2.22}$$

with functions b_ℓ determined by $|\lambda_j(t)|_{1,s,M,\ell'}$, $|A(t)|_{1/s,s,M,\ell'}$, $\ell' = \ell'(\ell, n)$, that can be chosen independent of $b(t)$ provided that

$$\int_0^t \left(\varphi(\tau,\xi) + b(\tau)\langle\xi\rangle^{1/s}\right) d\tau \le \lambda\langle\xi\rangle^{1/s}, \;\; 0 \le \lambda \le \lambda_0.$$

To the operator $\varphi(t, D_x) + A$ we can apply the sharp Gårding inequality whereas, taking

$$b(t) = b_{\ell_0}(t), \;\; \ell_0 = \ell_0(n),$$

we have

$$|\langle BU(t), U(t)\rangle| \le b(t)\langle\langle D_x\rangle^{1/s}U(t), U(t)\rangle. \tag{2.23}$$

So, from (2.21), we obtain

$$\|U(t)\|_0^2 \le C_0\left(\|U(0)\|_0^2 + \int_0^t \|L_w U(\tau)\|_0^2 d\tau\right) \tag{2.24}$$

by Gronwall's method.

In order to generalize (2.24) to the case $\mu > 0$, we only need to notice that for each μ

$$\langle D_x\rangle^\mu L_w \langle D_x\rangle^{-\mu} = L_w + R_\mu, \;\; R_\mu \in L^1([0,T]; S^0).$$

So we have also

$$\|U(t)\|_\mu^2 \le C_\mu\left(\|U(0)\|_\mu^2 + \int_0^t \|L_w U(\tau)\|_\mu^2 d\tau\right) \tag{2.25}$$

which gives (2.19) for the operator L. □

3. Weakly Hyperbolic Equations

In this section we consider the Cauchy problem

$$\begin{cases} P(t,x,D_t,D_x)u(t,x) = 0, & (t,x) \in [0,T] \times \mathbf{R}^n, \\ u(0,x) = u_0(x), \;\; \partial_t u(0,x) = u_1(x), \end{cases} \tag{3.1}$$

for the second order operator

$$P = D_t^2 - a(t)Q(x, D_x) + b(t,x,D_x) + c(t,x) \tag{3.2}$$

where

$$\begin{cases} \alpha \in C^\infty([0,T];\mathbf{R}), \\ Q(x,\xi) = \displaystyle\sum_{i,j=1}^{n} q_{ij}(x)\xi_i\xi_j, \quad q_{ij} \in \mathcal{B}^\infty(\mathbf{R}^n;\mathbf{R}), \\ b(t,x,\xi) = \displaystyle\sum_{j=1}^{n} b_j(t,x)\xi_j, \quad b_j \in C([0,T];\mathcal{B}^\infty(\mathbf{R}^n)). \end{cases} \tag{3.3}$$

Here the weak hyperbolicity of P is expressed by

$$\alpha(t) \geq 0, \quad Q(x,\xi) \geq q_0|\xi|^2, \; q_0 > 0, \tag{3.4}$$

and, following [9], we are going to impose an intermediate condition between the effective hyperbolicity and the Levi condition. Notice that for this class of operators, the effective hyperbolicity is equivalent to

$$\alpha(t) = 0 \Rightarrow \alpha''(t) > 0$$

that can be expressed also as follows

$$\sum_{h=0}^{2} |\alpha^{(h)}(t)| \neq 0.$$

We have the following result concerning C^∞ well-posedness:

Theorem 3.1. *Let us consider the operator* (3.2) *under condition* (3.4) *and let us assume that for an integer $k \geq 2$ and $\gamma \geq 0$*

$$\sum_{h=0}^{k} |\alpha^{(h)}(t)| \neq 0, \quad t \in [0,T], \tag{3.5}$$

$$\begin{cases} |\partial_x^\beta b_j(t,x)| \leq C_\beta \alpha^\gamma(t), \quad \gamma + \dfrac{1}{k} \geq \dfrac{1}{2}, \\ t \in [0,T], x \in \mathbf{R}^n, \; j = 1,\ldots,n. \end{cases} \tag{3.6}$$

Then the Cauchy problem (3.1) *is well-posed in C^∞.*

Proof. Our aim is to reduce the scalar equation $Pu = f$ to an equivalent system $LU = F$ with L that fulfills all the assumptions in Theorem 2.1.

The first step is to factorize the principal part of P by means of the approximated characteristic root

$$\tilde{\lambda}(t,x,\xi) = \sqrt{\alpha(t) + \langle\xi\rangle^{-2}}\sqrt{Q(x,\xi)}. \tag{3.7}$$

We have

$$\begin{aligned} D_t^2 - \alpha(t)Q(x,D_x) = \; & (D_t - \tilde{\lambda}(t,x,D_x))(D_t + \tilde{\lambda}(t,x,D_x)) \\ & + S(t,x,D_x) + R(t,x,D_x), \end{aligned} \tag{3.8}$$

where

$$S(t,x,\xi)=\frac{i\alpha'(t)}{2\sqrt{\alpha(t)+\langle\xi\rangle^{-2}}}\sqrt{Q(x,\xi)}+\frac{\alpha(t)}{4Q(x,\xi)}\sum_{j=1}^{n}\partial_{\xi_j}Q(x,\xi)D_{x_j}Q(x,\xi),$$

$R(t,x,\xi)\in C([0,T];S^0)$.

Now, in order to reduce the scalar operator (3.2) to an equivalent system operator of the first order, we define

$$\omega(t,\xi)=\sqrt{1+\alpha(t)\langle\xi\rangle^2} \tag{3.9}$$

and

$$\begin{cases} u_0=\omega(t,D_x)u \\ u_1=(D_t+\tilde\lambda(t,x,D_x))u. \end{cases} \tag{3.10}$$

Notice that

$$\omega\in C([0,T];S^1),\quad \omega^{-1}\in C([0,T];S^0),\quad \sqrt{\alpha}\omega^{-1}\in C([0,T];S^{-1}).$$

We have

$$(D_t+\tilde\lambda)u_0=\omega u_1+i2^{-1}\alpha'(t)\langle D_x\rangle^2\omega^{-2}u_0+[\tilde\lambda,\omega]\omega^{-1}u_0, \tag{3.11}$$

where the operator $[\tilde\lambda,\omega]\omega^{-1}$ is of order 0 because

$$[\tilde\lambda,\omega]\omega^{-1}=[\sigma,\omega],\quad \sigma(x,\xi)=\langle\xi\rangle^{-1}\sqrt{Q(x,\xi)}.$$

Then, from (3.3), (3.8), (3.10), (3.11) and (3.9), the problem (3.1) for the operator (3.2) is equivalent to the Cauchy problem

$$L_1U=0,\quad U(0,x)=U_0, \tag{3.12}$$

with

$$U=(u_0,u_1)$$

for the first order system

$$L_1=\partial_t+\begin{pmatrix} i\tilde\lambda(t,x,D_x) & -i\omega(t,D_x) \\ 0 & -i\tilde\lambda(t,x,D_x)\end{pmatrix}$$

$$+A_1(t,x,D_x)\alpha'(t)\langle D_x\rangle^2\omega^{-2}(t,D_x)+B_1(t,x,D_x)b(t,x,D_x)\omega^{-1}(t,D_x)$$

$$+R_1(t,x,D_x),$$

where $A_1(t,x,\xi),B_1(t,x,\xi),R_1(t,x,\xi)$ are 2×2 matrices such that

$$A_1,B_1,R_1\in C^0([0,T];S^0).$$

The matrix

$$\begin{pmatrix} i\tilde\lambda(t,x,\xi) & -i\omega(t,\xi) \\ 0 & -i\tilde\lambda(t,x,\xi)\end{pmatrix}$$

can be diagonalized by

$$M(x,\xi) = \begin{pmatrix} 1 & \dfrac{\langle\xi\rangle}{2\sqrt{Q(x,\xi)}} \\ 0 & 1 \end{pmatrix}$$

which is elliptic of order zero, so problem (3.12) is equivalent to

$$\begin{cases} LU = 0 \\ U(0,x) = \tilde{U}_0, \end{cases} \tag{3.13}$$

where

$$L = \partial_t + \begin{pmatrix} i\tilde{\lambda}(t,x,D_x) & 0 \\ 0 & -i\tilde{\lambda}(t,x,D_x) \end{pmatrix}$$

$$+A(t,x,D_x)\alpha'(t)\langle D_x\rangle^2\omega^{-2}(t,D_x) + B(t,x,D_x)b(t,x,D_x)\omega^{-1}(t,D_x)$$

$$+R(t,x,D_x),$$

with new 2×2 matrices

$$A, B, R \in C^0([0,T]; S^0).$$

We have that

$$\alpha'(t)\langle\xi\rangle^2\omega^{-2}(t,\xi) = \frac{\alpha'(t)}{(\alpha(t)+\langle\xi\rangle^{-2})^{1-1/N}} \cdot \frac{1}{(\alpha(t)+\langle\xi\rangle^{-2})^{1/N}}$$

belongs to $\in L^1([0,T]; S^{2/N})$ for every $N \geq 2$, so in particular to $\in L^1([0,T]; S^1)$, since $\alpha^{1/N}$ is an absolutely continuous function in view of Lemma 1 in [10].

Then we notice that from (3.6) with $\gamma = 1/2 - 1/k$

$$\begin{aligned} b(t,x,\xi)\omega^{-1}(t,\xi) &= \frac{b(t,x,\xi)}{\langle\xi\rangle\sqrt{\alpha(t)+\langle\xi\rangle^{-2}}} \\ &= \frac{b(t,x,\xi)}{\langle\xi\rangle(\alpha(t)+\langle\xi\rangle^{-2})^{\gamma}} \cdot \frac{1}{(\alpha(t)+\langle\xi\rangle^{-2})^{1/k}} \end{aligned}$$

so, in particular, $b(t,x,\xi)\omega^{-1}(t,\xi) \in C([0,T]; S^{2/k})$, $k \geq 2$, because (3.6) implies

$$\frac{b(t,x,\xi)}{\langle\xi\rangle(\alpha(t)+\langle\xi\rangle^{-2})^{\gamma}} \in C([0,T]; S^0).$$

Now, in order to apply Theorem 2.1 to the operator L , we need to check that

$$\varphi(t,\xi) = C\frac{|\alpha'(t)|}{\alpha(t)+\langle\xi\rangle^{-2}} + C\frac{1}{(\alpha(t)+\langle\xi\rangle^{-2})^{1/k}},$$

C a sufficiently large constant, satisfies the last condition in (2.3).

As simple particular cases of Lemma 1 and Lemma 2 of [9] we have

$$\begin{cases} \displaystyle\int_0^T \frac{|\alpha'(t)|}{\alpha(t)+\langle\xi\rangle^{-2}}dt \le c_0 + \delta\log\langle\xi\rangle, \\ \displaystyle\int_0^T \frac{1}{(\alpha(t)+\langle\xi\rangle^{-2})^{1/k}}dt \le c_0 + \delta\log\langle\xi\rangle, \end{cases} \tag{3.14}$$

for some positive constants c_0, δ. For instance, to estimate the second integral, one uses that α has isolated zeros of order less or equal to k and in a neighborhood of such a zero one just takes into account that

$$\begin{aligned} \int_0^T \frac{1}{(t^k+\langle\xi\rangle^{-2})^{1/k}}dt &\le \int_0^{\langle\xi\rangle^{-2/k}} \frac{1}{\langle\xi\rangle^{-2/k}}dt + \int_{\langle\xi\rangle^{-2/k}}^T \frac{1}{t}dt \\ &= 1 + \log\frac{T}{\langle\xi\rangle^{-2/k}}. \end{aligned}$$

The same arguments can be applied to all the derivatives $\partial_\xi^\beta\varphi$ after having noticed that

$$\partial_\xi^\beta\varphi(t,\xi) = q_1^{(\beta)}(t,\xi)\frac{|\alpha'(t)|}{\alpha(t)+\langle\xi\rangle^{-2}} + q_2^{(\beta)}(t,\xi)\frac{1}{(\alpha(t)+\langle\xi\rangle^{-2})^{1/k}},$$

$$q_1^{(\beta)}, q_2^{(\beta)} \in C([0,T];S^{-|\beta|}),$$

thus problem (3.13) satisfies all the assumptions (2.1), (2.2), (2.3).

The proof is complete in view of Theorem 2.1 and the equivalence between (3.13) and (3.1). □

We can also prove Gevrey well-posedness. Let us take Gevrey regularity

$$q_{ij} \in \gamma^s(\mathbf{R}^n;\mathbf{R}),\ b_j \in C([0,T];\gamma^s(\mathbf{R}^n)). \tag{3.15}$$

in (3.3). We have:

Theorem 3.2. *Let us consider the operator* (3.2) *under conditions* (3.4), (3.15) *and* (3.5) *with* $k>2$. *If for* $\gamma \ge 0$

$$\begin{cases} |\partial_x^\beta b_j(t,x)| \le CM^{|\beta|}\beta!^s\alpha^\gamma(t), \quad \gamma + \dfrac{1}{k} < \dfrac{1}{2}, \\ t\in[0,T], x\in\mathbf{R}^n,\ j=1,\dots,n, \end{cases} \tag{3.16}$$

then the Cauchy problem (3.1) *is well-posed in* γ^s *for* $1 < s < \dfrac{1-\gamma}{\frac{1}{2}-(\gamma+\frac{1}{k})}$.

Proof. Here the scalar equation $Pu=f$ is reduced to an equivalent system $LU=F$ with L that fulfills all the assumptions in Theorem 2.3. After having defined

$$\tilde\lambda(t,x,\xi) = \sqrt{\alpha(t)+\langle\xi\rangle^{-1/(1-\gamma)}}\sqrt{Q(x,\xi)} \tag{3.17}$$

and

$$\omega(t,\xi) = \langle\xi\rangle\sqrt{\alpha(t) + \langle\xi\rangle^{-1/(1-\gamma)}} \tag{3.18}$$

(see (3.7) and (3.9)), one uses the same arguments of the proof of Theorem 3.1 in the frame of the calculus of Gevrey symbols $S^{m,s}$.

Then one applies again Lemma 1 and Lemma 2 of [9] to have

$$\begin{cases} \displaystyle\int_0^T \frac{|\alpha'(t)|}{\alpha(t) + \langle\xi\rangle^{-1/(1-\gamma)}}dt \le c_0 + \delta\log\langle\xi\rangle, \\ \displaystyle\int_0^T \frac{1}{\left(\alpha(t) + \langle\xi\rangle^{-1/(1-\gamma)}\right)^{1/2-\gamma}}dt \le \lambda\langle\xi\rangle^{1/s_0}, \\ \displaystyle\langle\xi\rangle^{-\gamma/(1-\gamma)}\int_0^T \frac{1}{\sqrt{\alpha(t) + \langle\xi\rangle^{-1/(1-\gamma)}}}dt \le \lambda\langle\xi\rangle^{1/s_0}, \end{cases} \tag{3.19}$$

with

$$s_0 = \frac{1-\gamma}{\frac{1}{2} - \left(\gamma + \frac{1}{k}\right)}.$$

Theorem 2.3 gives the γ^s well-posedness of the Cauchy problem (3.1). □

4. Strictly Hyperbolic Equations

In this section we consider the Cauchy problem

$$\begin{cases} P(t,x,D_t,D_x)u(t,x) = 0, \quad (t,x) \in [0,T]\times\mathbf{R}^n, \\ u(0,x) = u_0(x), \quad \partial_t u(0,x) = u_1(x), \end{cases} \tag{4.1}$$

for the general second order operator

$$P = D_t^2 + 2Q_1(t,x,D_x)D_t - Q_2(t,x,D_x) \tag{4.2}$$

with symbols

$$Q_1 \in C([0,T];S^1),\ Q_2 \in C([0,T];S^2) \tag{4.3}$$

under the assumption of strict hyperbolicity

$$Q_{1,p}^2(t,x,\xi) + Q_{2,p}(t,x,\xi) \ge c_0|\xi|^2,\ c_0 > 0,\ |\xi| \text{ large}, \tag{4.4}$$

where $Q_{j,p}$ denotes the principal symbol of Q_j, $j = 1,2$.

Here the degeneracy of the problem (4.1) comes from the low regularity of the symbols with respect to t. Precisely, we assume

$$t^q\partial_t Q_{j,p} \in B^0([0,T];S^j),\ q \ge 1,\ j = 1,2. \tag{4.5}$$

Following [4] and [5], we show that Theorem 2.1 implies the C^∞ well-posedness of problem (4.1) in the case $q = 1$, whereas Theorem 2.3 gives the γ^s well-posedness of problem (4.1) for $q > 1$, $s < q/(q-1)$.

Theorem 4.1. *Let us consider the operator P given by (4.2) and (4.3) under conditions (4.4) and (4.5) with $q = 1$.*
Then the Cauchy problem (4.1) is well-posed in C^∞.

Proof. Also here we reduce the scalar equation $Pu = f$ to an equivalent system $LU = F$ with L that fulfills all the assumptions in Theorem 2.1.

The first step is again to factorize the principal part of P by means of approximated characteristic roots. Let us denote

$$\lambda_j(t, x, \xi), \; j = 1, 2, \tag{4.6}$$

the roots of the principal symbol of P and let us introduce the mollified roots

$$\tilde{\lambda}_j(t, x, \xi) = \int \lambda_j(\tau, x, \xi)\varrho((t-\tau)\langle\xi\rangle)\langle\xi\rangle d\tau \tag{4.7}$$

with $\varrho \in C_0^\infty(\mathbf{R})$, $0 \leq \varrho \leq 1$, $\int \varrho(\tau)d\tau = 1, \lambda_j(\tau, x, \xi) = \lambda_j(T, x, \xi)$, for $\tau \geq T$, $\lambda_j(\tau, x, \xi) = \lambda_j(0, x, \xi)$, for $\tau \leq 0$.
Under condition (4.5) with $q = 1$, the symbols

$$A_j^{(0)} = \lambda_j - \tilde{\lambda}_j, \;\; A_j^{(1)} = \langle\xi\rangle^{-1}\partial_t\tilde{\lambda}_j \tag{4.8}$$

fulfill all the conditions in (2.3). In fact we have both

$$A_j^{(k)} \in C([0, T]; S^1)$$

and

$$tA_j^{(k)} \in C([0, T]; S^0)$$

thus we have $|A_j^{(k)}(t, x, \xi)| \leq \varphi(t, \xi)$ choosing

$$\varphi(t, \xi) = \psi(t\langle\xi\rangle)\delta\langle\xi\rangle + (1 - \psi(t\langle\xi\rangle))\delta/t,$$

$$\psi(y) = 1 \text{ for } |y| \leq 1, \quad \psi(y) = 0 \text{ for } |y| \geq 2$$

with a large $\delta > 0$ and a smooth function ψ, $0 \leq \psi \leq 1$. Also the other conditions in (2.3) are fulfilled, in particular

$$\int_0^T \varphi(t, \xi)dt \leq \delta \int_0^{2\langle\xi\rangle^{-1}} \langle\xi\rangle dt + \delta \int_{2\langle\xi\rangle^{-1}}^T (1/t)dt \leq c_0 + \delta \log\langle\xi\rangle.$$

So we can factorize P as follows

$$\begin{aligned} P(t, x, D_t, D_x) = \;\; & (D_t - \tilde{\lambda}_2(t, x, D_x))(D_t - \tilde{\lambda}_1(t, x, D_x)) \\ & + R_0(t, x, D_x)\langle D_x\rangle + R_1(t, x, D_x)D_t \end{aligned}$$

with R_j satisfying (2.3) for $j = 0, 1$.
Then, given a scalar function $u(t, x)$, we define

$$u_0 = \langle D_x\rangle u, \;\; u_1 = (D_t - \tilde{\lambda}_1(t, x, D_x))u, \tag{4.9}$$

so that the problem (4.1) for the operator (4.2) is equivalent to the Cauchy problem

$$L_1 U = 0, \quad U(0,x) = U_0, \tag{4.10}$$

with

$$U = (u_0, u_1)$$

for the first order system

$$L_1 = \partial_t - \begin{pmatrix} i\tilde{\lambda}_1(t,x,D_x) & i\langle D_x\rangle \\ 0 & i\tilde{\lambda}_2(t,x,D_x) \end{pmatrix} + A_1(t,x,D_x),$$

where $A_1(t,x,\xi)$ is a 2×2 matrix that satisfies (2.3).

The triangular matrix

$$\begin{pmatrix} \tilde{\lambda}_1(t,x,\xi) & \langle\xi\rangle \\ 0 & \tilde{\lambda}_2(t,x,\xi) \end{pmatrix}$$

can be diagonalized by

$$M(t,x,\xi) = \begin{pmatrix} 1 & \dfrac{\langle\xi\rangle}{\tilde{\lambda}_2(t,x,\xi) - \tilde{\lambda}_1(t,x,\xi)} \\ 0 & 1 \end{pmatrix}$$

which is elliptic of order zero.
Notice that also $\partial_t M$ and $\partial_t M^{-1}$ fulfill (2.3), so problem (4.10) is equivalent to

$$LU = 0, \quad U(0,x) = \tilde{U}_0, \tag{4.11}$$

where

$$L = \partial_t - \begin{pmatrix} i\tilde{\lambda}_1(t,x,D_x) & 0 \\ 0 & i\tilde{\lambda}_2(t,x,D_x) \end{pmatrix} + A(t,x,D_x) \tag{4.12}$$

with a new 2×2 matrix $A(t,x,\xi)$ that still satisfies (2.3).

The proof is complete in view of Theorem 2.1 and the equivalence between (4.11) and (4.1). □

We give now a result of Gevrey well-posedness under the assumptions

$$Q_1 \in C([0,T]; S^{1,s}), \; Q_2 \in C([0,T]; S^{2,s}), \tag{4.13}$$

$$t^q \partial_t Q_{j,p} \in B^0([0,T]; S^{j,s}), \; q > 1, \; j = 1,2. \tag{4.14}$$

Theorem 4.2. *Let us consider the operator given by* (4.2) *and* (4.13) *under conditions* (4.4) *and* (4.14).
Then the Cauchy problem (4.1) *is well-posed in* γ^s *for* $s < q/(q-1)$.

Proof. Here we reduce the scalar equation $Pu = f$ to an equivalent system $LU = F$ with L that fulfills all the assumptions in Theorem 2.3.

With the same notation of the proof of Theorem 4.1, for the symbols

$$A_j^{(0)} = \lambda_j - \tilde{\lambda}_j, \quad A_j^{(1)} = \langle\xi\rangle^{-1}\partial_t\tilde{\lambda}_j$$

we have both

$$A_j^{(k)} \in C([0,T]; S^{1,s}) \tag{4.15}$$

and

$$t^q A_j^{(k)} \in C([0,T]; S^{0,s}) \tag{4.16}$$

thus we obtain

$$t^{q(1-1/s)} A_j^{(k)} \in C([0,T]; S^{1/s,s})$$

using (4.15) for $t^q\langle\xi\rangle \leq 1$ and (4.16) for $t^q\langle\xi\rangle \geq 1$. Since

$$q(1-1/s) < 1,$$

this gives

$$A_j^{(k)} \in L^1([0,T]; S^{1/s,s}),$$

in particular the symbols $A_j^{(k)}$ satisfy (2.17).

We can now perform the same reduction and diagonalization procedure as in the proof of Theorem 4.1, here in the frame of the symbolic calculus in classes $S^{m,s}$. Then, Theorem 2.3 gives the γ^s well-posedness of the Cauchy problem (4.1) for $s < q/(q-1)$. □

References

[1] R. Agliardi and M. Cicognani, The Cauchy problem for a class of Kovalevskian pseudo-differential operators. *Proc. Amer. Math. Soc.* **132** (2004), 841–845.

[2] A. Ascanelli and M. Cicognani, Energy estimate and fundamental solution for degenerate hyperbolic Cauchy problems, *J. Diff. Equat.*, to appear.

[3] M. D. Bronšteĭn, The Cauchy problem for hyperbolic operators with characteristics of variable multiplicity, *Trudy Moskov. Mat. Obshch.* **41** (1980), 83–99.

[4] M. Cicognani, The Cauchy problem for strictly hyperbolic operators with non-absolutely continuous coefficients, *Tsukuba J. Math.* **27** (2003), 1–12.

[5] M. Cicognani, Coefficients with unbounded derivatives in hyperbolic equations, *Math. Nachr.* **277** (2004), 1–16.

[6] F. Colombini, E. De Giorgi and S. Spagnolo, Sur les équations hyperboliques avec des coefficients qui ne dépendent que du temps, *Ann. Scuola Norm. Sup. Pisa* **6** (1979), 511–559.

[7] F. Colombini, D. Del Santo and T. Kinoshita, Well-posedness of the Cauchy problem for a hyperbolic equation with non-Lipschitz coefficients, *Ann. Scuola Norm. Sup. Pisa* **Cl. Sci. 1** (2002), 327–358.

[8] F. Colombini, D. Del Santo and M. Reissig, On the optimal regularity coefficients in hyperbolic Cauchy problem, *Bull. Sci. Math.* **127** (2003), 328–347.

[9] F. Colombini, H. Ishida and N. Orru, On the Cauchy problem for finitely degenerate hyperbolic equations of second order, *Ark. Mat.* **38** (2000), 223–230.

[10] F. Colombini, E. Jannelli and S. Spagnolo, Well-posedness in Gevrey classes of the Cauchy problem for a non strictly hyperbolic equation with coefficients depending on time, *Ann. Scuola Norm. Sup. Pisa* **Cl. Sci. 10** (1983), 291–312.

[11] F. Colombini and N. Lerner, Hyperbolic operators with non-Lipschitz coefficients, *Duke Math. J.* **77** (1995), 657–698.

[12] F. Colombini and T. Nishitani, On finitely degenerate hyperbolic operators of second order, *Osaka J. Math.* **41** (2004), 933–947.

[13] F. Hirosawa, Loss of regularity for second order hyperbolic equations with singular coefficients, Preprint.

[14] F. Hirosawa and M. Reissig, About the optimality of oscillations in non-Lipschitz coefficients for strictly hyperbolic equations, *Ann. Scuola Normale Sup. Pisa* **8** (2004), 589–608.

[15] V. Ja. Ivriĭ, Conditions for correctness in Gevrey classes of the Cauchy problem for hyperbolic operators with characteristics of variable multiplicity, (Russian) *Sibirsk. Mat. Ž.* **17** (1976), 1256–1270.

[16] K. Kajitani, Cauchy problem for nonstrictly hyperbolic systems in Gevrey classes, *J. Math. Kyoto Univ.* **23** (1983), 599–616.

[17] H. Kumano-go, *Pseudo-Differential Operators*, MIT Press, Cambridge, Massachusetts, and London, England, 1981.

[18] A. Kubo and M. Reissig, Construction of parametrix for hyperbolic equations with fast oscillations in non-Lipschitz coefficients, *Comm. Partial Differential Equations* **28** (2003), 1471–1502.

[19] T. Nishitani, The Cauchy problem for weakly hyperbolic equations of second order, *Comm. Partial Differential Equations* **5** (1980), 1273-1296.

[20] M. Reissig, A refined diagonalization procedure to handle fast oscillations in degenerate hyperbolic problems, in *Proc. Conference Hyperbolic problems and related topics - Cortona 2002*, Editors: F. Colombini and T. Nishitani, International Press, Somerville, 2003.

Alessia Ascanelli
Dipartimento di Matematica, Università di Ferrara
Via Machiavelli 35, 44100 Ferrara, Italy
e-mail: alessia.ascanelli@unife.it

Massimo Cicognani
Dipartimento di Matematica, Università di Bologna
Piazza di Porta S. Donato 5, 40127 Bologna, Italy
and
Facoltà di Ingegneria II, Via Genova 181, 47023 Cesena, Italy
e-mail: cicognani@dm.unibo.it

Operator Theory:
Advances and Applications, Vol. 164, 43–52

Quasilinear Hyperbolic Equations with SG-Coefficients

Marco Cappiello and Luisa Zanghirati

Abstract. We study some classes of quasilinear symmetric hyperbolic systems with coefficients of SG type. We obtain results of existence and uniqueness for the solution in some weighted Sobolev spaces.

Mathematics Subject Classification (2000). Primary 35L60; Secondary 35S05.

Keywords. Quasilinear hyperbolic equations, pseudo-differential operators.

1. Introduction

The aim of this work is to investigate existence, uniqueness and regularity for the solutions of a class of quasilinear hyperbolic systems globally defined in the space variables. Quasilinear hyperbolic systems appear in the literature in a very large number of works with many applications in mathematical physics. Here we consider systems of the form

$$\begin{cases} \frac{\partial u}{\partial t} = K(t,x,u,D)u + f(t,x,u) & (t,x) \in [-T,T] \times \mathbb{R}^n \\ u(0,x) = g(x) & x \in \mathbb{R}^n \end{cases} \tag{1.1}$$

where

$$K(t,x,u,D) = \sum_{j=1}^{n} A_j(t,x,u) \frac{\partial}{\partial x_j}$$

and A_j are $n \times n$ matrices, smooth in their arguments. We will assume (1.1) to be symmetric hyperbolic imposing that A_j are symmetric:

$$A_j = A_j^{\star}. \tag{1.2}$$

Similar systems have been considered in [6], [7] assuming the coefficients uniformly bounded in $\mathbb{R}^n$ with respect to the space variables x. Here we are interested to the case in which (1.1) is a SG-hyperbolic system of order $(1,1)$, cf. Section 2.

Namely, we assume that, for every $(t,u) \in [-T,T] \times \mathbb{R}^n$, the entries of A_j satisfy an estimate of the form

$$\left|D_x^\beta D_u^\gamma A_j(t,x,u)\right| \le C_{\beta\gamma}\langle x\rangle^{1-|\beta|} \tag{1.3}$$

for all $\beta,\gamma \in \mathbb{Z}_+^n$ and for some positive constant $C_{\beta\gamma}$ smoothly depending on (t,u), with the notation $\langle x\rangle = (1+|x|^2)^{\frac{1}{2}}$. Linear symmetrizable SG-hyperbolic systems have been studied by H.O. Cordes [2] who proved the well-posedness in the Schwartz spaces $\mathcal{S}(\mathbb{R}^n), \mathcal{S}'(\mathbb{R}^n)$ and in the weighted Sobolev spaces $H_s, s = (s_1,s_2) \in \mathbb{R}^2$ defined as follows:

$$H_s = \{u \in \mathcal{S}'(\mathbb{R}^n) : \Pi_s u \in L^2(\mathbb{R}^n)\} \tag{1.4}$$

where Π_s denotes the pseudo-differential operator with symbol $\langle\xi\rangle^{s_1}\langle x\rangle^{s_2}$. Such spaces coincide with the usual Sobolev spaces for $s_2 = 0$. The results above have been extended to weakly hyperbolic systems with Levi conditions by S. Coriasco [3] and by S. Coriasco and L. Rodino [4]. In a recent work, we have considered quasilinear symmetrizable SG-hyperbolic pseudo-differential systems with initial data in H_s, for s_1, s_2 sufficiently large, proving existence and uniqueness of the solution. To do this, we have combined the techniques of the SG calculus with a method developed in [10] to study similar systems defined for x in a compact manifold. Unfortunately, these techniques and the general pseudo-differential setting do not allow to compute exactly the smallest values of s_1, s_2 for which a solution exists, cf. [10] and provide a meaningful result only for smooth data. In many applications and concrete physical models, one is led to consider differential systems of the form (1.1), for which more precise results are needed and can be obtained using different pseudo-differential techniques. In this work, assuming that

$$f \in C^\infty\left([-T,T]\times\mathbb{R}_u^n, H_s\right) \quad \textit{and} \quad g \in H_s \tag{1.5}$$

we prove existence and uniqueness for the solution of (1.1) in H_s, for $s_1 > \frac{n}{2} + 1, s_2 > 1$. To obtain our results, we combine the arguments used in [11] for systems locally defined in x and the techniques coming from the SG calculus and involving in particular the use of SG mollifiers defined in [2].

2. SG-Pseudo-Differential Operators and Weighted Sobolev Spaces

In this section, we introduce the classes SG of pseudo-differential operators and recall some basic properties of the spaces (1.4) that we will use in the next sections. For proofs and details, we address the reader to [8], [2], [9], [5].
In the following, given $m = (m_1,m_2), m' = (m_1',m_2')$, we will write $m \ge m'$ if $m_j \ge m_j', j = 1,2$. We will further denote $e_1 = (1,0), e_2 = (0,1), e = (1,1)$.

Definition 2.1. For any $m = (m_1,m_2) \in \mathbb{R}^2$, we shall denote by SG^m the space of all functions $p(x,\xi) \in C^\infty(\mathbb{R}^{2n})$ such that

$$\sup_{(x,\xi)\in\mathbb{R}^{2n}} \langle\xi\rangle^{-m_1+|\alpha|}\langle x\rangle^{-m_2+|\beta|}\left|D_\xi^\alpha D_x^\beta p(x,\xi)\right| < +\infty$$

for all $\alpha, \beta \in \mathbb{Z}_+^n$.

Proposition 2.2. i) *If* $m' \geq m$, *then* $SG^m \subseteq SG^{m'}$.
ii) *If* $p \in SG^m$, *then* $D_\xi^\alpha D_x^\beta p \in SG^{m-|\alpha|e_1-|\beta|e_2}$.
iii) $\bigcap_{m\in\mathbb{R}^2} SG^m = \mathcal{S}(\mathbb{R}^{2n})$.

Given $p \in SG^m$, the pseudo-differential operator $P = p(x, D)$ with symbol p is defined as standard by

$$Pu(x) = (2\pi)^{-n} \int_{\mathbb{R}^n} e^{i\langle x,\xi\rangle} p(x, \xi)\hat{u}(\xi)d\xi, \qquad u \in \mathcal{S}(\mathbb{R}^n), \tag{2.1}$$

where $\hat{u}$ denotes the Fourier transform of u. We denote by LG^m the space of all operators (2.1) with symbol in SG^m and by $\mathcal{K}$ the space of all operators (2.1) with symbol in $\mathcal{S}(\mathbb{R}^{2n})$.

Proposition 2.3. *Given* $p \in SG^m$, *the operator* P *is linear and continuous from* $\mathcal{S}(\mathbb{R}^n)$ *to* $\mathcal{S}(\mathbb{R}^n)$ *and it extends to a linear continuous map from* $\mathcal{S}'(\mathbb{R}^n)$ *to itself.*

Proposition 2.4. *Every* $P \in \mathcal{K}$ *can be extended to a linear continuous map from* $\mathcal{S}'(\mathbb{R}^n)$ *to* $\mathcal{S}(\mathbb{R}^n)$.

Proposition 2.5. *Let* $p \in SG^m, q \in SG^{m'}$. *Then, the following statements hold:*

i) *There exists* $s \in SG^{m+m'}$ *such that* $p(x, D)q(x, D) = s(x, D) + K$ *for some* $K \in \mathcal{K}$.
ii) *Denoting by* R *the commutator* $[p(x, D), q(x, D)]$, *we have* $R = r(x, D) + K'$ *for some* $r \in SG^{m+m'-e}, K' \in \mathcal{K}$.
iii) *Denoting by* $P^\star$ *the* L^2-*adjoint of* P, *we have* $P^\star = p'(x, D) + K''$ *for some* $p' \in SG^m, K'' \in \mathcal{K}$.

Proposition 2.6. i) *For every* $s = (s_1, s_2) \in \mathbb{R}^2, H_s$ *is a Hilbert space endowed with the inner product*

$$(u, v)_s = (\Pi_s u, \Pi_s v)_{L^2}.$$

An equivalent norm is given by $\|\Pi_s^\star u\|_{L^2}$, *where* $\Pi_s^\star = \langle D\rangle^{s_1}\langle x\rangle^{s_2}$.
ii) *If* $t_j \geq s_j, j = 1, 2$, *then* $H_t \subseteq H_s$. *Moreover, if* $t_j > s_j, j = 1, 2$ *then the embedding* $H_t \hookrightarrow H_s$ *is compact.*
iii)

$$\bigcap_{s\in\mathbb{R}^2} H_s = \mathcal{S}(\mathbb{R}^n), \qquad \bigcup_{s\in\mathbb{R}^2} H_s = \mathcal{S}'(\mathbb{R}^n).$$

iv) *If* $s_1 \in \mathbb{Z}_+$, *then* H_s *is the space of all functions* $u \in L^2(\mathbb{R}^n)$ *such that* $\langle x\rangle^{s_2}\partial_x^\alpha u(x)$ *is in* $L^2(\mathbb{R}^n)$ *for all* $|\alpha| \leq s_1$ *and an equivalent norm is given by*

$$\sum_{|\alpha|\leq s_1} \|\langle x\rangle^{s_2}\partial_x^\alpha u(x)\|_{L^2}.$$

v) *If* $s_1 > \frac{n}{2} + j$ *for some* $j \in \mathbb{Z}_+$, *then we have the compact embedding*

$$H_s \hookrightarrow C^j_{s_2,(o)}(\mathbb{R}^n)$$

where we denote by $C^j_{s_2,(o)}(\mathbb{R}^n)$ *the Banach space of all functions* $f \in C^j(\mathbb{R}^n)$ *such that* $\langle x\rangle^{s_2}\partial_x^\alpha u(x) \to 0$ *when* $|x| \to +\infty$ *for* $|\alpha| \le j$ *endowed with the norm*

$$\sup_{x\in\mathbb{R}^n} \langle x\rangle^{s_2} \sum_{|\alpha|\le j} |\partial_x^\alpha u(x)| .$$

We denote $C^j_{(o)}(\mathbb{R}^n) = C^j_{0,(o)}(\mathbb{R}^n)$.

Proposition 2.7. *Let* $m \in \mathbb{R}^2$ *and* $p \in SG^m$. *Then, for every* $s \in \mathbb{R}^2$, *the operator* P *defined by* (2.1) *is linear and continuous from* H_s *to* H_{s-m}.

3. Main Results

We can now study the problem (1.1) under the assumptions (1.2), (1.3) and (1.5). In this section, we will give three main results, the first concerning the existence of a solution for (1.1) in the weighted Sobolev spaces (1.4), the second proving uniqueness and stability of the solution, the third precising its regularity. We attempt to find a solution of (1.1) as a limit of solutions of a mollified problem. To this end, we need to introduce some SG-mollifiers, following [2].
Let ψ be a function in $C_0^\infty(\mathbb{R}^n), \psi \ge 0$ such that $supp(\psi) \subset \{x \in \mathbb{R}^n : |x| \le 1\}$ and $\int_{\mathbb{R}^n} \psi(x)dx = 1$. Let us set $\psi_\varepsilon(x) = \varepsilon^{-n}\psi\left(\frac{x}{\varepsilon}\right)$ for $\varepsilon \in (0,1]$ and denote by J_ε the convolution operator

$$J_\varepsilon v(x) = \psi_\varepsilon * v(x) = \hat{\psi}(\varepsilon D)v(x).$$

Set $J_0 = I$, where I is the identity operator. Furthermore, denote by L_ε the multiplication operator by $\hat{\psi}(\varepsilon x)$ and consider the problem

$$\begin{cases} \frac{\partial u_\varepsilon}{\partial t} = L_\varepsilon K_\varepsilon J_\varepsilon u_\varepsilon + f_\varepsilon & (t,x) \in [-T,T]\times\mathbb{R}^n \\ u_\varepsilon(0,x) = g(x) & x \in \mathbb{R}^n \end{cases} \tag{3.1}$$

where $K_\varepsilon = \sum_{j=1}^n A_j(t,x,J_\varepsilon u_\varepsilon)\frac{\partial}{\partial x_j}$ and $f_\varepsilon = L_\varepsilon f(t,x,J_\varepsilon u_\varepsilon)$.
First of all, we observe that, for every fixed $\varepsilon \in (0,1]$ and for every $s \in \mathbb{R}^2$, the operator $L_\varepsilon K_\varepsilon J_\varepsilon$ is bounded on H_s since $\hat{\psi} \in \mathcal{S}(\mathbb{R}^n)$. Then, the problem (3.1) admits a unique solution $u_\varepsilon(t)$ defined for t in an interval $I_1 = [-T', T']$ for some $T' \in]0,T]$ and for $x \in \mathbb{R}^n$. Our aim is to prove that u_ε exists in an interval independent of ε and that it converges to a solution u of (1.1) when $\varepsilon \to 0$. Before doing this, let us give a proposition which collects some properties of the SG mollifiers. We refer to [2] for the proof.

Proposition 3.1. i) *The families* $\{J_\varepsilon : 0 \le \varepsilon \le 1\}$ *and* $\{L_\varepsilon : 0 \le \varepsilon \le 1\}$ *define bounded functions* $[0,1] \to \mathcal{L}(H_s)$ *for any* $s \in \mathbb{R}^2$, *which are norm continuous in* $(0,1]$ *and strongly continuous at* 0. *If* ψ *is an even function,* J_ε *is hermitian symmetric in* $L^2(\mathbb{R}^n)$ *for all* $\varepsilon \in [0,1]$.

ii) *The family* $\{[J_\varepsilon, L_\varepsilon] : 0 < \varepsilon \le 1\}$ *is bounded in* LG^{-e}.

iii) *The families* $\{[\Pi_s, J_\varepsilon]/\varepsilon : 0 < \varepsilon \le 1\}$ *and* $\{[\Pi_s, L_\varepsilon]/\varepsilon : 0 < \varepsilon \le 1\}$ *are bounded in* LG^{s-e} *for every* $s \in \mathbb{R}^2$.

Let us find an energy estimate for the H_s-norm of u_ε. Denoting $v_\varepsilon = \Pi_s u_\varepsilon$, we have

$$\begin{aligned}\frac{d}{dt}\|u_\varepsilon(t)\|_s^2 &= (\Pi_s L_\varepsilon K_\varepsilon J_\varepsilon u_\varepsilon, v_\varepsilon) + (v_\varepsilon, \Pi_s L_\varepsilon K_\varepsilon J_\varepsilon u_\varepsilon) + 2Re\,(v_\varepsilon, \Pi_s f_\varepsilon)\\ &= ((K_\varepsilon + K_\varepsilon^\star)v_\varepsilon, L_\varepsilon J_\varepsilon v_\varepsilon) + (v_\varepsilon, K_\varepsilon^\star[J_\varepsilon, L_\varepsilon]v_\varepsilon) + (v_\varepsilon, [L_\varepsilon, K_\varepsilon]J_\varepsilon v_\varepsilon)\\ &\quad + ([K_\varepsilon, J_\varepsilon]v_\varepsilon, L_\varepsilon v_\varepsilon) + 2Re\,([\Pi_s, L_\varepsilon K_\varepsilon J_\varepsilon]u_\varepsilon, v_\varepsilon) + 2Re\,(v_\varepsilon, \Pi_s f_\varepsilon),\end{aligned}$$

where $(\cdot,\cdot)$ denotes the inner product in $L^2(\mathbb{R}^n)$. Now, by the assumption (1.2), it follows that

$$\begin{aligned}(K_\varepsilon + K_\varepsilon^\star)w &= -\sum_{j=1}^n \left(\partial_{x_j} A_j(t,x,J_\varepsilon u_\varepsilon)\right) w\\ &= -\sum_{j=1}^n \left[(\partial_{x_j} A_j)(t,x,J_\varepsilon u_\varepsilon)w + \sum_{\ell=1}^n \partial_{x_j}(J_\varepsilon u_\varepsilon)_\ell \cdot (\partial_{u_\ell} A_j)(t,x,J_\varepsilon u_\varepsilon)\right] w.\end{aligned}$$

By v) of Proposition 2.6, the condition $s_1 > \frac{n}{2}+1, s_2 > 1$ and (1.3) imply that

$$\begin{aligned}((K_\varepsilon + K_\varepsilon^\star)v_\varepsilon, L_\varepsilon J_\varepsilon v_\varepsilon) &\le C_s \sum_{j=1}^n \|\partial_{x_j} A_j(t,x,J_\varepsilon u_\varepsilon)\|_{L^\infty} \cdot \|u_\varepsilon(t)\|_s^2\\ &\le C_s'(\|J_\varepsilon u_\varepsilon(t)\|_{C^1_{1,(o)}})\|u_\varepsilon(t)\|_s^2.\end{aligned} \tag{3.2}$$

To estimate the term $(v_\varepsilon, K_\varepsilon^\star[J_\varepsilon, L_\varepsilon]v_\varepsilon)$, we observe that

$$\begin{aligned}K_\varepsilon^\star[J_\varepsilon, L_\varepsilon]v_\varepsilon &= -\sum_{j=1}^n A_j(t,x,J_\varepsilon u_\varepsilon)\partial_{x_j}[J_\varepsilon, L_\varepsilon]v_\varepsilon\\ &\quad -\sum_{j=1}^n \left(\partial_{x_j} A_j(t,x,J_\varepsilon u_\varepsilon)\right)[J_\varepsilon, L_\varepsilon]v_\varepsilon.\end{aligned}$$

Applying ii) of Proposition 3.1 and arguing as before, we obtain

$$(v_\varepsilon, K_\varepsilon^\star[J_\varepsilon, L_\varepsilon]v_\varepsilon) \le C_s(\|J_\varepsilon u_\varepsilon(t)\|_{C^1_{1,(o)}})\|u_\varepsilon(t)\|_s^2.$$

Concerning the term $(v_\varepsilon, [L_\varepsilon, K_\varepsilon]J_\varepsilon v_\varepsilon)$, we have

$$[L_\varepsilon, K_\varepsilon]J_\varepsilon v_\varepsilon = -\sum_{j=1}^n A_j(t,x,J_\varepsilon u_\varepsilon)\left(\partial_{x_j}\hat{\psi}(\varepsilon x)\right) J_\varepsilon v_\varepsilon.$$

Then,

$$\begin{aligned}(v_\varepsilon, [L_\varepsilon, K_\varepsilon]J_\varepsilon v_\varepsilon) &\le C_s \sum_{j=1}^{n} \left\| A_j(t,x,J_\varepsilon u_\varepsilon)\partial_{x_j}\left(\hat{\psi}(\varepsilon x)\right)\right\|_{L^\infty} \|u_\varepsilon(t)\|_s^2 \\ &\le C'_s(\|J_\varepsilon u_\varepsilon\|_{L^\infty})\|u_\varepsilon(t)\|_s^2,\end{aligned}$$

being $\hat{\psi} \in \mathcal{S}(\mathbb{R}^n)$. For the term $([K_\varepsilon, J_\varepsilon]v_\varepsilon, L_\varepsilon v_\varepsilon)$ we observe that

$$[K_\varepsilon, J_\varepsilon]v_\varepsilon = \sum_{j=1}^{n}[A_j(t,x,J_\varepsilon u_\varepsilon), J_\varepsilon]\partial_{x_j} v_\varepsilon$$

Then, arguing as [12], (cf. Section 13.1) and using the properties of the commutators of SG operators, we get the estimate

$$([K_\varepsilon, J_\varepsilon]v_\varepsilon, L_\varepsilon v_\varepsilon) \le C_s(\|J_\varepsilon u_\varepsilon(t)\|_{C^1_{1,(o)}})\|u_\varepsilon(t)\|_s^2 \tag{3.3}$$

To estimate the term $(v_\varepsilon, [\Pi_s, L_\varepsilon K_\varepsilon J_\varepsilon]u_\varepsilon)$ we can split the commutator as follows

$$[\Pi_s, L_\varepsilon K_\varepsilon J_\varepsilon]u_\varepsilon = [\Pi_s, L_\varepsilon]K_\varepsilon J_\varepsilon u_\varepsilon + L_\varepsilon K_\varepsilon[\Pi_s, J_\varepsilon]u_\varepsilon + L_\varepsilon[\Pi_s, K_\varepsilon]J_\varepsilon u_\varepsilon. \tag{3.4}$$

For the first two terms in (3.4), we apply iii) of Proposition 3.1 while the last term can be estimated directly obtaining the same upper bound as in (3.3). Finally, we have

$$\begin{aligned}(v_\varepsilon, \Pi_s f_\varepsilon) &\le \|u_\varepsilon(t)\|_s^2 + C_s\|\Pi_s^\star f_\varepsilon\|_{L^2}^2 = \|u_\varepsilon(t)\|_s^2 + C_s\|\langle D\rangle^{s_1}(\langle x\rangle^{s_2} f_\varepsilon)\|_{L^2}^2 \\ &\le \|u_\varepsilon(t)\|_s^2 + C'_s(\|J_\varepsilon u_\varepsilon(t)\|_{L^\infty})\left(1 + \|J_\varepsilon u_\varepsilon(t)\|_s^2\right) \\ &\le C''_s(\|J_\varepsilon u_\varepsilon(t)\|_{L^\infty})\left(1 + \|u_\varepsilon(t)\|_s^2\right).\end{aligned} \tag{3.5}$$

using the Moser type estimates as in [11]. From the estimates above, we deduce that

$$\frac{d}{dt}\|u_\varepsilon(t)\|_s^2 \le C_s\left(\|J_\varepsilon u_\varepsilon(t)\|_{C^1_{1,(o)}}\right)\left(1 + \|u_\varepsilon(t)\|_s^2\right). \tag{3.6}$$

Now, by the condition $s_1 > \frac{n}{2} + 1, s_2 > 1$, the space H_s is compactly embedded in $C^1_{1,(o)}(\mathbb{R}^n)$. Then, applying Gronwall's inequality backward and forward in time as in [11], it follows that there exists an interval $\tilde{I} = (-T_1, T_2)$, with $T_1, T_2 \in]0, T]$ independent of ε such that

$$\|u_\varepsilon(t)\|_s \le K(t) \tag{3.7}$$

for every $t \in \tilde{I}$ and for some $K(t) > 0$ independent of ε.

Theorem 3.2. *Under the assumptions* (1.2), (1.3) *and* (1.5), *if* $s_1 > \frac{n}{2} + 1, s_2 > 1$, *then the problem* (1.1) *admits a solution* u *satisfying*

$$u \in L^\infty(\tilde{I}, H_s) \cap Lip(\tilde{I}, H_{s-e}). \tag{3.8}$$

Proof. From (1.1) and (3.7), the family $\{u_\varepsilon : 0 < \varepsilon \le 1\}$ is bounded in $C(\tilde{I}, H_s) \cap C^1(\tilde{I}, H_{s-e})$. Then, it has a weak limit point u satisfying (3.8). Moreover, by ii) of Proposition 2.6, H_{s-e} is compactly embedded in H_s, then there exists a sequence u_{ε_j} converging to u in $C(\tilde{I}, H_{s-e})$. On the other hand, $\{u_\varepsilon : 0 < \varepsilon \le 1\}$ is also bounded in $C(\tilde{I}, H_{s-\sigma e})$ for $\sigma \in (0,1)$ and for σ sufficiently small, $H_{s-\sigma e}$ is compactly embedded in $C^1_{1,(o)}(\mathbb{R}^n)$ by v) of Proposition 2.6. Hence,

$$u_{\varepsilon_j} \to u \quad in \quad C(\tilde{I}, C^1_{1,(o)}(\mathbb{R}^n)) \quad when \quad \varepsilon_j \to 0.$$

Consequently,

$$L_{\varepsilon_j} K_{\varepsilon_j} J_{\varepsilon_j} u_{\varepsilon_j} + L_{\varepsilon_j} f_{\varepsilon_j} \to K(t,x,u,D)u + f(t,x,u)$$

in the space $\mathcal{BC}(\tilde{I} \times \mathbb{R}^n)$ of all bounded and continuous functions on $\tilde{I} \times \mathbb{R}^n$ endowed with the sup-norm. On the other hand, $\frac{\partial u_\varepsilon}{\partial t} \to \frac{\partial u}{\partial t}$ weakly in $\mathcal{BC}(\tilde{I} \times \mathbb{R}^n)$. This concludes the proof. □

The next result shows that every solution u of (1.1) can be obtained as a limit of solutions u_ε of (3.1) for $\varepsilon \to 0$. This implies in particular that the solution of (1.1) is unique.

Theorem 3.3. *For $s_1 > \frac{n}{2} + 1, s_2 > 1$, the solution u of (1.1) is unique. Moreover, given a solution u_ε of (3.1), the following estimate holds for all $\varepsilon \in (0,1]$:*

$$\|u(t) - u_\varepsilon(t)\|_{L^2} \leq K_1(t) B_\varepsilon \tag{3.9}$$

where $B_\varepsilon = \max\left\{\|I - L_\varepsilon\|_{\mathcal{L}(H_{s-e}, L^2)}, \|I - J_\varepsilon\|_{\mathcal{L}(H_s, H_e)}\right\}$.

Proof. Let u be a solution of (1.1) and let u_ε satisfy

$$\begin{cases} \frac{\partial u_\varepsilon}{\partial t} = L_\varepsilon K_\varepsilon J_\varepsilon u_\varepsilon + f_\varepsilon & (t,x) \in [-T,T] \times \mathbb{R}^n \\ u_\varepsilon(0,x) = h(x) & x \in \mathbb{R}^n \end{cases} \tag{3.10}$$

for some $h \in H_s$. Set $v_\varepsilon = u - u_\varepsilon$. From (1.1) and (3.10), we have

$$\frac{\partial v_\varepsilon}{\partial t} = K(u,D)v_\varepsilon + K(u,D)u_\varepsilon - L_\varepsilon K(J_\varepsilon u_\varepsilon, D) J_\varepsilon u_\varepsilon + f(u) - L_\varepsilon f(J_\varepsilon u_\varepsilon)$$

where we have suppressed the dependence on (t,x) to simplify the notation. Now, we can write

$$\begin{aligned} &K(u,D)u_\varepsilon - L_\varepsilon K(J_\varepsilon u_\varepsilon, D) J_\varepsilon u_\varepsilon \\ &= [K(u,D) - K(u_\varepsilon, D)]u_\varepsilon + (I - L_\varepsilon)K(u_\varepsilon, D)u_\varepsilon \\ &\quad + L_\varepsilon K(u_\varepsilon, D)(I - J_\varepsilon)u_\varepsilon + L_\varepsilon[K(u_\varepsilon, D) - K(J_\varepsilon u_\varepsilon, D)]J_\varepsilon u_\varepsilon \end{aligned}$$

and

$$f(u) - L_\varepsilon f(J_\varepsilon u_\varepsilon) = f(u) - f(u_\varepsilon) + (I - L_\varepsilon) f(u_\varepsilon) + L_\varepsilon[f(u_\varepsilon) - f(J_\varepsilon u_\varepsilon)].$$

Moreover,

$$f(u) - f(w) = F(u,w)(u - w),$$

with

$$F(u,w) = \int_0^1 f'(\tau u + (1-\tau)w) d\tau,$$

where f' denotes the Jacobian matrix of f as a function of u. Similarly, we can write

$$K(u,D) - K(w,D) = (u - w) \cdot M(u,w,D).$$

Then, we obtain

$$\frac{\partial v_\varepsilon}{\partial t} = K(u,D)v_\varepsilon + v_\varepsilon \cdot M(u,u_\varepsilon,D)u_\varepsilon + F(u,u_\varepsilon)v_\varepsilon + R_\varepsilon$$

where

$$\begin{aligned} R_\varepsilon &= (I-L_\varepsilon)K(u_\varepsilon,D)u_\varepsilon + L_\varepsilon K(u_\varepsilon,D)(I-J_\varepsilon)u_\varepsilon \\ &\quad + L_\varepsilon(I-J_\varepsilon)u_\varepsilon \cdot M(u_\varepsilon,J_\varepsilon u_\varepsilon,D)J_\varepsilon u_\varepsilon \\ &\quad + (I-L_\varepsilon)f(u_\varepsilon) + L_\varepsilon F(u_\varepsilon,J_\varepsilon u_\varepsilon)(I-J_\varepsilon)u_\varepsilon. \end{aligned}$$

Repeating the arguments used to prove (3.6), we deduce that

$$\frac{d}{dt}\|v_\varepsilon(t)\|^2_{L^2} \le C(t)\|v_\varepsilon(t)\|^2_{L^2} + \|R_\varepsilon(t)\|^2_{L^2}$$

where $C(t)$ is a positive constant depending on the $C^1_{1,(o)}$-norm of u and u_ε. Hence, possibly shrinking $\tilde{I}$, by Gronwall's inequality, we get

$$\|v_\varepsilon(t)\|^2_{L^2} \le e^{\int_0^t C(\tau)d\tau}\left(\|g-h\|^2_{L^2} + \int_0^t \|R_\varepsilon(\tau)\|^2_{L^2} \cdot e^{-\int_0^\tau C(\sigma)d\sigma}d\tau\right). \tag{3.11}$$

Finally, observe that, by (1.3) and (1.5), it is easy to prove that

$$\|R_\varepsilon(t)\|^2_{L^2} \le C'(t)B^2_\varepsilon. \tag{3.12}$$

Applying (3.12) in (3.11), we obtain (3.9) for $g = h$. □

Proposition 3.4. *Under the assumptions* (1.2), (1.3) *and* (1.5), *for every* $g \in H_s$, *with* $s_1 > \frac{n}{2}+1, s_2 > 1$, *the solution* u *of* (1.1) *is in* $C(\tilde{I}, H_s)$.

Proof. To prove the proposition, it is sufficient to show that the H_s-norm of $u(t,\cdot)$ is continuous with respect to t. We want to estimate the rate of change of $\|u(t)\|_s$ by finding an estimate for its derivative. Indeed, since $K\Pi_s u \notin L^2$ in general, we need to use mollifiers again. We have

$$\begin{aligned} \frac{d}{dt}\|L_\varepsilon J_\varepsilon u(t)\|^2_s &= (L_\varepsilon J_\varepsilon K u, L_\varepsilon J_\varepsilon u)_s + (L_\varepsilon J_\varepsilon u, L_\varepsilon J_\varepsilon K u)_s \\ &\quad + 2Re\,(L_\varepsilon J_\varepsilon f(u), L_\varepsilon J_\varepsilon u)_s \\ &= (K\Pi_s L_\varepsilon J_\varepsilon u, \Pi_s L_\varepsilon J_\varepsilon u) + 2Re\,([\Pi_s L_\varepsilon J_\varepsilon, K]u, \Pi_s L_\varepsilon J_\varepsilon u) \\ &\quad + (\Pi_s L_\varepsilon J_\varepsilon u, K\Pi_s L_\varepsilon J_\varepsilon u) + 2Re\,(\Pi_s L_\varepsilon J_\varepsilon f(u), \Pi_s L_\varepsilon J_\varepsilon u) \\ &= ((K+K^\star)\Pi_s L_\varepsilon J_\varepsilon u, \Pi_s L_\varepsilon J_\varepsilon u) \\ &\quad + 2Re\,([\Pi_s L_\varepsilon J_\varepsilon, K]u, \Pi_s L_\varepsilon J_\varepsilon u) \\ &\quad + 2Re\,(\Pi_s L_\varepsilon J_\varepsilon f(u), \Pi_s L_\varepsilon J_\varepsilon u). \end{aligned}$$

Arguing as for (3.2), (3.4), (3.5), we obtain

$$\frac{d}{dt}\|L_\varepsilon J_\varepsilon u(t)\|^2_s \le C_s(\|u(t)\|_{C^1_{1,(o)}})\|u(t)\|^2_s. \tag{3.13}$$

for some positive constant $C_s\left(\|u(t)\|_{C^1_{1,(o)}}\right)$ independent of ε. By v) of Proposition 2.6 and by Theorem 3.2, the right-hand side of (3.13) is bounded in $\tilde{I}$ uniformly with respect to ε. Then, $\|L_\varepsilon J_\varepsilon u(t)\|^2_s$ is a Lipschitz function of t uniformly with respect to ε and the same holds for $\|u(t)\|_s^2$, since $\|L_\varepsilon J_\varepsilon u(t)\|_s^2 \to \|u(t)\|_s^2$ when $\varepsilon \to 0$. This concludes the proof. □

Remark 3.5. Repeating the arguments in [11], it is easy to prove that, under the assumptions on s_1, s_2, we can extend the solution of (1.1) to an open interval $(-T_1', T_2')$ independent of s, preserving the uniqueness.

From Remark 3.5 and iii) of Proposition 2.6 , we deduce the following result.

Corollary 3.6. *If $g \in \mathcal{S}(\mathbb{R}^n)$ and $f \in C^\infty([-T,T] \times \mathbb{R}^n_u, \mathcal{S}(\mathbb{R}^n))$, then the problem* (1.1) *admits a unique solution $u \in C^\infty(I_2, \mathcal{S}(\mathbb{R}^n))$ defined in an open interval $I_2 = (-T_1', T_2')$ for some $T_1', T_2' \in]0, T]$.*

References

[1] M. Cappiello and L. Zanghirati, The Cauchy problem for quasilinear SG-hyperbolic systems, Math. Nachr., to appear.

[2] H.O. Cordes, *The Technique of Pseudodifferential Operators.* Cambridge University Press, 1995.

[3] S. Coriasco, Fourier integral operators in SG classes.II. Application to SG hyperbolic Cauchy problems, *Ann. Univ. Ferrara Sez VII* **44** (1998), 81–122.

[4] S. Coriasco and L. Rodino, Cauchy problem for SG-hyperbolic equations with constant multiplicities, *Ricerche di Matematica*, Suppl. Vol. **XLVIII** (1999), 25–43.

[5] Y.V. Egorov and B.-W. Schulze, *Pseudo-Differential Operators, Singularities, Applications*, Operator Theory: Advances and Applications, **93**, Birkhäuser Verlag, Basel, 1997.

[6] A.E. Fischer and J.E. Marsden, The Einstein evolution equations as a first order quasi-linear symmetric hyperbolic system, *Comm. Math. Phys.* **28** (1972), 1–38 .

[7] T. Kato, The Cauchy problem for quasi-linear symmetric hyperbolic systems, *Arch. Rat. Mech. Anal.* **58** 3 (1975),181–205.

[8] C. Parenti, Operatori pseudodifferenziali in $\mathbb{R}^n$ e applicazioni, *Ann. Mat. Pura Appl.* **93** (1972), 359–389.

[9] E. Schrohe, Spaces of weighted symbols and weighted Sobolev spaces on manifolds, in Pseudo-Differential Operators, Proc. Oberwolfach 1986, Editors: H. O Cordes, B Gramsch and H. Widom, *Lecture Notes in Mathematics* **1256**, Springer, New York, 1987, 360–377.

[10] M. Taylor, *Pseudodifferential Operators*, Princeton Mathematical Series **34**, Princeton University Press, Princeton, N.J., 1981.

[11] M. Taylor, *Pseudodifferential Operators and Nonlinear PDE*, Progress in Mathematics, **100**, Birkhäuser Boston, Inc., Boston, MA, 1991.

[12] M. Taylor, *Partial Differential Equations III: Nonlinear equations*, Applied Math. Sci. **117**, Springer-Verlag, New York, 1997.

Marco Cappiello and Luisa Zanghirati
Department of Mathematics
University of Ferrara
Via Machiavelli, 35
44100 Ferrara, Italy
e-mail: cppmrc@unife.it
zan@unife.it

Operator Theory:
Advances and Applications, Vol. 164, 53–63

Representation of Solutions and Regularity Properties for Weakly Hyperbolic Systems

Ilia Kamotski and Michael Ruzhansky

Abstract. Regularity properties of generic hyperbolic systems with diagonalizable principal part will be established in L^p and other function spaces. Sharp regularity of solutions will be discussed. Applications will be given to solutions of scalar weakly hyperbolic equations with non-involutive characteristics. Established representation of solutions and its properties allow to derive spectral asymptotics for elliptic systems with diagonalizable principal part.

Mathematics Subject Classification (2000). Primary 35S30, 35L45, 58J40; Secondary 35L30, 35C20.

Keywords. Hyperbolic systems, elliptic systems, spectral asymptotics, regularity of solutions.

1. Introduction

Let X be a smooth manifold without boundary of dimension $n \geq 3$. Let P be an elliptic self-adjoint pseudo-differential operator of order one acting on half-densities on m-dimensional cross-sections of vector bundles on X. Since we will be mostly interested in local properties of solutions, we may already assume that P acts on functions, and can think of it as an $m \times m$ matrix of pseudo-differential operators of order one. We consider the following Cauchy problem for $u = u(t,x)$

$$\begin{cases} iu' - Pu = 0, & (t,x) \in [0,T] \times X, \\ u|_{t=0} = u_0, \end{cases} \tag{1.1}$$

and we think of u_0 as of an m-vector. It is well known that if equation (1.1) is strictly hyperbolic, the system can be diagonalized and its solution can be given as a sum of Fourier integral operators applied to Cauchy data. Our question here is to study what happens when P has multiple characteristics.

This work was supported by EPSRC grant GR/R67583/01.

Everywhere in this paper $\Psi^\mu = \Psi^\mu_{1,0}(X)$ will denote the space of classical pseudo-differential operators of order μ of type $(1,0)$. The space of Fourier integral operators of order μ with amplitudes of type $(1,0)$ will be denoted by I^μ. All Fourier integral operators in the sequel will be non-degenerate, which means that its canonical relation satisfies the local graph condition, i.e. it is locally a graph of a symplectic diffeomorphism from $T^*X\backslash 0$ to itself.

Let $A(x,\xi)$ denote the principal symbol of P. If A is a diagonal matrix, properties of system (1.1) have been studied by many authors. For example, Kumano-go 1981 used the calculus of Fourier integral operators with multiphases to show that the Cauchy problem (1.1) is well-posed in L^2 and Sobolev spaces H^s. Systems with symmetric principal part A have been extensively studied as well (e.g. Ivrii 1998, Kucherenko 1974, etc.) In a generic situation, such systems have double characteristics, and their normal forms have been found by Braam and Duistermaat 1993. Recently, Colin de Verdiere 2003 used these representations to derive some asymptotic properties of such systems. Polarization properties of similar systems were studied in Dencker 1992.

More elaborate analysis of system (1.1) becomes possible if we assume that the principal symbol matrix $A(x,\xi)$ is smoothly (microlocally) diagonalizable with smooth eigenvalues $a_j(x,\xi)$ and smooth eigenspaces. Then, as it was pointed out in Rozenblum 1980, there exists a finite dimensional cover $\tilde{X}$ of X such that A lifted to $\tilde{X}$ is globally diagonalized. In this situation, Rozenblum showed that the Picard series for this problem gives an expansion with respect to smoothness under the condition that at all points of multiplicity, bicharacteristics intersect characteristics transversally. In other words, if $a_j(x,\xi) = a_k(x,\xi)$ for $j \neq k$, then the Poisson bracket $\{a_j, a_k\}(x,\xi) \neq 0$. However, this condition is non-generic even for diagonal systems. For example, it is clear that if one of characteristics has a fold, there may be a point where this transversality condition fails, and it is not possible to remove it by small perturbations.

The purpose of this paper is to present results removing the transversal intersection condition and to investigate regularity properties of system (1.1) in a generic setting. Apart from this, our results will cover systems arising from some scalar weakly hyperbolic equations. Now we will formulate our main assumption.

Let us define operator

$$H_{a_j} f = \{a_j, f\}, j = 1, \ldots, m,$$

where

$$H_g(f) = \{g, f\} = \sum_{k=1}^{n} \left(\frac{\partial g}{\partial \xi_k} \frac{\partial f}{\partial x_k} - \frac{\partial g}{\partial x_k} \frac{\partial f}{\partial \xi_k} \right)$$

is the usual Poisson bracket.

Our assumption is that at points of multiplicity $a_j = a_k$ (assuming that they are not identically the same), bicharacteristics of a_j intersect level sets $\{a_k = 1\}$ with finite order, i.e. $H^M_{a_j} a_k \neq 0$ for some M at points where $a_j = a_k$. In other words, for $j \neq k$,

$$
\begin{aligned}
& a_j(x,\xi) = a_k(x,\xi) \Longrightarrow \\
& H_{a_j}^M a_k(x,\xi) = \{a_j, \{a_j, \cdots \{a_j, a_k \overbrace{\}\}\ldots\}}^{M} (x,\xi) \neq 0,
\end{aligned}
\tag{1.2}
$$

for some number M. This number M may depend on the point (x,ξ). While the function $M = M(x,\xi)$ is locally bounded, it is allowed to grow at infinity. We note here that the transversal situation considered by Rozenblum requires (1.2) to hold with $M = 1$. In the strictly hyperbolic case we may set $M = 0$. The case $M = 2$ corresponds to the case when a_j and a_k define glancing hypersurfaces (as considered by Melrose). Here we also assume that a_j and a_k are not identically the same at (x,ξ).

Let us now give some examples where this property holds while the tranversality condition ($M = 1$) fails.

Example 1. In scalar equations with Levi conditions studied by Chazarain 1974, Mizohata-Ohya 1971, Zeman, one assumed that $\{a_j, a_k\} = C_{jk}(a_j - a_k)$. It is clear that in this situation $a_j(x,\xi) = a_k(x,\xi)$ implies $\{a_j, a_k\}(x,\xi) = 0$. However, in a general case when $C_{jk}(x,\xi)$ is non-constant, condition (1.2) may be satisfied.

Example 2. Let L be a scalar operator with involutive characteristics. More precisely, let us denote $\partial_j = D_t + a_j(t,x,D_x)$. Let

$$L = \partial_1 \cdots \partial_m + \sum_{k<m} b_{j_1,\cdots,j_k} \partial_{j_1} \cdots \partial_{j_k} + c,$$

where $b(t,x,D_x), c(t,x,D_x) \in \Psi^0$ are pseudo-differential operators of order zero for all t is some bounded interval. Let us assume that operator L has involutive characteristics, i.e. that

$$[\partial_j, \partial_k] \equiv \partial_j \partial_k - \partial_k \partial_j = \alpha_{jk} \partial_j + \beta_{jk} \partial_k + \gamma_{jk},$$

where $\alpha_{jk}, \beta_{jk}, \gamma_{jk} \in \Psi^0$ are pseudo-differential operators of order zero. Then it was shown in Morimoto 1979 that the Cauchy problem for the equation $Lu = f$ is diagonalizable (with $1 + \sum_{j=1}^{m-1} m!/j!$ components). Even in the simplest case of characteristics not depending on t, we have

$$\{a_j, a_k\} = \alpha_{jk}(x,\xi)(a_j - a_k) + \gamma_{jk},$$

similar to Example 1.

Propagation of singularities of systems with vanishing Poisson brackets has been also studied in these situations. For example, Iwasaki and Morimoto (1982, 1984) studied propagation of singularities of 3×3 systems, where the second Poisson bracket vanish. Also, Ichinose 1982 studied 2×2 systems with vanishing second Poisson brackets.

Regularity properties of solutions of hyperbolic equations have been extensively studied over the years. In the case of scalar higher order strictly hyperbolic equations, this reduces to the corresponding questions for Fourier integral operators. Seeger, Sogge and Stein 1991 showed that Fourier integral operators of order $-(n-1)|1/p-1/2|$ are bounded from L^p_{comp} to L^p_{loc}, $1<p<\infty$. This implies the loss of $(n-1)|1/p-1/2|$ derivatives in Cauchy problems for strictly hyperbolic equations. We will extend this result to the weakly hyperbolic case of equations or systems satisfying condition (1.2). Let us now give an example of second order weakly hyperbolic equations where this condition is satisfied.

Example 3. Let us consider the second order equation

$$u''+b(x,D_x)u'+c(x,D_x)u=0,$$

where $b\in\Psi^1$, $c\in\Psi^2$. Let us denote $\langle x\rangle=(1+x^2)^{1/2}$. Introducing $v=\begin{pmatrix}\langle D_x\rangle u\\ u'\end{pmatrix}$, the matrix form of this equation is given by

$$v'=\begin{pmatrix}0 & \langle D_x\rangle\\ -\langle D_x\rangle^{-1}c & -b\end{pmatrix}v.$$

Let b_1 and c_2 be principal symbols of b and c. The equation is hyperbolic if $b_1^2\geq 4c_2$, with multiple roots at $b_1^2=4c_2$. Assume that $b_1^2-4c_2=\mu^2$, with $\mu\in S^1$ being a symbol of order one. Then characteristics a_1,a_2 satisfy $\{a_1,a_2\}=\frac{1}{2}\{b_1,\mu\}$, which may vanish.

Let the Cauchy data be $u(0)=f_0, u'(0)=f_1$. Let $\alpha=(n-1)|1/p-1/2|$ and $1<p<\infty$. If μ is elliptic, it is known that if $f_j\in L^p_{\alpha-j}$, then $u(t,\cdot)\in L^p_{loc}$. Our result of Theorem 2.4 will imply that the same is true for any $\mu\in S^1$.

As before, we assume that the principal symbol A of operator P may be (micro)locally diagonalized. As we already mentioned, we may then lift A to a finite dimensional cover of X to separate the characteristic roots and eigenspaces.

Therefore, we may assume from now on that A is globally diagonalized, that is

$$P=A+B, A=\operatorname{diag}\{A_1,\dots,A_m\},$$

where $A_j\in\Psi^1$ are scalar pseudo-differential operators with principal symbols $a_j(x,\xi)$ (eigenvalues of A). Here a_j's may be identically equal to each other or may intersect with any finite order, according to our assumption (1.2). Here B is an $m\times m$ matrix of classical pseudo-differential operators or order zero. Without loss of generality we may assume that $B_{jj}=0$ for $1\leq j\leq m$. Indeed, we can always add these terms to the corresponding entries of A. In the case of some of a_j's being identically equal to each other locally at some points, the construction is slightly different, but all the results remain valid. Substitution $U=e^{-iAt}V$ leads to the equation

$$\begin{cases} iV'=Z(t)V,\\ V|_{t=0}=I,\end{cases} \tag{1.3}$$

with $Z(t) = e^{iAt}Be^{-iAt}$. Writing the Picard series for problem (1.3), we obtain the expansion

$$V(t) = I + \int_0^t Z(t_1)dt_1 + \int_0^t \int_0^{t_1} Z(t_1)Z(t_2)dt_2dt_1 + \cdots \tag{1.4}$$

A general term of this series is

$$Q_l = \overbrace{\int_0^t \int_0^{t_1} \cdots \int_0^{t_{l-1}}}^{l} e^{iA_1t_1}b_{12}e^{iA_2(t_2-t_1)} \cdots dt_l \ldots dt_1.$$

It is easy to see that $||Q_l||_{L^2\to L^2} \leq C/l!$ and that series (1.4) converges in L^2 and in H^s. Using the notion of a multiphase for Fourier integral operators, Kumano-go 1981 studied propagation of singularities of Q_l. Instead of introducing multiphases for Q_l, we will analyze operators Q_l in more detail and will show its smoothing properties in Sobolev spaces under assumption (1.2). Here, contrary to the transversal case ($M = 1$) considered by Rozenblum 1980, we do not have good control on the singular supports of integral kernels of operators Q_l, so more elaborate geometric analysis is required.

In fact, our Theorem 2.1 asserts that $Q_l(t)$ maps L^2 to some Sobolev space $H^{p(l)}$, and that $p(l) \to \infty$ as $l \to \infty$. Then we will show that this allows to treat the Picard expansion as a series with finitely many terms, with many conclusions.

2. Results

Here we will formulate our results concerning operators Q_l and solutions to systems (1.1) and (1.3).

Our first main result will be Theorem 2.1 on the smoothing properties of terms Q_l. The other two important results will be Theorem 2.4 on the L^p-regularity of solutions to system (1.1) and Theorem 2.3 on the spectral asymptotics for elliptic operator P satisfying our assumptions. To obtain L^p-estimates, we use Theorem 2.5, which we regard as a general principle behind regularity estimates for general Cauchy problems based on several natural properties of the right hand side operators $Z(t)$. We will illustrate its use in several situations in Corollaries 1-3. Finally, Theorem 2.2 is a statement on the propagation of singularities on operators Q_l or, more generally, of Fourier integral operators where the frequency integration is performed over a cone rather than the whole space. Note that these results will still hold for microlocally smoothly block-diagonalizable operators with any (finite) geometry of characteristics, i.e. characteristics satisfying our assumption (1.2). Proof of Theorems 2.1–2.4 can be found in our paper [11].

The following Theorem establishes a smoothing property of operators Q_l under our assumption (1.2).

Theorem 2.1. *Let condition* (1.2) *be satisfied, that is, assume that for any* $(x,\xi) \in T^*X$ *and any* $1 \leq j,k \leq m$, $j \neq k$, *with* $a_j(x,\xi) = a_k(x,\xi)$, *we have*

$$\{a_j, \overbrace{\{a_j, \cdots \{a_j, a_k \}\} \ldots\}}^{M} (x,\xi) \neq 0$$

for some number M. *Then* Q_l *is a bounded linear operator from* L^2_{comp} *to* $H^{p(l)}_{loc}$, *with* $p(l) \sim \frac{l}{6M+2}$, *as* $l \to \infty$.

In fact, a more precise statement here is to say that Q_l is a bounded linear operator from L^2_{comp} to $H^{p(l)}_{loc}$, with $p(l) = ((-3n/2-2)(3[l/2]-1-n)+([l/2]-n-1)(l/(2M)-n-1))(3[l/2]-2n-2+l/(2M))^{-1}$. In turn, this order can be improved further. However, in our applications it will be important that $p(l) \to \infty$ as $l \to \infty$.

This theorem implies, in particular, that the series

$$V(t) = I + Q_1(t) + Q_2(t) + \cdots$$

is a series with respect to smoothness. This fact allows one to refine the study of propagation of singularities and regularity properties of solutions to systems (1.1) and (1.3).

Let us now discuss the propagation of singularities for operators Q_l. This result is essentially a reformulation of Rozenblum's result in our case of finite geometry under condition (1.2). It is clear (also from multi-phase analysis) that singularities propagate along broken Hamiltonian flows. Let

$$J = \{j_1, \ldots, j_{l+1}\},\ 1 \leq j_k \leq m,\ j_k \neq j_{k+1}.$$

Let $\Phi_J(t,x,\xi)$ be the corresponding broken Hamiltonian flow. It means that points follow bicharacteristics of a_{j_1} until meeting the characteristic of a_{j_2}, and then continue along the bicharacteristic of a_{j_2}, etc. Note that singularities may accumulate if wave front sets for different broken trajectories project to the same point of X.

We can write

$$Q_l = \int_\Delta I(\bar{t})d\bar{t},$$

where $\bar{t} = (t_1,\ldots,t_l)$, Δ is a symplex in $\mathbf{R}^l$ and $I(\bar{t}) = Z(t_1) \circ \ldots \circ Z(t_l)$. It is possible to treat it as a standard Fourier integral operator with the change of variables $\bar{t} = \zeta|\xi|^{-1}$. Let K be a cone in $\mathbf{R}^N = \mathbf{R}^{n+l}$. Let

$$Iu(x) = \int_K \int_Y e^{i\phi(x,y,\theta)} a(x,y,\theta)u(y)dyd\theta$$

be a Fourier integral operator with integration over the cone K with respect to θ. Let K_j be K or a face of K. Let $\phi_j(x,y,\theta_j) = \phi|_{K_j}, \theta_j \in K_j$. Let $\Lambda_j \subset T^*X \times T^*Y$ be a Lagrangian manifold with boundary:

$$\Lambda_j = \{(x, \frac{\partial \phi_j}{\partial x}, y, -\frac{\partial \phi_j}{\partial y}) : \frac{\partial \phi_j}{\partial \theta_j} = 0\}.$$

For $G \subset T^*Y$, let $\Lambda_j(G) = \{z \in T^*X : \exists\zeta \in G : (z, \zeta) \in \Lambda_j\}$. Then we have the following statement on the propagation of singularities.

Theorem 2.2. *Let $u \in \mathcal{D}'(Y)$. Then $WF(Iu) \subset \cup_j \Lambda_j(WF(u))$.*

From this, we can deduce first and second terms of the spectral asymptotic of operator P (see also [10]). Let us call T a period of symbol $A(x, \xi)$ if there exists J such that $j_1 = j_{l+1}$, and trajectory of Φ_J is closed: $\Phi_J(T, x, \xi) = (x, \xi)$. Then we have the following extension of well-known results of Hörmander 1968, Duistermaat–Guillemin 1975, Safarov–Vassiliev 1997, and Rozenblum 1980.

Theorem 2.3. *Assume that X is compact and assume that condition (1.2) holds. Let D be the set of $(x, \xi) \in T^*X$ such that there exist T and J such that $\Phi_J(T, x, \xi) = (x, \xi)$. Assume that the measure of D is zero. Then for the spectrum of P the following Weyl formula holds:*

$$N(\lambda) = \sharp\{j : \lambda_j < \lambda\} = c_n \lambda^n + c_n' \lambda^{n-1} + o(\lambda^{n-1}),$$

where λ_j are eigenvalues of P.

It turns out that smoothing properties of Q_l in Theorem 2.1 can be used to establish local L^p properties of solutions to systems with multiplicities (1.1). For strictly hyperbolic equations such estimates have been established by Seeger, Sogge and Stein 1999, and some optimal estimates were given by Ruzhansky 2000.

Theorem 2.4. *Assume that condition (1.2) holds. Let $1 < p < \infty$ and let $\alpha = (n-1)|1/p - 1/2|$. Let $u_0 \in L^p_\alpha$ be compactly supported. Let $u = u(t, x)$ be the solution of the Cauchy problem (1.1). Then $u(t, \cdot) \in L^p_{loc}$ for small t. We also have*

$$\sup_{t \in [0,T]} ||u(t, \cdot)||_{L^p} \le C ||u_0||_{L^p_\alpha},$$

uniformly for all u_0 supported in some fixed compact subset of X.

As a consequence, if the Cauchy data u_0 is compactly supported, we obtain local estimates in other spaces as well:

- $u_0 \in L^p_{s+\alpha}$ implies $u(t, \cdot) \in L^p_s$, $s \in \mathbf{R}$.
- $u_0 \in \mathrm{Lip}(s + (n-1)/2)$ implies $u(t, \cdot) \in \mathrm{Lip}(s)$.
- Let $1 < p \le q \le 2$. Then $u_0 \in L^p_{s-1/q+n/p-(n-1)/2}$ implies $u(t, \cdot) \in L^q_s$. Dual result holds for $2 \le p \le q < \infty$.

The proof of Theorem 2.4 relies on the smoothing properties of operators Q_l and the following general principle for solutions to differential functional equations.

Theorem 2.5. *Let $W_0, W_1 \hookrightarrow W$ be linear subspaces of $\mathcal{D}'(X)$. Let $Z(t)$, $t \in [0, T]$, be a family of $L^2(X)$–bounded operators such that $\sup_{t \in [0,T]} ||Z(t)||_{L^2 \to L^2} < \infty$. Assume also that*

(i) **(Boundedness)** *$Z(t)$ extends to a bounded linear operator from W_1 to W, for all $t \in [0, T]$.*
(ii) **(Calculus)** *$Z(t_1) \circ \cdots \circ Z(t_l) : W_1 \to W$ are bounded for all l, and for all $t_1, \ldots, t_l \in [0, T]$.*

(iii) **(Smoothing)** *There exists l such that*

$$Z(t)\int_0^t\int_0^{t_1}\cdots\int_0^{t_l} Z(t_1)\cdots Z(t_l)dt_l\cdots dt_1$$

is bounded from W to W_0, for all $t\in[0,T]$.

(iv) **(Zero Cauchy data)** *The Cauchy problem*

$$\begin{cases} v'-Z(t)v=r, & r(t)\in W_0,\\ v(0)=0 \end{cases} \tag{2.1}$$

has a unique solution $v(t)=v(t,\cdot)\in L^2$ which also satisfies $v(t,\cdot)\in W$ for all $t\in[0,T]$.

Then the solution $u=u(t,x)$ of the Cauchy problem

$$\begin{cases} u'-Z(t)u=r, & r(t)\in W_0,\\ u(0)\in W_1. \end{cases} \tag{2.2}$$

satisfies $u(t,\cdot)\in W$ for all $t\in[0,T]$.

Moreover, if W_0, W_1, W are normed spaces and if solutions v to (2.1) satisfy $||v(t,\cdot)||_W\le C||r(t)||_{W_0}$ for all $t\in[0,T]$, then also

$$||u(t,\cdot)||_W\le C(||u(0)||_{W_1}+||r(t)||_{W_0}),$$

for all $t\in[0,T]$.

Let us now give some examples of the use of Theorem 2.5 in different situations. Below we will always assume the well-posedness, namely we will always assume that $Z(t)$ and $r(t)$ are such that system (2.1) has a unique solution $v(t,\cdot)\in L^2$, at least locally.

Corollary 1. Let $Z(t)$ be a family of pseudo-differential operators. Let us choose $W=W_1=W_0=L^p_{comp}$ with $1<p<\infty$. Let $Z(t)\in\Psi^0$ and assume that $\sup_{t\in[0,T]}(||Z(t)||_{L^2\to L^2}, ||Z(t)||_{L^p\to L^p})<\infty$ (note that this condition always holds pointwise in t). Then the solution $u=u(t,x)$ of the Cauchy problem

$$\begin{cases} u'-Z(t)u=r, & r(t)\in L^p_{comp},\\ u(0)\in L^p_{comp}. \end{cases}$$

satisfies $u(t,\cdot)\in L^p_{loc}$ and $u'(t,\cdot)\in L^p_{loc}$ for all $t\in(0,T]$.

Here properties (i) and (ii) of Theorem 2.5 follow from local properties of pseudo-differential operators of order zero. Property (iii) follows from the fact that $Z(t)$ are locally bounded in L^p. Finally, property (iv) follows from Duhamel principle and smoothing properties of the Picard series.

Corollary 2. Let $Z(t)$ be a family of Fourier integral operators. Let us choose $W=L^2\cap L^p_{comp}$, $W_1=(L^p_\alpha)_{comp}$ with $\alpha=(n-1)|1/p-1/2|$, and let $W_0=H^s_{comp}\subset L^p_{comp}$ with some sufficiently large s (for example, with $s=(pn-2n)/2p$ when $p\ge 2$). Let $Z\in I^{-\epsilon}$, $\epsilon>0$, be a family of non-degenerate Fourier integral operators of negative order $-\epsilon$ such that the composition of two such operators is again a non-degenerate Fourier integral operator. Assume also that $\sup_{t\in[0,T]} ||Z(t)||_{H^s\to H^s} < \infty$ (note that this condition always holds pointwise in t). Then the solution $u = u(t,x)$ of the Cauchy problem

$$\begin{cases} u' - Z(t)u = r, & r(t) \in H^s_{comp}, \\ u(0) \in (L^p_\alpha)_{comp} \cap L^2. \end{cases}$$

satisfies $u(t,\cdot) \in L^p_{loc}$ for all $t \in (0,T]$.

Here property (i) of Theorem 2.5 follows from the fact that non-degenerate Fourier integral operators of order zero map $(L^p_\alpha)_{comp}$ to L^p_{loc}. Property (ii) follows from the calculus of non-degenerate Fourier integral operators of order zero. Smoothing condition (iii) follows from the fact that $Z(t)$ are operators of negative orders. Finally, property (iv) is again a consequence of Duhamel principle and smoothing properties of the Picard series.

The last corollary is the situation of equation (1.2), where $Z(t)$ is a Fourier integral operator of order zero of the form $e^{iAt}Be^{-iAt}$.

Corollary 3. Let $Z(t) = ie^{iAt}Be^{-iAt}$ be a Fourier integral operator of order zero, where A and B are as in the Introduction, and A satisfies condition (1.2). Let us choose spaces W, W_0, W_1 as in Corollary 2. The smoothing property (iii) of Theorem 2.5 now follows from Theorem 2.1. Therefore, a combination of Theorems 2.1 and 2.5 yields Theorem 2.4.

Before proving Theorem 2.5, let us make a remark.

Remark 2.6. *Under conditions of Theorem 2.5, let also $R(t)$ be a continuous linear operator from W_1 to W_0. Then the solution operator $U(t)$ of the Cauchy problem*

$$\begin{cases} U' - Z(t)U = R(t), \\ U(0) = I \end{cases} \tag{2.3}$$

satisfies $U(t)w \in W$ for all $w \in W_1$. Moreover, if W_0, W_1, W are normed spaces and the estimate for solutions of (2.1) in Theorem 2.5 holds, then there is a constant $C > 0$ such that

$$\|U(t)w\|_W \le C(\|w\|_{W_1} + \|R(t)w\|_{W_0}), \tag{2.4}$$

for all $w \in W_1$.

We will now prove Theorem 2.5 and Remark 2.6. We note immediately that the statement of this remark with $R(t) = 0$ and assumption (iv) of Theorem 2.5 imply the statement of the theorem. Therefore, it is sufficient to prove the remark.

Let $R(t)$ be a continuous linear operator from W_1 to W_0. Let $U_0(t)$ be some partial solution to the problem

$$\begin{cases} U_0' - Z(t)U_0(t) = R(t), \\ U_0(0) = 0. \end{cases}$$

Then the solution $U(t)$ of the Cauchy problem (2.3) satisfies

$$U(t) = U_0(t) + I + \int_0^t Z(t_1)dt_1 + \int_0^t \int_0^{t_1} Z(t_1)Z(t_2)dt_2dt_1 + \dots \qquad (2.5)$$

The convergence of this series can be understood in L^2. Indeed, because of assumption (i), the term of this series with k integrals can be estimated by $t^k \sup_t ||Z(t)||^k_{L^2\to L^2}/k!$ From this and the well-posedness assumption it also follows that $U(t)$ is a solution of (2.3) from L^2 to L^2. Let us now define

$$S_N(t) = I + \int_0^t Z(t_1)dt_1 + \int_0^t \int_0^{t_1} Z(t_1)Z(t_2)dt_2dt_1 + \dots$$
$$+ \int_0^t \int_0^{t_1} \dots \int_0^{t_{N-1}} Z(t_1)Z(t_2)\dots Z(t_N)dt_N \dots dt_2dt_1.$$

Let $V(t) = U(t) - S_N(t)$, it is equal to $U_0(t)$ plus the remainder of the series (2.5). Then we have

$$V'(t) - Z(t)V(t) = R(t) - Z(t) \int_0^t \int_0^{t_2} \dots \int_0^{t_{N-1}} Z(t_2)\dots Z(t_N)dt_2 \dots dt_N. \quad (2.6)$$

Choosing $N = l$, and renumbering t_l's, it follows from assumption (iii) of the theorem that the second term is continuous from W_1 to W_0. Since also $R(t)$ is continuous from W_1 to W_0, it follows that the right hand side is also a continuous linear operator from W_1 to W_0.

Let $w = u(0) \in W_1$ be the Cauchy data for (2.2). If we denote by $\rho(t)$ the value of the operator in the last line of (2.6) at w, we will have $\rho(t) \in W_0$. The value of $V(0)$ is

$$V(0) = U(0) - S_N(0) = 0.$$

It follows now that $V(t)w$ solves Cauchy problem in (iv), so it belongs to W by the assumption. Since $S_N(t)$ is continuous from W_1 to W by assumption (ii), and $V(t)w = U(t)w - S_N(t)w$ is in W, be obtain $u(t,\cdot) = U(t)w \in W$.

Moreover, suppose that we also have the estimate $\|v(t,\cdot)\|_W \le C\|\rho(t)\|_{W_0}$ in (iv). Then we also have

$$\begin{aligned}\|u(t,\cdot)\|_W &\le \|V(t)w\|_W + \|S_N(t)w\|_W \le C\|\rho(t)\|_{W_0} + C\|w\|_{W_1}\\ &\le C(\|w\|_{W_1} + \|R(t)w\|_{W_0}),\end{aligned}$$

which implies (2.4).

References

[1] P. J. Braam and J. J. Duistermaat, Normal forms of real symmetric systems with multiplicity, *Indag. Math. (N.S.)* **4** (4) (1993), 407–421.

[2] J. Chazarain, Propagation des singularites pour une classe d'operateurs a caracteristiques multiples et resolubilite locale, (French) *Ann. Inst. Fourier (Grenoble)* **24** (1) (1974), 203–223.

[3] Y. Colin de Verdière, The level crossing problem in semi-classical analysis, I: Symmetric case, II: The Hermitian case, Preprints.

[4] N. Dencker, On the propagation of singularities for pseudo-differential operators with characteristics of variable multiplicity, *Comm. Partial Differential Equations* **17** (9–10) (1992), 1709–1736.

[5] J.J. Duistermaat and V. Guillemin, The spectrum of positive elliptic operators and periodic bicharacteristics, *Invent. Math.* **29** (1975), 39–79.

[6] L. Hörmander, The spectral function of an elliptic operator, *Acta Math.* **121** (1968), 193–218.

[7] W. Ichinose, Propagation of singularities for a hyperbolic system with double characteristics, *Osaka J. Math.* **19** (1) (1982), 171–187.

[8] V. Ya. Ivrii, *Microlocal Analysis and Precise Spectral Asymptotics*, Springer Monographs in Mathematics. Springer-Verlag, Berlin, 1998.

[9] C. Iwasaki and Y. Morimoto, Propagation of singularities of solutions for a hyperbolic system with nilpotent characteristics, II, *Comm. Partial Differential Equations* **9** (15) (1984), 1407–1436.

[10] I. Kamotski and M. Ruzhansky, Representation of solutions and regularity properties for weakly hyperbolic systems, *Funct. Anal. Applic.*, to appear.

[11] I. Kamotski and M. Ruzhansky, Regularity properties, representation of solutions and spectral asymptotics of systems with multiplicities, Preprint, arXiv:math.AP/0402203.

[12] V. V. Kucherenko, Asymptotic behavior of the solution of the system $A(x, -ih\partial/\partial x)u = 0$ as $h \to 0$ in the case of characteristics with variable multiplicity, (Russian) *Izv. Akad. Nauk SSSR Ser. Mat.* **38** (1974), 625–662.

[13] H. Kumano-go, *Pseudo-Differential operators.* MIT press, Cambridge, Mass. London, 1981.

[14] S. Mizohata and Y. Ohya, Sur la condition d'hyperbolicite pour les equations a caracteristiques multiples. II, (French) *Japan. J. Math.* **40** (1971), 63–104.

[15] Y. Morimoto, Fundamental solution for a hyperbolic equation with involutive characteristics of variable multiplicity, *Comm. Partial Differential Equations* **4** (6) (1979), 609–643.

[16] G. Rozenblum, Spectral asymptotic behavior of elliptic systems, (Russian) *Zap. LOMI* **96** (1980), 255–271, 311–312.

[17] M. Ruzhansky, Singularities of affine fibrations in the regularity theory of Fourier integral operators, *Russian Math. Surveys* **55** (2000), 99-170.

[18] Yu. Safarov and D. Vassiliev, *The Asymptotic Distribution of Eigenvalues of Partial Differential Operators.* Translations of Mathematical Monographs **155** AMS, Providence, RI, 1997.

Ilia Kamotski and Michael Ruzhansky
Department of Mathematics, Imperial College
180 Queen's Gate, London SW7 2BZ
UK
e-mail: `kamotski@ic.ac.uk`
`ruzh@ic.ac.uk`

Operator Theory:
Advances and Applications, Vol. 164, 65–78

Global Calculus of Fourier Integral Operators, Weighted Estimates, and Applications to Global Analysis of Hyperbolic Equations

Michael Ruzhansky and Mitsuru Sugimoto

Abstract. The aim of this paper is to present certain global regularity properties of hyperbolic equations. In particular, it will be determined in what way the global decay of Cauchy data implies the global decay of solutions. For this purpose, global weighted estimates in Sobolev spaces for Fourier integral operators will be reviewed. We will also present elements of the global calculus under minimal decay assumptions on phases and amplitudes.

Mathematics Subject Classification (2000). Primary 35L40; Secondary 35B65.

Keywords. Fourier integral operators, hyperbolic equations, global estimates.

1. Introduction

The aim of this paper is to present some results on the global regularity properties of hyperbolic equations. In particular, we will establish how the global decay of Cauchy data implies the global decay of solutions. For example, let $u(t,x)$ be the solution of the Cauchy problem

$$\begin{cases} i\partial_t u - a(D_x)u = 0, \\ u(0,x) = u_0(x), \ x \in \mathbb{R}^n, \end{cases} \tag{1.1}$$

where $a(D_x)$ is a pseudo-differential operator of order 1 with a real valued symbol satisfying the dispersive condition. We will show that solutions to (1.1) satisfy estimates of the form

$$||t^k \langle x\rangle^{-k} \langle x\rangle^{-s} u||_{L^2(\mathbb{R}_t\times\mathbb{R}^n_x)} \leq C_k ||\langle x\rangle^k u_0||_{L^2(\mathbb{R}^n_x)},$$

for all $k = 0, 1, 2, \ldots$ and $s > 1/2$. (See Section 6 for the precise statement). We will use the notation $\langle x\rangle = (1 + |x|^2)^{1/2}$. These estimates are an extended hyperbolic

This work was completed with the aid of UK-Japan Joint Project Grant by The Royal Society and Japan Society for the Promotion of Science.

version of global smoothing estimates for equations of Schrödinger type presented in [17]. Our proof of these estimates will rely on the global weighted estimates and the global calculus for a class of Fourier integral operators. For these purpose we will first present such a calculus making minimal decay assumptions on phases and amplitudes.

Let T be an operator of the form

$$Tu(x) = \int_{\mathbb{R}^n}\int_{\mathbb{R}^n} e^{i(x\cdot\xi+\varphi(y,\xi))} a(x,y,\xi)u(y) đ\xi\, dy, \tag{1.2}$$

with a real valued phase function $\varphi \in C^\infty(\mathbb{R}^n \times \mathbb{R}^n)$ and with a smooth amplitude $a(x,y,\xi)$, satisfying

$$|\partial_x^\alpha \partial_y^\beta \partial_\xi^\gamma a(x,y,\xi)| \le C_{\alpha\beta\gamma} \langle x\rangle^{m_1} \langle y\rangle^{m_2} \langle \xi\rangle^{m_3}, \quad x,y,\xi \in \mathbb{R}^n,$$

for all α,β,γ. As usual, we understand these operators as limits in $\mathcal{S}(\mathbb{R}^n)$ as $\epsilon \to 0$, of operators with amplitudes $a_\epsilon(x,y,\xi) = a(x,y,\xi)\gamma(\epsilon y, \epsilon\xi)$, where $\gamma \in C_0^\infty(\mathbb{R}^n \times \mathbb{R}^n)$ equals one near the origin. These operators extend to $\mathcal{S}'(\mathbb{R}^n)$.

As usual, we will denote $D^\alpha = i^{-|\alpha|}\partial^\alpha$. We denote $đ\xi = (2\pi)^{-n} d\xi$. We will also use the standard notation $a = a(x,y,\xi) \in S^m_{1,0}(\mathbb{R}^n \times \mathbb{R}^n \times \mathbb{R}^n)$ if

$$|\partial_x^\alpha \partial_y^\beta \partial_\xi^\gamma a(x,y,\xi)| \le C_{\alpha\beta\gamma} \langle \xi\rangle^{m-|\gamma|},$$

for all multi-indices α,β,γ and all $x,y,\xi \in \mathbb{R}^n$.

We note that Fourier integral operators assume the form (1.2) microlocally while there are topological obstructions to such parametrizations globally if one is using real valued phase functions. However, operators in the form (1.2) appear naturally in applications to hyperbolic and Schrödinger equations, which we will discuss later.

By a slight abuse of terminology, we will still call operators (1.2) Fourier integral operators. We will also present results for adjoint operators. In particular, this includes operators of the form

$$Su(x) = \int_{\mathbb{R}^n} e^{i\varphi(x,\xi)} a(x,\xi)\hat{u}(\xi) đ\xi,$$

which appear as parametrices for strictly hyperbolic equations.

We will be interested in global L^2 and Sobolev properties of operators (1.2). Local properties of these operators are quite well understood. For example, if φ is a non-degenerate phase function and $a \in S^0_{1,0}$, then operators (1.2) are continuous from $L^2_{comp}(\mathbb{R}^n)$ to $L^2_{loc}(\mathbb{R}^n)$. General local properties of these operators in L^p and other function spaces have been establishes by Seeger, Sogge and Stein in [19], and a survey of related results and their dependence on geometric properties of the corresponding wave fronts can be found in [13].

Global L^2 properties of operators (1.2) have been thoroughly investigated in the case of pseudo-differential operators when $\varphi(y,\xi) = -y\cdot\xi$. Global L^2–boundedness of some classes of Fourier integral operators was shown by Asada-Fujiwara ([1]) and Kumano-go ([10]) under conditions on infinitely many derivatives of the phase φ. Some results with conditions on phase and amplitude in

Sobolev-Kato spaces were obtained by Boulkhemair in [3]. However, in all these results one required that $\partial_\xi \partial_\xi \varphi$ is globally bounded, which fails in many important situations. For example, in applications to global smoothing problems one often has $\varphi(y,\xi) = y \cdot \psi(\xi)$, in which case the assumption above fails.

An important case of operators in (1.2) are SG–operators. We say that $a \in SG^{m_1,m_2,m_3}(\mathbb{R}^n \times \mathbb{R}^n \times \mathbb{R}^n)$ if

$$|\partial_x^\alpha \partial_y^\beta \partial_\xi^\gamma a(x,y,\xi)| \le C_{\alpha\beta\gamma} \langle x\rangle^{m_1-|\alpha|} \langle y\rangle^{m_2-|\beta|} \langle \xi\rangle^{m_3-|\gamma|}$$

holds for all α,β,γ and all $x,y,\xi \in \mathbb{R}^n$. The calculus of pseudo-differential operators with SG–symbols is described in [7]. Assuming that the phase φ in (1.2) satisfies $\varphi \in SG^{1,1}$ and

$$C_1\langle y\rangle \le \langle \partial_\xi \varphi(y,\xi)\rangle \le C_2\langle y\rangle,\ C_1\langle \xi\rangle \le \langle \partial_y \varphi(y,\xi)\rangle \le C_2\langle \xi\rangle,$$

Coriasco showed in [8] that operators T with amplitudes $a \in SG^{0,0,0}$ are bounded in $L^2(\mathbb{R}^n)$.

In [2], Boggiato, Buzano, and Rodino studied global problems with phase and amplitude from some polynomial classes associated to the Newton polygon of the hyperbolic operator. They obtained the $L^2(\mathbb{R}^n)$-boundedness of corresponding operators under similar decay assumptions on φ and a.

As an application, one derives weighted estimates in Sobolev spaces for the solution u of Cauchy problem

$$\partial_t u + ia(t,x,D_x)u = 0,\ u(0) = u_0.$$

with $a \in SG^{1,1}$ (as in [8]) or in some similar classes (as in [2]).

However, in this paper we will present results which show that assumptions on the phase and amplitude can be considerably weakened for operators to be still bounded in weighted Sobolev L^2 spaces, and for the calculus.

A natural question is, what are the minimal requirements for global L^2, Sobolev, weighted Sobolev estimates and for the global calculus to hold? In the following sections we will give some answers to these questions. In the last sections we will use this to investigate global properties of solutions to hyperbolic equations. In particular, in Section 6 we will prove a decay estimate for solutions of problem (1.1).

2. L^2–Boundedness

In [15] we proved the following result for operators given by (1.2). Assume that on the support of the amplitude a we have

$$\text{(C1)} \qquad |\det \partial_y \partial_\xi \varphi(y,\xi)| \ge C > 0, \forall (y,\xi) \in \mathbb{R}^n \times \mathbb{R}^n,$$

$$\text{(C2)} \qquad |\partial_y^\alpha \partial_\xi \varphi(y,\xi)| \le C_\alpha, |\partial_y \partial_\xi^\beta \varphi(y,\xi)| \le C_\beta, \forall (y,\xi), 1 \le |\alpha|, |\beta| \le 2n+2.$$

In this situation we have shown that operators T in (1.2) are bounded in $L^2(\mathbb{R}^n)$. Note that condition (C1) is a global version of the "local graph condition", which is necessary even for local L^2-boundedness of operators with amplitudes $a \in S^0_{1,0}$.

Note also that (C1) and (C2) are satisfied in many applications. For example, they are satisfied when $\varphi(y,\xi) = \psi(\xi)\cdot y$, ψ is homogeneous of order one for large ξ and $|\det D\psi(\xi)| \geq C > 0$. Such phases appear in the global smoothing problems for hyperbolic and Schrödinger equation (see [16]). These conditions are also satisfied in the SG-setting.

Let us first review results in $L^2(\mathbb{R}^n)$ for operators in (1.2) and their adjoints. Consider first operators in the form

$$Tu(x) = \int_{\mathbb{R}^n}\int_{\mathbb{R}^n} e^{i(x\cdot\xi+\varphi(y,\xi))} a(x,\xi)u(y) đ\xi\, dy,$$

which is a special case of (1.2) with amplitudes independent of y.

Theorem 2.1. *Let $\varphi(y,\xi)$ satisfy conditions* (C1) *and* (C2). *Let $a(x,\xi)$ satisfy one of the following conditions:*

(1) $\partial_x^\alpha \partial_\xi^\beta a(x,\xi) \in L^\infty(\mathbb{R}^n_x \times \mathbb{R}^n_\xi)$, $\alpha,\beta \in \{0,1\}^n$.
(2) $\partial_x^\alpha \partial_\xi^\beta a(x,\xi) \in L^\infty(\mathbb{R}^n_x \times \mathbb{R}^n_\xi)$, $|\alpha|,|\beta| \leq [n/2]+1$.
(3) $\exists \lambda,\lambda' > n/2 : (1-\Delta_x)^{\lambda/2}(1-\Delta_\xi)^{\lambda'/2} a(x,\xi) \in L^\infty(\mathbb{R}^n_x \times \mathbb{R}^n_\xi)$.
(4) $\partial_x^\alpha \partial_\xi^\beta a(x,\xi) \in L^\infty(\mathbb{R}^n_x \times \mathbb{R}^n_\xi)$, $|\alpha| \leq [n/2]+1, \beta \in \{0,1\}^n$.
(5) *there exists p with $2 \leq p < \infty$ such that* $\partial_x^\alpha \partial_\xi^\beta a(x,\xi) \in L^p(\mathbb{R}^n_x \times \mathbb{R}^n_\xi)$, $|\alpha| \leq [n(1/2-1/p)]+1, |\beta| \leq 2n$.

Then T is $L^2(\mathbb{R}^n)$–bounded.

This theorem follows from a more general result for operators with symbols in Besov spaces which was proved by the authors in [15]. Other conditions for the $L^2(\mathbb{R}^n)$-boundedness of operators of this form were announced in [14] and proved in [15]. Note that in the case of pseudo-differential operators, (1) was proved by Calderon and Vaillancourt ([4]) for $\alpha,\beta \in \{0,1,2,3\}^n$, (2) and (3) were proved by Cordes in [6], (4) and (5) were proved by Coifman and Meyer ([5]) for $\beta \subset \{0,1,2\}^n$. Results for pseudo-differential operators with symbols in Besov spaces can be found in [21].

Slightly different results hold for amplitudes independent of x. Let T now be defined by

$$Tu(x) = \int_{\mathbb{R}^n}\int_{\mathbb{R}^n} e^{i(x\cdot\xi+\varphi(y,\xi))} a(y,\xi)u(y) đ\xi\, dy,$$

so phase and amplitude depend on the same set of variables. The same result holds for the adjoint operators

$$T^*u(x) = \int_{\mathbb{R}^n}\int_{\mathbb{R}^n} e^{i(\varphi(x,\xi)-y\cdot\xi)} a(x,\xi)u(y) đ\xi\, dy = \int_{\mathbb{R}^n} e^{i\varphi(x,\xi)} a(x,\xi)\hat{u}(\xi) đ\xi.$$

Theorem 2.2. *Assume that*

$$|\partial_y^\alpha \partial_\xi^\beta a(y,\xi)| \leq C_{\alpha\beta},$$

for $|\alpha|,|\beta| \leq 2n+1$. Also assume (C1), (C2), *i.e. that on* $\operatorname{supp} a(y,\xi)$ *we have*

$$|\det \partial_y \partial_\xi \varphi(y,\xi)| \geq C > 0$$

and

$$|\partial_y^\alpha \partial_\xi \varphi(y,\xi)| \le C_\alpha, \qquad |\partial_y \partial_\xi^\beta \varphi(y,\xi)| \le C_\beta$$

for $1 \le |\alpha|, |\beta| \le 2n+2$. *Then the operator* T *is* $L^2(\mathbb{R}^n)$*-bounded, and satisfies*

$$||T||_{L^2\to L^2} \le C \sup_{|\alpha|,|\beta|\le 2n+1} ||\partial_y^\alpha \partial_\xi^\beta a(y,\xi)||_{L^\infty(\mathbb{R}^n_y\times\mathbb{R}^n_\xi)}.$$

Now let us consider the case of general amplitudes. Here we will use a slightly different, but equivalent (after taking adjoints) representation of T. We will now allow the amplitude to depend on all of x, y, ξ.

Theorem 2.3. *Let* T *be defined by*

$$Tu(x) = \int_{\mathbb{R}^n}\int_{\mathbb{R}^n} e^{i(\varphi(x,\xi)-y\cdot\xi)} a(x,y,\xi)u(y)\, dy\, đ\xi.$$

Let the phase $\varphi(x,\xi) \in C^\infty$ *for some positive constants and all* $|\alpha|, |\beta| \ge 1$ *satisfy*

$$|\det \partial_x \partial_\xi \varphi(x,\xi)| \ge C_0 > 0,$$
$$\exists x_\beta : |\partial_\xi^\beta \varphi(x_\beta,\xi)| \le C_\beta,$$
$$|\partial_x^\alpha \partial_\xi^\beta \varphi(x,\xi)| \le C_{\alpha\beta}.$$

Let amplitude $a = a(x,y,\xi) \in C^\infty$ *for some* $m \in \mathbb{R}$ *satisfy*

$$|\partial_x^\alpha \partial_y^\beta \partial_\xi^\gamma a(x,y,\xi)| \le C_{\alpha\beta\gamma} \langle x\rangle^m \langle y\rangle^{-m-|\beta|}.$$

for all α, β, γ *and all* $x, y, \xi \in \mathbb{R}^n$. *Then* T *is* $L^2(\mathbb{R}^n)$*–bounded.*

Proofs of theorems in this section can be found in [15].

3. Weighted Estimates in Sobolev Spaces

We will say that $f \in H^{s_1,s_2}(\mathbb{R}^n)$ if $f \in \mathcal{S}'(\mathbb{R}^n)$ and $\Pi_{s_1,s_2} f \in L^2(\mathbb{R}^n)$, where Π_{s_1,s_2} is a pseudo-differential operator with symbol $\pi_{s_1,s_2}(x,\xi) = \langle x\rangle^{s_1}\langle\xi\rangle^{s_2}$. We can also define operators $\tilde\Pi_{s_1,s_2}$ with symbols $\tilde\pi_{s_1,s_2}(y,\xi) = \langle y\rangle^{s_1}\langle\xi\rangle^{s_2}$. Then it can be shown (e.g. [7]) that $\tilde\Pi_{-s_1,-s_2} = \Pi^{-1}_{s_1,s_2}$. Also, Sobolev space defined by $\tilde\Pi_{s_1,s_2}$, i.e. the space of all $f \in \mathcal{S}'(\mathbb{R}^n)$ with $\tilde\Pi_{s_1,s_2} f \in L^2(\mathbb{R}^n)$, coincide with $H^{s_1,s_2}(\mathbb{R}^n)$.

We will be studying operators

$$Tu(x) = \int_{\mathbb{R}^n}\int_{\mathbb{R}^n} e^{i(\varphi(x,\xi)-y\cdot\xi)} a(x,y,\xi)u(y)\, dy\, đ\xi$$

with real-valued phase $\varphi = \varphi(x,\xi) \in C^\infty(\mathbb{R}^n\times\mathbb{R}^n)$ and amplitude $a = a(x,y,\xi) \in C^\infty(\mathbb{R}^n\times\mathbb{R}^n\times\mathbb{R}^n)$.

Theorem 3.1. *Let operator* T *be defined by*

$$Tu(x) = \int_{\mathbb{R}^n}\int_{\mathbb{R}^n} e^{i(\varphi(x,\xi)-y\cdot\xi)} a(x,y,\xi)u(y)\, dy\, đ\xi.$$

Let the phase $\varphi = \varphi(x,\xi) \in C^\infty$ *for all* $|\alpha|, |\beta| \geq 1$ *satisfy* $|\det \partial_x \partial_\xi \varphi(x,\xi)| \geq C_0 > 0$, *and*

$$|\partial_x^\alpha \varphi(x,\xi)| \leq C_\alpha \langle \xi \rangle, |\partial_x^\alpha \partial_\xi^\beta \varphi(x,\xi)| \leq C_{\alpha\beta}.$$

Assume one of the following:

(1) *For all* α, β, *and* γ,

$$|\partial_x^\alpha \partial_y^\beta \partial_\xi^\gamma a(x,y,\xi)| \leq C_{\alpha\beta\gamma} \langle x \rangle^{m_1} \langle y \rangle^{m_2 - |\beta|} \langle \xi \rangle^{m_3},$$

and for all $|\beta| \geq 1$,

$$|\partial_\xi^\beta \varphi(x,\xi)| \leq C_\beta \langle x \rangle.$$

(2) *For all* α, β, *and* γ,

$$|\partial_x^\alpha \partial_y^\beta \partial_\xi^\gamma a(x,y,\xi)| \leq C_{\alpha\beta\gamma} \langle x \rangle^{m_1 - |\alpha|} \langle y \rangle^{m_2} \langle \xi \rangle^{m_3},$$

and for all α *and* $|\beta| \geq 1$,

$$|\partial_x^\alpha \partial_\xi^\beta \varphi(x,\xi)| \leq C_{\alpha\beta} \langle x \rangle^{1-|\alpha|}.$$

Then T *is bounded from* $H^{s_1,s_2}(\mathbb{R}^n)$ *to* $H^{s_1-m_1-m_2, s_2-m_3}(\mathbb{R}^n)$, *for all* $s_1, s_2 \in \mathbb{R}^n$.

The proof of this theorem relies on the composition formulae that we will describe in the next section.

4. Compositions with Pseudo-Differential Operators

Some global calculus of operators (1.2) and their adjoints can be constructed already under quite mild assumptions on phases and amplitudes. Let operator T be globally defined by

$$Tu(x) = \int_{\mathbb{R}^n} \int_{\mathbb{R}^n} e^{i(\varphi(x,\xi) - y\cdot\xi)} a(x,y,\xi) u(y) dy đ\xi,$$

with a smooth amplitude $a(x,y,\xi)$, satisfying

$$|\partial_x^\alpha \partial_y^\beta \partial_\xi^\gamma a(x,y,\xi)| \leq C_{\alpha\beta\gamma} \langle x \rangle^{m_1} \langle y \rangle^{m_2} \langle \xi \rangle^{m_3}$$

for all α, β, γ and all $x, y, \xi \in \mathbb{R}^n$. Pseudo-differential operators are defined by

$$Pu(x) = \int_{\mathbb{R}^n} \int_{\mathbb{R}^n} e^{i(x-y)\cdot\xi} p(x,\xi) u(y) dy đ\xi$$

with $p(x,\xi)$ satisfying similar estimates. We will now describe the main features of compositions of T and P:

• If $|\partial_x^\alpha \partial_\xi^\beta p(x,\xi)| \leq C_{\alpha\beta} \langle x \rangle^{t_1} \langle \xi \rangle^{t_2 - |\beta|}$ for all α, β and all $x, \xi \in \mathbb{R}^n$, then the amplitude $c = c(x,y,\xi)$ of the composition TP satisfies

$$|\partial_x^\alpha \partial_y^\beta \partial_\xi^\gamma c(x,y,\xi)| \leq C_{\alpha\beta\gamma} \langle x \rangle^{m_1} \langle y \rangle^{m_2 + t_1} \langle \xi \rangle^{m_3 + t_2},$$

with no conditions on the phase. See Theorems 4.2 and 4.3 below for two alternative descriptions.

• Similar for PT with some conditions on the phase. See Theorem 4.1 below.

• We have global asymptotic expansions for amplitudes. For example, the amplitude of TP has an expansion improving in ξ:

$$c(x,z,\xi) \sim \sum_\alpha \frac{i^{-|\alpha|}}{\alpha!} \partial_y^\alpha \left[a(x,y,\xi) \partial_\xi^\alpha p(y,\xi) \right] |_{y=z}.$$

In this paper, when we say that the asymptotic expansion $a \sim \sum_{j=1}^\infty a_j$ is improving in ξ, it means that for every α, β, γ and M there are N and C such that

$$\left| \partial_x^\alpha \partial_y^\beta \partial_\xi^\gamma \left(a - \sum_{j=1}^N a_j \right) \right| \le C \langle x \rangle^{m_1} \langle y \rangle^{m_2} \langle \xi \rangle^{m_3 - M}.$$

Similarly, we can define expansions improving in x or in y.

If a and p have additional decay properties, similar property can be extracted for c. For example, if a, p are SG-functions, so is c. If they satisfy conditions of Boggiatto-Buzano-Rodino in [2], so does c.

Let us write formal compositions first. The composition PT is of the form

$$\begin{aligned}
(PTu)(x) &= \iint e^{i(x-y)\cdot\eta} p(x,\eta) Tu(y) dy đ\eta \\
&= \iint \iint e^{i((x-y)\cdot\eta + \varphi(y,\xi) - z\cdot\xi)} p(x,\eta) a(y,z,\xi) u(z) dz đ\xi dy đ\eta \\
&= \iint e^{i(\varphi(x,\xi) - z\cdot\xi)} c_1(x,z,\xi) u(z) dz đ\xi,
\end{aligned}$$

where

$$c_1(x,z,\xi) = \iint e^{i(\varphi(y,\xi) - \varphi(x,\xi) + (x-y)\cdot\eta)} a(y,z,\xi) p(x,\eta) dy đ\eta.$$

We have the following theorem on composition $P(x,D) \circ T$.

Theorem 4.1 (Composition PT). *Let operator T be defined by*

$$Tu(x) = \int_{\mathbb{R}^n} \int_{\mathbb{R}^n} e^{i(\varphi(x,\xi) - y\cdot\xi)} a(x,y,\xi) u(y) dy đ\xi.$$

Let the phase $\varphi = \varphi(x,\xi) \in C^\infty(\mathbb{R}^n \times \mathbb{R}^n)$ be such that

$$C_1 \langle \xi \rangle \le \langle \nabla_x \varphi(x,\xi) \rangle \le C_2 \langle \xi \rangle, \quad x, \xi \in \mathbb{R}^n,$$

for some $C_1, C_2 > 0$, and such that for all $|\alpha|, |\beta| \ge 1$ we have

$$|\partial_x^\alpha \varphi(x,\xi)| \le C_\alpha \langle \xi \rangle, \quad |\partial_x^\alpha \partial_\xi^\beta \varphi(x,\xi)| \le C_{\alpha\beta}, \quad x, \xi \in \mathbb{R}^n.$$

Let $a = a(x,y,\xi) \in C^\infty(\mathbb{R}^{3n})$ satisfy

$$|\partial_x^\alpha \partial_y^\beta \partial_\xi^\gamma a(x,y,\xi)| \le C_{\alpha\beta\gamma} \langle x \rangle^{m_1} \langle y \rangle^{m_2} \langle \xi \rangle^{m_3},$$

for all α, β, γ, and all $x, y, \xi \in \mathbb{R}^n$. Let $p = p(x,\xi) \in C^\infty(\mathbb{R}^n \times \mathbb{R}^n)$ for all α, β satisfy

$$|\partial_x^\alpha \partial_\xi^\beta p(x,\xi)| \le C_{\alpha\beta} \langle x \rangle^{t_1} \langle \xi \rangle^{t_2 - |\beta|}, \quad x, \xi \in \mathbb{R}^n.$$

Then the composition $B = P(x, D) \circ T$ is an operator of the form

$$Bu(x) = \int_{\mathbb{R}^n} e^{i(\varphi(x,\xi) - z\cdot\xi)} c(x,z,\xi) u(z) dz đ\xi,$$

with amplitude $c(x, z, \xi)$ satisfying

$$|\partial_x^\alpha \partial_z^\beta \partial_\xi^\gamma c(x,z,\xi)| \le C_{\alpha\beta\gamma} \langle x\rangle^{m_1+t_1} \langle z\rangle^{m_2} \langle \xi\rangle^{m_3+t_2},$$

for all α, β, γ, and all $x, z, \xi \in \mathbb{R}^n$. Moreover, we have the asymptotic expansion, improving in ξ:

$$c(x,z,\xi) \sim \sum_\alpha \frac{i^{-|\alpha|}}{\alpha!} \partial_\xi^\alpha p(x, \nabla_x\varphi(x,\xi)) \partial_y^\alpha \left[e^{i\Psi(x,y,\xi)} a(y,z,\xi)\right] |_{y=x},$$

where $\Psi(x, y, \xi) = \varphi(y, \xi) - \varphi(x, \xi) + (x - y) \cdot \nabla_x\varphi(x, \xi)$.

The composition TP is of the form

$$\begin{aligned}(TPu)(x) &= \iint e^{i(\varphi(x,\xi) - y\cdot\xi)} a(x,y,\xi) Pu(y) dy đ\xi \\ &= \iint \iint e^{i(\varphi(x,\xi) - y\cdot\xi + (y-z)\cdot\eta)} a(x,y,\xi) p(y,\eta) u(z) dz đ\eta dy đ\xi \qquad (4.1) \\ &= \iint e^{i(\varphi(x,\eta) - z\cdot\eta)} c_2(x,z,\eta) u(z) dz đ\eta,\end{aligned}$$

where

$$c_2(x,z,\eta) = c_2(x,\eta) = \iint e^{i(\varphi(x,\xi) - \varphi(x,\eta) + y\cdot(\eta-\xi))} a(x,y,\xi) p(y,\eta) dy đ\xi. \qquad (4.2)$$

Note that this representation of TP makes the amplitude dependent on only two variables. This will be used later in Corollary 4.4.

Let us now describe another representation of TP to obtain a slightly different estimate for its amplitude. We can represent TP as

$$(TPu)(x) = \int_{\mathbb{R}^n} \int_{\mathbb{R}^n} e^{i(\varphi(x,\xi) - z\cdot\xi)} c(x,z,\xi) u(z) dz đ\xi, \qquad (4.3)$$

with amplitude

$$c(x,z,\xi) = \int_{\mathbb{R}^n} \int_{\mathbb{R}^n} e^{i(y-z)\cdot(\eta-\xi)} a(x,y,\xi) p(y,\eta) dy đ\eta. \qquad (4.4)$$

Note that in comparison with representation (4.1)-(4.2), the amplitude here depends on three variables, but there is no entry of the phase. This allows to treat more general phases and amplitudes. The following two theorems describe the composition $T \circ P(x, D)$.

Theorem 4.2 (Composition TP). *Let operator T be defined by*

$$Tu(x) = \int_{\mathbb{R}^n} \int_{\mathbb{R}^n} e^{i(\varphi(x,\xi) - y\cdot\xi)} a(x,y,\xi) u(y)\, dy\, đ\xi,$$

where amplitude $a = a(x,y,\xi) \in C^\infty(\mathbb{R}^{3n})$ *satisfies*

$$|\partial_x^\alpha \partial_y^\beta \partial_\xi^\gamma a(x,y,\xi)| \le C_{\alpha\beta\gamma} \langle x\rangle^{m_1} \langle y\rangle^{m_2} \langle \xi\rangle^{m_3}, \quad x,y,\xi \in \mathbb{R}^n,$$

for all α, β, γ. *Let the phase* $\varphi(x,\xi)$ *be any function. Let* $p \in C^\infty(\mathbb{R}^n \times \mathbb{R}^n)$ *for all* α, β *satisfy the estimate*

$$|\partial_y^\alpha \partial_\eta^\beta p(y,\eta)| \le C_{\alpha\beta} \langle y\rangle^{t_1} \langle \eta\rangle^{t_2-|\beta|}, \quad y,\eta \in \mathbb{R}^n.$$

Then for all α, β, γ, *the amplitude* $c(x,z,\xi)$ *of* $T \circ P$ *in* (4.4) *satisfies*

$$|\partial_x^\alpha \partial_z^\beta \partial_\xi^\gamma c(x,z,\xi)| \le C_{\alpha\beta\gamma} \langle x\rangle^{m_1} \langle z\rangle^{m_2+t_1} \langle \xi\rangle^{m_3+t_2}, \quad x,z,\xi \in \mathbb{R}^n.$$

Moreover, we have an asymptotic expansion, improving in ξ:

$$c(x,z,\xi) \sim \sum_\alpha \frac{i^{-|\alpha|}}{\alpha!} \partial_y^\alpha \left[a(x,y,\xi) \partial_\xi^\alpha p(y,\xi) \right] |_{y=z}.$$

The composition formula for TP is given in Theorem 4.2 under no conditions on the phase. However, the following theorem shows that if the phase satisfies natural non-degeneracy assumptions and the amplitude $a(x,y,\xi)$ has decay properties in y, the amplitude of the composition TP can be made dependent on two variables only. In this way we obtain another representation for the composition $T \circ P(x,D)$.

Theorem 4.3 (Composition TP**).** *Let operator* T *be defined by*

$$Tu(x) = \int_{\mathbb{R}^n} \int_{\mathbb{R}^n} e^{i(\varphi(x,\xi) - y\cdot\xi)} a(x,y,\xi) u(y) dy đ\xi.$$

Let the phase $\varphi = \varphi(x,\xi) \in C^\infty(\mathbb{R}^n \times \mathbb{R}^n)$ *be such that*

$$C_1 \langle x\rangle \le \langle \nabla_\xi \varphi(x,\xi)\rangle \le C_2 \langle x\rangle, \quad x,\xi \in \mathbb{R}^n, \tag{4.5}$$

for some $C_1, C_2 > 0$, *and such that for all* $|\alpha|, |\beta| \ge 1$ *we have*

$$|\partial_\xi^\beta \varphi(x,\xi)| \le C_\beta \langle x\rangle, \quad |\partial_x^\alpha \partial_\xi^\beta \varphi(x,\xi)| \le C_{\alpha\beta}, \quad x,\xi \in \mathbb{R}^n. \tag{4.6}$$

Let $a = a(x,y,\xi) \in C^\infty(\mathbb{R}^{3n})$ *satisfy*

$$|\partial_x^\alpha \partial_y^\beta \partial_\xi^\gamma a(x,y,\xi)| \le C_{\alpha\beta\gamma} \langle x\rangle^{m_1} \langle y\rangle^{m_2-|\beta|} \langle \xi\rangle^{m_3}, \tag{4.7}$$

for all α, β, γ, *and all* $x, y, \xi \in \mathbb{R}^n$. *Let* $p = p(x,\xi) \in C^\infty(\mathbb{R}^n \times \mathbb{R}^n$ *for all* α, β *satisfy*

$$|\partial_x^\alpha \partial_\xi^\beta p(x,\xi)| \le C_{\alpha\beta} \langle x\rangle^{t_1-|\alpha|} \langle \xi\rangle^{t_2}, \quad x,\xi \in \mathbb{R}^n.$$

Then the composition $B = T \circ P(x,D)$ *is an operator of the form*

$$Bu(x) = \int_{\mathbb{R}^n} e^{i\varphi(x,\xi)} c(x,\xi) \hat{u}(\xi) đ\xi,$$

with amplitude $c(x,\xi)$ *satisfying*

$$|\partial_x^\alpha \partial_\xi^\beta c(x,\xi)| \le C_{\alpha\beta} \langle x\rangle^{m_1+m_2+t_1} \langle \xi\rangle^{m_3+t_2},$$

for all α, β, γ, *and all* $x, z, \xi \in \mathbb{R}^n$.

Moreover, we have the asymptotic expansion, improving in x:

$$c(x,\xi)\sim\sum_{\alpha,\beta}\frac{i^{-(|\alpha|+|\beta|)}}{\alpha!\beta!}\partial_x^\alpha p(\nabla_\xi\varphi(x,\xi),\xi)\partial_\eta^{\alpha+\beta}\left[e^{i\Psi(\eta,\xi,x)}\partial_y^\beta a(x,\nabla_\xi\varphi(x,\xi),\eta)\right]|_{\eta=\xi},$$

where $\Psi(\eta,\xi,x)=\varphi(x,\xi)-\varphi(x,\eta)+(\eta-\xi)\cdot\nabla_\xi\varphi(x,\xi)$.

If we apply this Theorem with $p\equiv 1$, we can reduce the amplitude to functions dependent on two variables only. Operators of such type are the reduced forms of Fourier integral operators and the local versions are described by, for example, Kumano-go [11], Stein [20], and many other authors.

Corollary 4.4. *Let T be the operator from Theorem 4.3 with phase φ satisfying* (4.5), (4.6), *and amplitude satisfying* (4.7). *Then T can be written in the form*

$$Tu(x)=\int_{\mathbb{R}^n}e^{i\varphi(x,\xi)}c(x,\xi)\hat{u}(\xi)đ\xi,$$

with amplitude $c(x,\xi)$ satisfying

$$|\partial_x^\alpha\partial_\xi^\beta c(x,\xi)|\le C_{\alpha\beta}\langle x\rangle^{m_1+m_2}\langle\xi\rangle^{m_3},$$

for all α,β,γ, and all $x,z,\xi\in\mathbb{R}^n$. Moreover,

$$c(x,\xi)\sim\sum_\beta\frac{i^{-|\beta|}}{\beta!}\partial_\eta^\beta\left[e^{i\Psi(\eta,\xi,x)}\partial_y^\beta a(x,\nabla_\xi\varphi(x,\xi),\eta)\right]|_{\eta=\xi},$$

where Ψ is as in Theorem 4.3.

We can apply Corollary 4.4 to pseudo-differential operators to obtain a "normal form" for pseudo-differential operators with decay in only one of the variables. In the SG-setting a similar reduction is described by Cordes in [7].

Corollary 4.5. *Let T be a pseudo-differential operator of the form*

$$Tu(x)=\int_{\mathbb{R}^n}\int_{\mathbb{R}^n}e^{i(x-y)\cdot\xi}a(x,y,\xi)u(y)\,dy\,đ\xi,$$

with amplitude $a=a(x,y,\xi)\in C^\infty(\mathbb{R}^{3n})$ satisfying

$$|\partial_x^\alpha\partial_y^\beta\partial_\xi^\gamma a(x,y,\xi)|\le C_{\alpha\beta\gamma}\langle x\rangle^{m_1}\langle y\rangle^{m_2-|\beta|}\langle\xi\rangle^{m_3},\ \forall\alpha,\beta,\gamma,\ \ x,y,\xi\in\mathbb{R}^n.$$

Then T can be written in the form

$$Tu(x)=\int_{\mathbb{R}^n}e^{ix\cdot\xi}c(x,\xi)\hat{u}(\xi)đ\xi,$$

with amplitude $c(x,\xi)$ satisfying

$$|\partial_x^\alpha\partial_\xi^\beta c(x,\xi)|\le C_{\alpha\beta}\langle x\rangle^{m_1+m_2}\langle\xi\rangle^{m_3},\ \forall\alpha,\beta,\ \ x,\xi\in\mathbb{R}^n.$$

Moreover, we have the asymptotic expansion

$$c(x,\xi)\sim\sum_\beta\frac{i^{-|\beta|}}{\beta!}\partial_\xi^\beta\partial_y^\beta a(x,y,\xi)|_{y=x}.$$

Note that from the asymptotic expansions in these theorems it is clear that if a has additional decay with respect to some variables, so does the new amplitude. Proofs of theorems of this section will appear in [18].

5. Hyperbolic Equations

By estimates of the preceding sections we can obtain weighted estimates for solutions of the Cauchy problem

$$\begin{cases} i\partial_t u - a(t,x,D_x)u = 0, \\ u(0,x) = u_0(x), \end{cases}$$

where $a \in C^\infty((-T,T), S^0_{1,0})$ is real-valued, and estimate

$$|\partial_t^\alpha \partial_x^\beta \partial_\xi^\gamma a(t,x,\xi)| \leq C_{\alpha\beta\gamma}(1+|\xi|)^{1-|\gamma|}$$

holds for all multiindices α, β, γ and all $t \in (-T,T), x, \xi \in \mathbb{R}^n$. Then we have global weighted estimates for solutions:

$$||u(t,\cdot)||_{H^{s_1,s_2}} \leq C||u_0||_{H^{s_1,s_2}}.$$

Let now P be an $m \times m$ matrix of pseudo-differential operators of order one on $\mathbb{R}^n$. Let $u = u(t,x)$ be the solution of Cauchy problem

$$\begin{cases} i\partial_t u - Pu = 0, & (t,x) \in [0,T] \times \mathbb{R}^n, \\ u|_{t=0} = u_0. \end{cases}$$

Let $A(x,\xi)$ be the principal symbol of P and assume that it is smoothly diagonalizable. In this case we may allow $A(x,\xi)$ to have characteristics of variable multiplicities. Local L^p and other properties of solutions to such equations have been analyzed in [9]. Now we also have the global weighted estimate:

$$||u(t,\cdot)||_{H^{s_1,s_2}} \leq C||u_0||_{H^{s_1,s_2}}.$$

6. Global Smoothing for Hyperbolic Equations

Let us consider the following Cauchy problem

$$\begin{cases} i\partial_t u - a(D_x)u = 0, \\ u(0,x) = u_0(x). \end{cases} \tag{6.1}$$

Let us assume that operator $a(D_x)$ has a real valued symbol $a(\xi) \in C^\infty(\mathbb{R}^n)$ satisfying $\nabla a(\xi) \neq 0$ for all $\xi \in \mathbb{R}^n$. Assume also that $a(\xi) = a_1(\xi) + a_0(\xi)$ for large ξ, where $a_1(\xi) \in C^\infty(\mathbb{R}^n \setminus 0)$ is a positively homogeneous function of order 1 satisfying $\nabla a_1(\xi) \neq 0$ for all $\xi \neq 0$, and $a_0(\xi) \in S^0$. In the case $n = 1$, we can see that the following estimate for the solution $u = u(t,x)$ of (6.1) is true for $s > 1/2$:

$$||\langle x \rangle^{-s} u||_{L^2_{t,x}(\mathbb{R}\times\mathbb{R}^n)} \leq C||u_0||_{L^2_x(\mathbb{R}^n)}. \tag{6.2}$$

Note that $u(t,x)$ is just a translation of u_0 (if $a(\xi)$ is homogeneous). In [17] we have presented a result which implies that the estimate above is also true for

$n \geq 2$. Analysis of previous sections allows us to apply this to obtain further global smoothing properties of solutions to (6.1).

Theorem 6.1. *Let $u(t,x)$ be the solution of the Cauchy problem (6.1). Then for $s > 1/2$ we have the estimate*

$$||t\langle x\rangle^{-1}\langle x\rangle^{-s}u||_{L^2(\mathbb{R}_t\times\mathbb{R}^n_x)} \leq C||\langle x\rangle u_0||_{L^2(\mathbb{R}^n_x)}. \tag{6.3}$$

Moreover, for any $k = 0, 1, 2, \ldots$ we also have

$$||t^k\langle x\rangle^{-k}\langle x\rangle^{-s}u||_{L^2(\mathbb{R}_t\times\mathbb{R}^n_x)} \leq C||\langle x\rangle^k u_0||_{L^2(\mathbb{R}^n_x)}. \tag{6.4}$$

Proof. The solution $u = u(t,x)$ to (6.1) is given by

$$u(t,x) = T_t u_0(x) = \int_{\mathbb{R}^n}\int_{\mathbb{R}^n} e^{i[(x-y)\cdot\xi + ta(\xi)]}u_0(y)dy đ\xi.$$

Let $\sigma(X,D)$ be a pseudo-differential operator with symbol $\sigma(x,\xi) = x\cdot\gamma(\xi)$, where γ is a symbol to be chosen later. Let $\tau(x,\xi) = \sigma(x + t\nabla_\xi a(\xi),\xi) = (x + t\nabla_\xi a(\xi))\cdot\gamma(\xi)$.

If Ψ is the function defined in Theorem 4.3, it is straightforward to see that

$$\Psi(\eta,\xi,x) = x\cdot(\xi-\eta) + t(a(\xi) - a(\eta)) + (\eta - \xi)\cdot(x + t\nabla_\xi a(\xi)),$$

if we take $\phi(x,\xi) = x\cdot\xi + ta(\xi)$. We can observe that $\partial_{\eta_j} e^{i\Psi(\eta,\xi,x)}|_{\eta=\xi} = 0$. We can also observe that the corresponding function $\Psi(x,y,\xi)$ in Theorem 4.1 is identically zero. Since σ is linear in x, it follows from Theorems 4.1 and 4.3 that $\tau(X,D)\circ T_t = T_t\circ\sigma(X,D)$.

Taking $\sigma(X,D)u_0$ as the Cauchy data in (6.1) and using (6.2) we get

$$||\langle x\rangle^{-s}T_t\sigma(X,D)u_0||_{L^2(\mathbb{R}_t\times\mathbb{R}^n_x)} \leq C||\sigma(X,D)u_0||_{L^2(\mathbb{R}^n_x)},$$

which means that we have the estimate

$$||\langle x\rangle^{-s}(x + t\nabla a(D))\cdot\gamma(D)u||_{L^2(\mathbb{R}_t\times\mathbb{R}^n_x)} \leq C||x\cdot\gamma(D)u_0||_{L^2(\mathbb{R}^n_x)}$$

for $s > 1/2$. If γ is a pseudo-differential operator of order zero, then

$$||x\cdot\gamma(D)u_0||_{L^2_x} \leq ||\langle x\rangle\langle\gamma(D)\rangle u_0||_{L^2_x} \leq C||\langle x\rangle u_0||_{L^2_x}.$$

Moreover, it follows that

$$\begin{aligned}
&||\langle x\rangle^{-s}t\,\nabla a(D)\cdot\gamma(D)u||_{L^2_{t,x}} \\
&\leq ||\langle x\rangle^{-s}(x + t\nabla a(D))\cdot\gamma(D)u||_{L^2_{t,x}} + ||\langle x\rangle^{-s}x\cdot\gamma(D)u||_{L^2_{t,x}} \\
&\leq C(||\langle x\rangle u_0||_{L^2_x} + ||\langle x\rangle^{-s+1}u||_{L^2_{t,x}}).
\end{aligned}$$

Now, if we take $\gamma(D) = \nabla a(D)$ and use the $L^2(\mathbb{R}^n)$-boundedness of the operator $\langle x\rangle^{-s}|\nabla a(D)|^{-2}\langle x\rangle^s$ (Theorem 3.1), we get

$$\begin{aligned}
||\langle x\rangle^{-s}tu||_{L^2_{t,x}} &\leq C||\langle x\rangle^{-s}t\,\nabla a(D)\cdot\gamma(D)u||_{L^2_{t,x}} \\
&\leq C(||\langle x\rangle u_0||_{L^2_x} + ||\langle x\rangle^{-s+1}u||_{L^2_{t,x}}) \\
&\leq C||\langle x\rangle u_0||_{L^2_x},
\end{aligned}$$

since $||\langle x\rangle^{-s+1}u||_{L^2_{t,x}} \leq C||u_0||_{L^2_x}$ for $s > 3/2$. This completes the proof of (6.3). Formula (6.4) follows from it by an iteration argument. □

We remark that if $a(\xi)$ itself is homogeneous of order one with nonzero gradient and a possible singularity at the origin, Theorem 6.1 still holds for $n \geq 4$. To justify it, we use the $L^2(\mathbb{R}^n)$-boundedness of the operator $\langle x\rangle^{-s}|\nabla a(D)|^{-2}\langle x\rangle^s$ for $-n/2 < s < n/2$ (see [12] for example). We finally remark that estimate (6.3) is sharp in the sense that in general it does not hold for $s = 1/2$. For example, its failure can be easily seen for $n = 1$ and $a(\xi) = \xi$.

References

[1] K. Asada and D. Fujiwara, On some oscillatory integral transformations in $L^2(\mathbb{R}^n)$, *Japan. J. Math. (N.S.)* **4** (1978), 299–361.

[2] P. Boggiato, E. Buzano and L. Rodino, *Global Hypoellipticity and Spectral Theory,* Akademie Verlag, Berlin, 1996.

[3] A. Boulkhemair, Estimations L^2 precisees pour des integrales oscillantes, *Comm. Partial Differential Equations* **22** (1997), 165–184.

[4] A. P. Calderón and R. Vaillancourt, On the boundedness of pseudo-differential operators, *J. Math. Soc. Japan* **23** (1971), 374–378.

[5] R. R. Coifman and Y. Meyer, *Au-delà des opérateurs pseudo-différentiels*, Astérisque **57** (1978).

[6] H. O. Cordes, On compactness of commutators of multiplications and convolutions, and boundedness of pseudodifferential operators, *J. Funct. Anal.* **18** (1975), 115–131.

[7] H. O. Cordes, *The Technique of Pseudodifferential Operators*, Cambridge University Press 1995.

[8] S. Coriasco, Fourier integral operators in SG classes I: composition theorems and action on SG Sobolev spaces, *Rend. Sem. Mat. Univ. Pol. Torino* **57** (1999), 249–302.

[9] I. Kamotski and M. Ruzhansky, Regularity properties, representation of solutions and spectral asymptotics of systems with multiplicities, Preprint, arXiv:math.AP/0402203.

[10] H. Kumano-go, A calculus of Fourier integral operators on $\mathbb{R}^n$ and the fundamental solution for an operator of hyperbolic type, *Comm. Partial Differential Equations* **1** (1976), 1–44.

[11] H. Kumano-go, *Pseudo-Differential Operators*, MIT Press, 1981.

[12] D. S. Kurtz and R. L. Wheeden, Results on weighted norm inequalities for multipliers, *Trans. Amer. Math. Soc.* **255** (1979), 343–362.

[13] M. Ruzhansky, Singularities of affine fibrations in the regularity theory of Fourier integral operators, *Russian Math. Surveys* **55**, 99–170 (2000).

[14] M. Ruzhansky and M. Sugimoto, Global L^2 estimates for a class of Fourier integral operators with symbols in Besov spaces, *Russian Math. Surveys* **58** (2003), 201–202.

[15] M. Ruzhansky and M. Sugimoto, Global L^2-boundedness theorems for a class of Fourier integral operators, *Comm. Partial Differential Equations*, to appear.

[16] M. Ruzhansky and M. Sugimoto, A smoothing property of Schrödinger equations in the critical case, *Math. Ann.*, to appear.

[17] M. Ruzhansky and M. Sugimoto, A new proof of global smoothing estimates for dispersive equations, in *Advances in Pseudo-Differential Operators*, Editors: R. Ashino, P. Boggiatto and M. W. Wong, Birkhäuser, 2004, 65–75.

[18] M. Ruzhansky and M. Sugimoto, Weighted L^2 estimates for a class of Fourier integral operators, Preprint.

[19] A. Seeger, C. D. Sogger and E. M. Stein, Regularity properties of Fourier integral operators, Ann. Math. **134** (1991), 231–251.

[20] E. M. Stein, *Harmonic Analysis: Real-Variable Methods, Orthogonality, and Oscillatory Integrals*, Princeton University Press, Princeton, 1993.

[21] M. Sugimoto, L^2-boundedness of pseudo-differential operators satisfying Besov estimates I, *J. Math. Soc. Japan* **40** (1988), 105–122.

Michael Ruzhansky
Department of Mathematics
Imperial College
180 Queen's Gate, London SW7 2BZ
UK
e-mail: ruzh@ic.ac.uk

Mitsuru Sugimoto
Department of Mathematics
Graduate School of Science
Osaka University
Machikaneyama-cho 1-16, Toyonaka, Osaka 560-0043
Japan
e-mail: sugimoto@math.wani.osaka-u.ac.jp

Operator Theory:
Advances and Applications, Vol. 164, 79–94

L^p-Continuity for Pseudo-Differential Operators

Gianluca Garello and Alessandro Morando

Abstract. The authors give a short survey about the L^p-continuity of pseudo-differential operators both with smooth and non-smooth symbols

Mathematics Subject Classification (2000). Primary 35S05, 35A17.

Keywords. Pseudo-differential operators, L^p spaces.

1. Introduction

Let us fix the attention on the classical pseudo-differential operators in Op $S^m_{\rho,\delta}$ defined for u in the class of the rapidly decreasing functions $\mathcal{S}(\mathbb{R}^n)$ by:

$$a(x,D)u(x) = (2\pi)^{-n} \int e^{ix\cdot\xi} a(x,\xi)\hat{u}(\xi)\, d\xi. \tag{1.1}$$

Here $\hat{u}(\xi)$ is the Fourier transform of u and the smooth symbol $a(x,\xi)$ belongs to the Hörmander symbol classes $S^m_{\rho,\delta}$ which are characterized, for $m \in \mathbb{R}$ and $0 \leq \delta \leq \rho \leq 1$ by the inequality:

$$|\partial_\xi^\alpha \partial_x^\beta a(x,\xi)| \leq c_{\alpha,\beta}\,(1+|\xi|)^{m-\rho|\alpha|+\delta|\beta|}, \quad x,\xi \in \mathbb{R}^n. \tag{1.2}$$

It is well known, see e.g. Hörmander [14], that for $0 \leq \delta < \rho \leq 1$ the pseudo-differential operators of order $m = 0$ are L^2 bounded and the same is when $\delta = \rho \neq 1$. In the case $\delta = \rho = 1$ a counterexample of Ching, [3] 1972, showed that the L^2 continuity is not in general assured; at the same time an unpublished result of Stein proved the continuity of such operators on the Sobolev spaces H^s of strictly positive order s, given by all the Schwartz distributions $u \in \mathcal{S}'$ such that $(1+|\xi|)^s\hat{u}(\xi) \in L^2$, equipped with natural norm $\|u\|_s = \|(1+|\xi|)^s\hat{u}(\xi)\|_{L^2}$. In order to find general conditions for the L^2 continuity of pseudo-differential operators in Op $S^m_{1,1}$ the reader can consider the theory of paradifferential operators well summarized in [18].

The authors are supported by F.I.R.B. grant of Italian Government.

In this paper the aim is to give a brief survey about L^p-continuity of classical pseudo-differential operators for $p \neq 2$.
It is well known, see e.g. Taylor [26], Wong [32], that the operator class Op $S^0_{1,0}$ maps continuously L^p to itself, for any $1 < p < \infty$ and the same is true when $\rho = 1$ and $0 < \delta < 1$, see Taylor [26] or the next estimate (2.3).
The matter is completely different when $\rho \neq 1$. We begin by considering in §2 the very basic results of Hörmander [14] which clarify that pseudo-differential operators of 0 order are not in general L^p bounded, but one needs operators of suitable negative order which strictly depends on ρ itself. The matter will be then completely clarified by Feffermann [5].
In §3 we consider pseudo-differential operators with non smooth symbols, which gain more and more interest in the last twenty five years because of their applications to the general theory of non linear partial differential equations. Of particular interest we quote here the works of Bourdaud [2], Nagase [20], [21].
In the last section we change mind and following the layout of Taylor [26] we consider pseudo-differential operators of 0 order which satisfy an additional property, which allows us to prove directly their boundedness on L^p, $1 < p < \infty$. The difference with respect to the preceding cases is that now we can apply the Marcinkiewicz Lemma on the L^p-continuity of Fourier multipliers.
In this framework we quote here some results of the authors about pseudo-differential operators both with smooth and non smooth weighted symbols. Some applications to the study of the regularity for semilinear partial differential equations are also given.

2. Classical Estimates

Let us consider now the problem of L^p-continuity for pseudo-differential operators in $\mathrm{Op}S^m_{\rho,\delta}$.
We begin by quoting the results of Hörmander [14], 1965. The matter is to determine when the inequalities (1.2) imply for p, q from 1 to ∞ and some $C > 0$ the estimate:

$$\|a(x,D)u\|_{L^p} \leq C\|u\|_{L^q}, \quad u \in \mathcal{S}(\mathbb{R}^n). \tag{2.1}$$

When the symbol is independent of x then $a(x, D) = a(D)$ is a translation invariant operator that is $\tau_h a(D) = a(D)\tau_h$, where $\tau_h u(x) = u(x - h)$, $h \in \mathbb{R}^n$. Denoting now by L^∞_0 the set of the L^∞ functions which tend to 0 at the infinitive, we can then apply the following

Theorem 2.1. [Theorem 1.1, Hörmander [13]] *If A is a bounded translation invariant operator from L^q to L^p and $q > p$ we have $A = 0$ if $q < \infty$ and if $q = \infty$ the restriction of A to L^∞_0 is 0.*

Then we can conclude that in (2.1) it must be $q \leq p$ unless $a(x, D) = 0$; from now on we can then assume that $1 \leq q \leq p \leq \infty$.

The following result is proved in [14] as a quite easy consequence of the L^2 continuity of the pseudo-differential operators in $\mathrm{Op}S^m_{\rho,\delta}$, $0 \le \delta < \rho \le 1$.

Theorem 2.2. *If $q \le 2 \le p$ the estimate (2.1) is valid for any $a(x, \xi)$ satisfying (1.2) if and only if*

$$m \le -n\left(\frac{1}{q} - \frac{1}{p}\right), \tag{2.2}$$

with strict inequality if $q = 1$ or $p = \infty$.

The cases where $q \le p \le 2$ or $2 \le q \le p$ are more interesting since now the result depends on the choice of $0 < \rho \le 1$. It is enough to consider only one of the preceding two cases since (2.1) is equivalent to the same estimate with $a(x, D)$ replaced by ${}^t a(x, D)$ and q, p replaced by q', p' such that $\frac{1}{q} + \frac{1}{q'} = \frac{1}{p} + \frac{1}{p'} = 1$.
In order to obtain (2.1) for any $a(x, D)$ satisfying (1.2), also in the case when the symbol is independent of x, it is essentially proved in Wainger [31], 1965, that we must have:

$$m \le -n\left\{\frac{1}{q} - \frac{1}{p} + (1-\rho)\max\left(\frac{1}{2} - \frac{1}{q}, \frac{1}{p} - \frac{1}{2}, 0\right)\right\}, \tag{2.3}$$

with strict inequality if $q = 1$ or $p = \infty$. The inequality (2.3) clearly agree with (2.2) when $q \le 2 \le p$.
When $p = \infty$ and $q > 2$ the sufficiency of the strict inequality (2.3) is also essentially contained in Wainger [31].
By an interpolation theorem of Stein [22] the sufficiency of the strict inequality (2.3) follows in general since we have already shown it for $q \le 2 \le p$.
At the end Hörmander [14] does not clarify if (2.1) is satisfied by any $a(x, D) \in \mathrm{Op}S^m_{\rho,\delta}$, $0 \le \delta < \rho < 1$ when there is equality in (2.3) and then Theorem 2.1 is not applicable.
If we consider the case $1 \le p = q \le \infty$, $p \ne 2$, then (2.3) reduces to

$$m \le -n(1-\rho)\left|\frac{1}{p} - \frac{1}{2}\right|, \tag{2.4}$$

which is completely proved by the following

Theorem 2.3. [Feffermann [5], 1973]

a) *let $a(x, \xi)$ be a symbol in $S^{-m}_{\rho,\delta}$ with $0 \le \delta < \rho < 1$ and $m < \frac{n}{2}(1-\rho)$. Then the operator $a(x, D)$ is bounded on L^p, $1 < p < \infty$, provided that:*

$$\left|\frac{1}{p} - \frac{1}{2}\right| \le \frac{m}{n}\frac{1}{1-\rho}. \tag{2.5}$$

b) *If $\left|\frac{1}{p} - \frac{1}{2}\right| > \frac{m}{n}\frac{1}{1-\rho}$ then the symbol*

$$a_{\rho,m}(\xi) = \frac{e^{i|\xi|^{1-\rho}}}{1 + |\xi|^m} \tag{2.6}$$

belongs to the class $S^{-m}_{\rho,0}$ and provides an operator $a_{\rho,m}(D)$ unbounded on L^p.

c) *Let $a(x,\xi)$ belong to $S^{-\frac{n}{2}(1-\rho)}_{\rho,\delta}$ so that the critical L^p space is L^1. Although $a(x,\xi)$ is unbounded on L^1, it is bounded on the Hardy space H^1.*

Part b) of the previous theorem is exactly the counter-example of Wainger [31], and part a) must be proved in the critical case when the inequality is not strict. Feffermann observe that a) may be derived from c) by a non trivial interpolation. Moreover the proof of a) and c) requires a technique discovered some times before by E.M. Stein and himself, [6], essentially based on the class of functions with bounded mean oscillation (B.M.O.), introduced by John and Nirenberg in [15]. More precisely a real-valued function f on $\mathbb{R}^n$ belongs to B.M.O. if the norm $\|f\|_{B.M.O.} = \sup_Q \frac{1}{|Q|}\int_Q |f(x) - av_Q f|\,dx$ is finite. Here Q denotes an arbitrary cube in $\mathbb{R}^n$ and $av_Q f = \frac{1}{|Q|}\int_Q f(x)\,dx$. At the end in order to prove the results Feffermann [5] observes that, using the techniques described in [6], a) and c) reduce to the following

Proposition 2.4. *Let $a(x,\xi)$ belong to $S^{-\frac{n}{2}(1-\rho)}_{\rho,\delta}$, for $0 \le \delta < \rho < 1$. Then $a(x,D)$ is a bounded operator from L^∞ to B.M.O..*

3. Symbols with Limited Smoothness

In this section we consider pseudo-differential operators with non regular symbols. As already said in the Introduction their importance in the literature is now increasing, because of their applications to linear partial differential equations with non smooth coefficients and non linear equations, as well as to applicative problems of different nature: signal theory, quantization, ...

We may quote as a basic example pseudo-differential operators with symbols which are differentiable with respect to the ξ variable a finite number of times and which belong to some function class with respect to x, for example Sobolev, Hölder or Besov spaces.

In this framework worthy to be quoted are two results of G. Bourdaud [2]. The first one deals with symbols $a(x,\xi)$ which are x-Hölderian and satisfy, with respect to ξ, the Hörmander's $S^0_{1,\delta}$-estimates. Some results of Kumano-go, Nagase [19] and Meyer [18] are thus generalized. In the second result the Hölder condition are suitably relaxed.

The method used for the proof, introduced by Coifman-Meyer [4], is similar to that used by ourselves in proving the results described in the last section. It is very essentially the following:

- one characterizes the functions by means of their Littlewood-Paley expansions, [18], [29];
- one reduces the problem to "elementary symbols" of the following type:

$$a(x,\xi) = \sum_{j\ge 0} M_j\left(2^{j\delta}x\right)\psi\left(2^{-j}\xi\right), \tag{3.1}$$

Theorem 3.3. *Assume that* $2 \le p < \infty$ *and the symbol* $a(x,\xi)$ *satisfies* (3.12) *and* (3.13) *with* $m_p = n(1-\rho)(\frac{1}{2}-\frac{1}{p})$, $\kappa = [\frac{n}{2}]+1$, $\mu_p = \kappa\frac{n(1-\rho)}{p\kappa(\rho-\delta)+n(1-\rho)}$; *then the operator* $a(x,D)$ *is* L^p *bounded.*

Theorem 3.4. *Assume that* $1 < p \le 2$ *and the symbol* $a(x,\xi)$ *satisfies* (3.12) *and* (3.13) *with* $m_p = n(1-\rho)(\frac{1}{p}-\frac{1}{2})$, $\kappa = n+1$, $\mu_p = \mu_2 = \frac{n(1-\rho)}{2\kappa(\rho-\delta)+n(1-\rho)}$; *then the operator* $a(x,D)$ *is* L^p *bounded.*

At the end of this section let us quote some interesting results of Sugimoto, where the symbols are entirely considered in weighted Besov spaces, both with respect x and ξ. Let us consider at first a partition of unity similar to that introduced by Bourdaud, that is, let $\mathcal{A}(\mathbb{R}^n)$ be the collection of all the systems $\Theta = \{\Theta_j(y)\}_{j=0}^{\infty} \subset \mathcal{S}(\mathbb{R}^n)$ that satisfy the following conditions:

i) $\mathrm{supp}\Theta_0 \subset \{y; |y| \le 2\}$, $\mathrm{supp}\Theta_j \subset \left\{y; 2^{j-1} \le |y| \le 2^{j+1}\right\}$, for j=1,2,...;
ii) $\sum_{j=0}^{\infty} \Theta_j(y) = 1$;
iii) for every multi-index $\alpha \in \mathbb{Z}_+^n$ there exists a positive constant C_α such that $2^{j|\alpha|}|\partial^\alpha\Theta_j(y)| \le C_\alpha$, for every $j = 0,1,\dots$ and all $y \in \mathbb{R}^n$.

In order to introduce the symbol classes we consider Besov classes on product spaces of dimension $2n$. We will take in the following $n = n_1 + \cdots + n_N = n_1' + \cdots + n_{N'}'$, where $n_r, n_s' \in \mathbb{N}$, for $r = 1,\dots,N$, $s = 1,\dots,N'$.
Let us consider then $p,q \in]0,\infty]$, $\bar\lambda = (\lambda_1,\dots,\lambda_N) \in \mathbb{R}^N$, $\bar\lambda' = (\lambda_1',\dots,\lambda_{N'}') \in \mathbb{R}^{N'}$, $\bar n = (n_1,\dots,n_N)$ and $\bar n' = (n_1',\dots,n_{N'}')$. We define now the symbol class $B_{p,q}^{\bar\lambda,\bar\lambda'} = B_{p,q}^{\bar\lambda,\bar\lambda'}(\mathbb{R}^{(\bar n,\bar n')})$ as the set of all the distributions $a(x,\xi) \in \mathcal{S}'(\mathbb{R}^{2n})$ such that $\|a\|_{B_{p,q}^{(\bar\lambda,\bar\lambda')}} < \infty$, where

$$\begin{aligned}\|a\|_{B_{p,q}^{(\bar\lambda,\bar\lambda')}} &= \left\| \|2^{\bar j\cdot\bar\lambda+\bar k\cdot\bar\lambda'}\mathcal{F}^{-1}\Phi_{\bar j,\bar k}\mathcal{F}a\|_{L^p(\mathbb{R}^{2n})} \right\|_{l^q} \\ &= \left\{ \sum_{\bar j,\bar k} \left(\int |2^{\bar j\cdot\bar\lambda+\bar k\cdot\bar\lambda'}\mathcal{F}^{-1}\Phi_{\bar j,\bar k}\mathcal{F}a(x,\xi)|^p \, dx d\xi \right)^{q/p} \right\}^{1/q};\end{aligned} \tag{3.14}$$

Slight modifications are needed in the case of $p = \infty$ or $q = \infty$. Here $\bar j = (j_1,\dots,j_N)$, $\bar k = (k_1,\dots,k_{N'})$ are non negative integer vectors. Moreover if we set $\Phi_{\bar j}^{\bar n}(y) = \Phi_{\bar j}^{\bar n}(y_1,\dots,y_N) = \Theta_{j_1}(y_1)\dots\Theta_{j_N}(y_N)$, with $\{\Theta_{j_r}(y_r)\}_{j_r=0}^{\infty} \in \mathcal{A}(\mathbb{R}^{n_r})$, we set $\Phi_{\bar j,\bar k}(y,\eta) = \Phi_{\bar j}^{\bar n}(y)\Phi_{\bar k}^{\bar n'}(\eta)$. Furthermore we always decompose the variables y and η in such a way that $y = (y_1,\dots,y_N)$, $\eta = (\eta_1,\dots,\eta_{N'})$; $y_r \in \mathbb{R}^{n_r}$, $\eta_s \in \mathbb{R}^{n_s'}$, $r = 1,\dots,N$; $s = 1,\dots,N'$.
All the basic theory of the generalized Besov spaces which is used for the definition of the symbol classes $B_{p,q}^{\bar\lambda,\bar\lambda'}$ is developed in Sugimoto [24].
We shall give now the first result about L^p-continuity.

Theorem 3.5. [*Theorem* 2.1.1 *in* [24]] *Let* $1 \le p \le 2$, *then there exists a constant* C *such that the estimate*

$$\|a(x,D)f\|_{L^p(\mathbb{R}^n)} \le C\|a(x,\xi)\|_{B^{\bar{0}}_{p,1}} \|f\|_{L^p(\mathbb{R}^n)} \tag{3.15}$$

holds for any $a(x,\xi) \in B^{\bar{0}}_{p,1}$ *and any* $f \in \mathcal{S}(\mathbb{R}^n)$.

In order to introduce the last Sugimoto result about L^p-continuity we need the following weighted Besov symbols.
Let us use the preceding notation and set moreover, for $\bar{\rho} = (\rho_1, \dots, \rho_{N'}) \in \mathbb{R}^{N'}$: $\omega_{\bar{\rho}}(\xi) = (1+|\xi_1|)^{\rho_1} \dots (1+|\xi_{N'}|)^{\rho_{N'}}$.
For $0 < q \le \infty$ the Sugimoto's weighted Besov symbol classes are now given by:

$$B^{\bar{\lambda},\lambda'}_{\infty,q,(0,\bar{\rho})} = \{a(x,\xi) \in \mathcal{S}'(\mathbb{R}^{2n}); \|a\|^{(\bar{\lambda},\lambda')}_{q,\bar{\rho}} < \infty\}. \tag{3.16}$$

Notice that here $\lambda' \in \mathbb{R}$ and

$$\|a\|^{(\bar{\lambda},\lambda')}_{q,\bar{\rho}} = \left\{ \sum_{\substack{\bar{j} \ge 0 \\ k \ge 0}} \left(2^{\bar{j}\cdot\bar{\lambda}+k\lambda'} \left\| (\mathcal{F}^{-1}\Phi_{\bar{j},k}\mathcal{F}a)(x,\xi)\omega_{\bar{\rho}}(\xi) \right\|_{L^\infty(\mathbb{R}^{2n})} \right)^q \right\}^{\frac{1}{q}}. \tag{3.17}$$

The last Sugimoto's results that we quote reads as follows:

Theorem 3.6. [*Theorem* 6.1 *in* [25]] *Using the preceding notations let us consider* $2 \le p \le \infty$ *and* $\bar{\rho} > \frac{\bar{n}'}{2}$. *Then there exists a constant* C *such that the estimate*

$$\|a(x,D)f\|_{L^p(\mathbb{R}^n)} \le C\|a(x,\xi)\|^{(\bar{0},\frac{n}{2})}_{1,\bar{\rho}} \|f\|_{L^p(\mathbb{R}^n)} \tag{3.18}$$

holds for all $a(x,\xi) \in \mathcal{S}(\mathbb{R}^{2n})$ *and all* $f \in L^p(\mathbb{R}^n) \cap FL^1(\mathbb{R}^n)$. *Here* $FL^1(\mathbb{R}^n)$ *is the set of the Schwartz distribution whose Fourier transform belongs to* L^1.

More refined results are given in Sugimoto [25] involving also the action of pseudo-differential operators on Hardy spaces.
Worthy to be noticed, at the end, are the results of Marschall [17] which use techniques of the pseudo-differential calculus in [18] and in some way generalize the results of Nagase.

4. L^p-Continuity for Pseudo-Differential Operators of Zero Order

In the most part of the existent literature about the L^p-continuity of pseudo-differential operators, many efforts are devoted in evaluating as well as possible the negative order $m_p \sim -n\left|\frac{1}{p} - \frac{1}{2}\right|$ arising from the foregoing bounds (2.4), (2.5).
A different point of view was marked by Taylor: arguing on the proof of the L^p-continuity of pseudo-differential operators in $\mathrm{Op}S^0_{1,0}$, he realized that the latter may be adapted to prove the L^p-boundedness of some suitable subclasses of the 0 order operators with symbols in $S^0_{\rho,0}$, $\rho \in]0,1[$. This can be done by changing the

Mikhlin-Hörmander Lemma about Fourier multipliers with the following analogous result due to Marcinkiewicz and Lizorkin (see [16, 23]).

Lemma 4.1. *Let us set* $\mathbb{K} := \{0,1\}^n$. *Let* $a(\xi)$ *be a continuous function together with its derivatives* $\partial_\xi^\gamma a(\xi)$, *for any* $\gamma \in \mathbb{K}$. *If there is a constant* $B > 0$ *such that*

$$|\xi^\gamma \partial_\xi^\gamma a(\xi)| \leq B, \qquad \xi \in \mathbb{R}^n, \quad \gamma \in \mathbb{K}, \tag{4.1}$$

then for every $p \in]1, +\infty[$ *we can find a constant* $A_p > 0$, *depending only on* p, B *and the dimension* n, *such that*

$$\|a(D)u\|_p \leq A_p \|u\|_p, \qquad u \in \mathcal{S}.$$

The classes of Taylor are defined assuming on the derivatives of its symbols a "variable coefficients" version of the decay condition stated by the foregoing Lemma 4.1. In [8, 9], we introduced a weighted version of Taylor's symbol classes. A local version of the latter were employed in [9] to find some L^p-regularity results for a class of inhomogeneous partial differential (and pseudo-differential) operators, the main example of which being an operator like

$$a(x, D) = \sum_{\alpha \in \mathcal{P}} a_\alpha(x) D^\alpha, \quad (D_j := -i\partial_j, \; j = 1, \ldots, n). \tag{4.2}$$

In the right-hand side of (4.2), $\mathcal{P}$ is a *complete polyhedron* of $\mathbb{R}^n$ in the sense of Gindikin-Volevich [30] and, for every multi-index $\alpha \in \mathcal{P}$, a_α is a given function in $C^\infty(\Omega)$. For a deep analysis of complete polyhedra and their relevant properties, we address to the book [1]. We say that (4.2) is *multi-quasi-elliptic* in Ω if its symbol $a(x, \xi) = \sum_{\alpha \in \mathcal{P}} a_\alpha(x) \xi^\alpha$ fulfills the following lower bounds: for every compact set $K \subset \Omega$ there exist positive constants $C = C_K$ and $R = R_K$ such that

$$|a(x, \xi)| \geq C \Lambda_{\mathcal{P}}(\xi), \quad x \in K, \; |\xi| \geq R. \tag{4.3}$$

For a given complete polyhedron $\mathcal{P}$, the weight function $\Lambda_{\mathcal{P}}$ from the right-hand side of (4.3) is defined by

$$\Lambda_{\mathcal{P}}(\xi) := \sqrt{\sum_{\alpha \in V(\mathcal{P})} \xi^{2\alpha}}, \quad \xi \in \mathbb{R}^n, \tag{4.4}$$

being $V(\mathcal{P})$ the set of the vertices of $\mathcal{P}$; let us note that, by definition of $\mathcal{P}$ itself, we have $V(\mathcal{P}) \subset \mathbb{Z}_+^n$ and $\mathbf{0} := (0, \ldots, 0) \in V(\mathcal{P})$. When $\mathcal{P}$ is the convex hull of the set $\{\mathbf{0}, me_j; \; j = 1, \ldots, n\}$, with given $m \in \mathbb{Z}_+ \setminus \{0\}$ and $e_j := (0, \ldots, 0, 1, 0, \ldots, 0)$ for $j = 1, \ldots, n$ (in such a case $\Lambda_{\mathcal{P}}(\xi)$ is equivalent to the elliptic weight $(1 + |\xi|)^m$), the bounds in (4.3) just express the well known property of ellipticity; in view of conditions (2.2), (2.3), the Hörmander classes $S^m_{1,0}$ provide in this last case the appropriate environment for the study of L^p-properties of solutions to the related elliptic partial differential equations (cf. [26]); in the case of a general complete polyhedron $\mathcal{P}$, however, sharper L^p-regularity results can be obtained by considering a weighted algebra of pseudo-differential operators whose symbols are modeled on the inhomogeneous structure of $\mathcal{P}$, through the function $\Lambda_{\mathcal{P}}$.

Actually, all this theory can be developed in a more general setting; we start by introducing an appropriate notion of weight function.

Definition 4.2. We say that a positive function $\Lambda(\xi) \in C^\infty(\mathbb{R}^n)$ is a weight function if there exist some constants $C > 1$, $1 \leq \mu_0 \leq \mu_1 \leq \mu$ for which the following assumptions are satisfied.

a) $\frac{1}{C}(1+|\xi|)^{\mu_0} \leq \Lambda(\xi) \leq C(1+|\xi|)^{\mu_1}, \quad \xi \in \mathbb{R}^n$;

b) for every multi-index $\alpha \in \mathbb{Z}^n_+$ there exists a positive constant C_α such that:

$$|\xi^\gamma \partial^{\gamma+\alpha}\Lambda(\xi)| \leq C_\alpha \Lambda(\xi)^{1-\frac{1}{\mu}|\alpha|}, \quad \xi \in \mathbb{R}^n,\ \gamma \in \mathbb{K}. \tag{4.5}$$

Similar definitions of weight functions can be found in [1], [7]. However, the peculiarity of Definition 4.2 is the invariance of estimate (4.5) under the differential operator $\xi^\gamma \partial^\gamma$: it prescribes condition (4.1) (which is referred to as "(LM)-condition", later on) on the decay of the γ-derivatives of $\Lambda(\xi)$ as long as $\gamma \in \mathbb{K}$. The function $\Lambda_{\mathcal{P}}$, corresponding to a complete polyhedron $\mathcal{P}$ by formula (4.4), provides an example of weight function according to the above definition (cf. [9] Lemma 2.1); in this case the constants μ_0, μ_1 and μ turn to be the *minimum, maximum* and *formal orders* of $\mathcal{P}$ respectively (see [1] for their precise definitions). For $m \in \mathbb{R}$ and $\rho \in \left]0, \frac{1}{\mu}\right]$, we will denote by $S^m_{\rho,\Lambda}$ the class of functions $a(x,\xi) \in C^\infty(\mathbb{R}^{2n})$ whose derivatives fulfill the estimates

$$|\partial^\alpha_\xi \partial^\beta_x a(x,\xi)| \leq C_{\alpha,\beta}\Lambda(\xi)^{m-\rho|\alpha|}, \quad (x,\xi) \in \mathbb{R}^{2n}, \tag{4.6}$$

for any multi-indices α, β and suitable constants $C_{\alpha,\beta} > 0$. Assumptions a) and b) of Definition 4.2 yield that the weighted symbol classes $S^m_{\rho,\Lambda}$ are related to the Hörmander ones through $S^h_{\rho\mu_1,0} \subset S^m_{\rho,\Lambda} \subset S^k_{\rho\mu_0,0}$, where $h := \min\{m\mu_0, m\mu_1\}$ and $k := \{m\mu_0, m\mu_1\}$. Since we have $\rho\mu_0 \leq \mu_0/\mu \leq 1$ (unless trivial cases), the symbol classes $S^m_{\rho,\Lambda}$ provide in general unbounded operators in L^p spaces for $p \neq 2$. Therefore, agreeing with the layout of Taylor [26], we give the following

Definition 4.3. For $m \in \mathbb{R}$ and $\rho \in \left]0, \frac{1}{\mu}\right]$, $M^m_{\rho,\Lambda}$ is the class of all functions $a \in C^\infty(\mathbb{R}^n)$ such that for every $\gamma \in \mathbb{K}$

$$\xi^\gamma \partial^\gamma_\xi a(x,\xi) \in S^m_{\rho,\Lambda}. \tag{4.7}$$

It is worth noticing that the requirement (4.7) is equivalent to assume that for any multi-indices $\alpha, \beta \in \mathbb{Z}^n_+$ there is a constant $C_{\alpha,\beta} > 0$ such that

$$|\xi \partial^{\gamma+\alpha}_\xi \partial^\beta_x a(x,\xi)| \leq \Lambda(\xi)^{m-\rho|\alpha|}, \quad (x,\xi) \in \mathbb{R}^{2n} \tag{4.8}$$

(cf. [9] Proposition 3.1). The last inequalities just prescribe the (LM)-condition on the ξ-derivatives of symbols, for every multi-index γ in $\mathbb{K}$, in addition to the $S^m_{\rho,\Lambda}$-estimates (4.6). For $\Lambda(\xi) = \sqrt{1+|\xi|^2}$ the symbol classes $M^m_{\rho,\Lambda}$ reduce to the homogeneous classes M^m_ρ studied by Taylor in [26]. For a general weight function $\Lambda(\xi)$, the pseudo-differential operators with symbols in $M^m_{\rho,\Lambda}$ give a workable algebra of L^p-bounded operators, for any $p \in]1,+\infty[$; indeed, for 0 order operators it holds the following

Theorem 4.4. *Let $a(x,\xi)$ belong to $M^0_{\rho,\Lambda}$. Then for every $p \in]1,+\infty[$ the operator $a(x,D)$ is L^p-bounded.*

The L^p property given by Theorem 4.4 can be easily extended to pseudo-differential operators with symbols $M^m_{\rho,\Lambda}$ and arbitrary order $m \in \mathbb{R}$: this is done by introducing a suitable scale of weighted Sobolev spaces, naturally arising in connection to such a kind of symbols. More precisely, for a given weight function $\Lambda(\xi)$, $p \in]1,+\infty[$ and real s, the Sobolev space $H^{s,p}_\Lambda$ is defined by setting

$$H^{s,p}_\Lambda := \{u \in \mathcal{S}' : \ \Lambda(D)^s u \in L^p(\mathbb{R}^n)\}, \tag{4.9}$$

where $\Lambda(D)^s$ is defined to be the Fourier multiplier $\Lambda(D)^s u = \mathcal{F}^{-1}\Lambda(\xi)^s \mathcal{F} u$; slightly modifying the foregoing definition, we are able to define local Sobolev spaces $H^{s,p}_{\Lambda,\mathrm{loc}}(\Omega)$ and compactly supported spaces $H^{s,p}_{\Lambda,\mathrm{comp}}(\Omega)$ on an arbitrary open subset Ω of $\mathbb{R}^n$ (their definitions are exactly the same as in the standard Sobolev case; cf. for instance [26]). Therefore, by noticing that $\Lambda(\xi) \in M^1_{\frac{1}{\mu},\Lambda}$ and using a few symbolic calculus, we get the continuity property below.

Proposition 4.5. *Let $a(x,\xi)$ belong to $M^m_{\rho,\Lambda}$ with arbitrary m. Then, for every real s, the following operator*

$$a(x,D) : H^{s+m,p}_\Lambda \to H^{s,p}_\Lambda$$

is linear and continuous, whenever p belongs to $]1,+\infty[$.

As usual, for an arbitrary open subset Ω of $\mathbb{R}^n$ the local symbol class $M^m_{\rho,\Lambda}(\Omega)$ is defined to be the set of all functions $a \in C^\infty(\Omega\times\mathbb{R}^n)$ such that for every function $\phi \in C^\infty_0(\Omega)$

$$\phi(x)a(x,\xi) \in M^m_{\rho,\Lambda}.$$

A simple analysis shows that classes $M^m_{\rho,\Lambda}(\Omega)$ are stable under the usual operations of the symbolic calculus (cf. [9], propositions 3.2-3.4). Let us also mention the following crucial property concerning the pseudo-differential operators with weighted elliptic symbols in $M^m_{\rho,\Lambda}(\Omega)$.

Proposition 4.6. *Let $a(x,\xi) \in M^m_{\rho,\Lambda}(\Omega)$ be Λ elliptic of order m in Ω; this means that for every compact subset K of Ω there are positive constants $C = C_K$ and $R = R_K$ such that*

$$|a(x,\xi)| \geq C\Lambda(\xi)^m, \quad x \in K, \ |\xi| \geq R. \tag{4.10}$$

Then there exists a symbol $q(x,\xi) \in M^{-m}_{\rho,\Lambda}(\Omega)$ such that the pseudo-differential operator $q(x,D)$ is properly supported and the following equalities hold true

$$q(x,D)p(x,D) = I + S, \qquad p(x,D)q(x,D) = I + R,$$

where $R, S : \mathcal{E}'(\Omega) \to C^\infty(\Omega)$ are suitable linear continuous operators and I is the identity operator.

Combining the results of Theorem 4.4 and Proposition 4.6 we are thus able to prove the following L^p-regularity result for solution to weighted elliptic pseudo-differential equations.

Proposition 4.7. *Let us suppose $a(x,\xi) \in M^m_{\rho,\Lambda}(\Omega)$ to be Λ-elliptic of order m in Ω and $u \in \mathcal{E}'(\Omega)$. If $a(x,D)u \in H^{s,p}_{\Lambda,\mathrm{loc}}(\Omega)$ then $u \in H^{s+m,p}_{\Lambda,\mathrm{comp}}(\Omega)$ for any $s \in \mathbb{R}$ and $p \in]1,+\infty[$. Furthermore, if $a(x,D)$ is properly supported then $u \in \mathcal{D}'(\Omega)$ and $a(x,D)u \in H^{s,p}_{\Lambda,\mathrm{loc}}(\Omega)$ imply $u \in H^{s+m,p}_{\Lambda,\mathrm{loc}}(\Omega)$.*

The preceding L^p-regularity result applies in particular to the case where $\Lambda_{\mathcal{P}}(\xi)$ is the weight function related by (4.4) to a complete polyhedron $\mathcal{P}$ and $a(x,D)$ a multi-quasi-elliptic partial differential operator like (4.2); since the latter turns to be a $\Lambda_{\mathcal{P}}$-elliptic (properly supported) operator of order 1, Proposition 4.7 yields that any distribution $u \in \mathcal{D}'(\Omega)$ such that $a(x,D)u \in H^{s,p}_{\Lambda_{\mathcal{P}},\mathrm{loc}}(\Omega)$ actually belongs to $H^{s+1,p}_{\Lambda_{\mathcal{P}},\mathrm{loc}}(\Omega)$.

In [10, 11, 12], we studied the L^p-boundedness of a class of pseudo-differential operators with weighted non regular symbols in the spirit of the ones presented in §3. Following [27], at the very beginning we can consider symbols $a(x,\xi)$ for which the function $\xi \mapsto a(.,\xi)$ is valued in an arbitrary Banach space X (with norm $\|.\|_X$): as in [8, 9], the key point is to require that the ξ derivatives of $\xi \mapsto a(.,\xi)$ obey a kind of (LM) condition. Non regular symbol classes were defined with respect to a general family of weight functions, including functions (4.4) as a significant example. For the remaining part of the paper, by weight function we mean a positive function $\Lambda(\xi) \in C^\infty(\mathbb{R}^n)$ for which there exist some constants $C > 0$ and $\mu_0 \geq 1$ such that:

I. $\Lambda(\xi) \geq C(1+|\xi|)^{\mu_0}, \quad \xi \in \mathbb{R}^n$;

II. for every $\gamma \in \mathbb{Z}^n_+$ there exists $C_\gamma > 0$ such that

$$\prod_{j=1}^n (1+\xi_j^2)^{\frac{\gamma_j}{2}} |\partial^\gamma \Lambda(\xi)| \leq C_\gamma \Lambda(\xi), \quad \xi \in \mathbb{R}^n; \tag{4.11}$$

III. $\Lambda(t\xi) \leq C\Lambda(\xi), \quad t,\xi \in \mathbb{R}^n, \ \max\limits_{1\leq j\leq n} |t_j| \leq 1, \ t\xi := (t_1\xi_1, \dots, t_n\xi_n)$;

IV. there exists a constant $\delta \in]0,1[$ such that

$$\Lambda(\xi) \leq C\left(\Lambda(\eta) + \Lambda(\xi-\eta) + \Lambda(\eta)^\delta \Lambda(\xi-\eta)^\delta\right), \quad \xi,\eta \in \mathbb{R}^n. \tag{4.12}$$

Let us notice that in assumption II we demand (4.11) to be satisfied for all derivatives of $\Lambda(\xi)$. Assumption IV, which is referred to as *δ-condition*, also plays a crucial role in this context. It has been shown by Garello in [7] that functions (4.4) display property IV, as well as I, III. Our non regular symbol classes obey the following

Definition 4.8. Let X be a Banach space and $\Lambda(\xi)$ a given weight function. For a real m and a positive integer N, $XM^m_\Lambda(N)$ is the class of all measurable functions $a : \mathbb{R}^n \times \mathbb{R}^n \to \mathbb{C}$ such that there exists a positive constant C for which

1. $\prod\limits_{j=1}^n (1+\xi_j^2)^{\frac{\gamma_j}{2}} |\partial_\xi^\gamma a(x,\xi)| \leq C\Lambda(\xi)^m, \quad x,\xi \in \mathbb{R}^n,$
2. $\prod\limits_{j=1}^n (1+\xi_j^2)^{\frac{\gamma_j}{2}} \|\partial_\xi^\gamma a(.,\xi)\|_X \leq C\Lambda(\xi)^m, \quad \xi \in \mathbb{R}^n$

hold for all multi-indices γ with $|\gamma| \le N$.

In [10, 11], the L^p-behavior of classes XM_Λ^m was studied when X is a weighted Sobolev class $H_\Lambda^{s,p}$ as previously defined. We obtain then the following

Theorem 4.9. *Let $a(x,\xi)$ be a symbol in $H_\Lambda^{r,p}M_\Lambda^m(N)$, with $m \in \mathbb{R}$, $N \ge 2n+1$, $p \in]1,+\infty[$ and $r > \frac{n}{(1-\delta)\mu_0 p}$. Then*

$$a(x,D) : H_\Lambda^{s+m,p} \to H_\Lambda^{s,p}$$

is linear and continuous for every $s \in [0,r]$.

Following the approach of Coifman-Meyer [4] and Taylor [27], as it was summarized in §3, the key point in the study of non regular classes $XM_\Lambda^m(N)$, with general X and sufficiently large N, consists to show that any 0 order symbol $a(x,\xi)$ can be decomposed as a series of certain elementary symbols; the last ones turn to be defined by suitable series like (3.1). Then the proof of Theorem 4.9 exploits a vector valued counterpart of Lemma 4.1 (cf. [28] Theorem 2.4/2), combined with a characterization of weighted Sobolev spaces $H_\Lambda^{s,p}$ through a non-homogeneous smooth partition of unity, due to Triebel (cf. [28] again). As a consequence of this analysis, Theorem 4.9 yields that the Sobolev spaces $H_\Lambda^{r,p}$ enjoy a Banach algebra property for any $p \in]1,+\infty[$, provided that r is sufficiently large; hence, as it was announced in the Introduction, by an usual iterative procedure, we were able to extend the L^p-regularity result stated by Proposition 4.7 to an appropriate class of multi-quasi-elliptic semilinear partial differential operators (cf. [10]). Namely, we consider in [10] a class of equations like

$$a(x,D)u = F(x,\partial^\alpha u, f)_{|\alpha\in\mathcal{P}\setminus\mathcal{F}(\mathcal{P})}. \tag{4.13}$$

Here $\mathcal{P}$ is a given complete polyhedron of $\mathbb{R}^n$, $a(x,D)$ is a multi-quasi-elliptic pseudo-differential operator in Ω, fulfilling (4.2), (4.3), while $F(x,\zeta)$ is a non linear function of $x \in \Omega$ and $\zeta \in \mathbb{C}^N$, with $N := 1 + \sum_{\alpha\in\mathcal{P}\setminus\mathcal{F}(\mathcal{P})} 1$, and f is a given function. In the right-hand side of (4.13), $\mathcal{F}(\mathcal{P})$ denotes the *lateral boundary* of $\mathcal{P}$, that is the union of the faces of $\mathcal{P}$ which do not lie on the coordinate hyperplanes of $\mathbb{R}^n$ (see [1] for precise definitions): dropping multi-indices $\alpha \in \mathcal{F}(\mathcal{P})$ from the right-hand side of (4.13) amounts to require that the non linear term, provided by F, has lower order than the linear part $a(x,D)$ (measured with respect to the weight $\Lambda_\mathcal{P}(\xi)$). About the non linear term F, we make the following additional assumptions:

i. F is locally smooth with respect to x and entire analytic with respect to ζ, namely

$$F(x,\zeta) = \sum_{\beta\in\mathbb{Z}_+^N} c_\beta(x)\zeta^\beta, \quad c_\beta \in C^\infty(\Omega),\ \zeta \in \mathbb{C}^N;$$

ii. for any compact subset K of Ω, $\alpha \in \mathbb{Z}_+^n$, $\beta \in \mathbb{Z}_+^N$ there exist constants $c_{\alpha,\beta} > 0$, $\lambda_\beta > 0$ such that

$$\sup_{x\in K} |\partial^\alpha c_\beta(x)| \le c_{\alpha,\beta}\lambda_\beta$$

and $F_1(\zeta) := \sum_{\beta \in \mathbb{Z}^n_+} \lambda_\beta \zeta^\beta$ is entire analytic.

Exploiting the algebra property of the $H^{s,p}_\Lambda$ spaces, combined with the result of Proposition 4.7, gives the following

Proposition 4.10. *For a complete polyhedron $\mathcal{P}$ of $\mathbb{R}^n$ and $p \in]1, +\infty[$, let f belong to $H^{t,p}_{\Lambda_\mathcal{P},\mathrm{loc}}(\Omega)$ with $t > \frac{n}{(1-\delta)\mu_0 p} + \delta$ and δ defined by (4.12); let $u \in H^{s,p}_{\Lambda_\mathcal{P},\mathrm{loc}}(\Omega)$ be a solution to the equation (4.13), for such an f, with $t \geq s > \frac{n}{(1-\delta)\mu_0 p} + \delta$. If $a(x, D)$ is multi-quasi-elliptic in Ω then $u \in H^{t+1,p}_{\Lambda_\mathcal{P},\mathrm{loc}}(\Omega)$.*

In [12], an analogous of Theorem 4.9 has been proved, where the weighted Sobolev spaces $H^{s,p}_\Lambda$ are replaced by a suitable scale of non-homogeneous Besov spaces $B^{s,\Lambda}_{p,q}$, $s \in \mathbb{R}$, $1 < p, q < +\infty$; such a kind of spaces were introduced by Triebel in [28].

References

[1] P. Boggiatto, E. Buzano and L. Rodino, *Global Hypoellipticity and Spectral Theory*, Mathematical Research, Vol. **92**, Akademie Verlag, Berlin, New York, 1996.

[2] G. Bourdaud, L^p-estimates for certain non-regular pseudodifferential operators, *Comm. Partial Differential Equations* **7** (9) (1982), 1023–1033.

[3] C.-H. Ching, Pseudodifferential operators with non regular symbols, *J. Differential Equations* **11** (1972), 436–447.

[4] R. R. Coifman and Y. Meyer, *Au-delà des operateurs pseudo-différentiels*, Astérisque no. **57**, Soc. Math. France, 1978.

[5] C. Fefferman, L^p bounds for pseudo-differential operators, *Israel J. Math.* **14** (1973), 413–417.

[6] C. Fefferman and E. M. Stein, H^p spaces of several variables, *Acta Math.* **129** (1972), 137–193.

[7] G. Garello, Pseudodifferential operators with symbols in weighted Sobolev spaces and regularity for non linear partial differential equations, *Math. Nachr.* **239-240** (2001), 62–79.

[8] G. Garello and A. Morando, A class of L^p-bounded pseudodifferential operators, in *Progress in Analysis, Vol. I*, Editors: H. G. W. Begehr, R. P. Gilbert and M. W. Wong, World Scientific, 2003, 689-696.

[9] G. Garello and A. Morando, L^p-bounded pseudodifferential operators and regularity for multi-quasi-elliptic equations, *Integral Equations Operator Theory* **51** (2005), 501–517.

[10] G. Garello and A. Morando, L^p-boundedness for pseudodifferential operators with non-smooth symbols and applications, *Boll. Unione Mat. Ital. Sez. B Artic. Ric. Mat. (8)* **8** (2005), 461–503.

[11] G. Garello and A. Morando, Continuity in weighted Sobolev spaces of L^p type for pseudo-differential operators with completely nonsmooth symbols, in *Advances in Pseudo-Differential Operators*, Editors: R. Ashino, P. Boggiatto and M. W. Wong, Birkhäuser, 2004, 91–106.

[12] G. Garello and A. Morando, Continuity in weighted Besov spaces for pseudodifferential operators with non regular symbols, to appear.

[13] L. Hörmander, Translation invariant operators in L^p spaces, *Acta Math.* **104** (1960), 93–140

[14] L. Hörmander, Pseudo-differential operators and hypoelliptic equations, in *Proc. Symp. Pure math.* **X** 1965, 138-183.

[15] F. John and L. Nirenberg, On functions of bounded mean oscillation, *Comm. Pure Appl. Math.* **14** (1961) 415–426.

[16] P.I. Lizorkin, (L_p, L_q)-multipliers of Fourier integrals, *Dokl. Akad. Nauk SSSR* **152** (1963), 808–811.

[17] J. Marschall, Pseudo-differential operators with non regular symbols of the class $S^m_{\rho,\delta}$, *Comm. Partial Differential Equations* **12** (8) (1987), 921–965.

[18] Y. Meyer, Remarques sur un théorème de J. M. Bony, *Rendiconti Circ. Mat. Palermo* **II** 1 (1981), 1–20.

[19] M. Nagase, L^p-boundedness of pseudodifferential operators with non regular symbols, *Comm. Partial Differential Equations* **2** (10) (1977), 1045–1071.

[20] M. Nagase, On a class of L^p-bounded pseudodifferential operators, *Sci. Rep. College Gen. Ed. Osaka Univ.* **33** (4) (1985), 1–7.

[21] M. Nagase, On sufficient conditions for pseudodifferential operators to be L^p-bounded, *Pseudodifferential operators, Oberwolfach, 1986*, Lecture Notes in Mathematics **1256**, Springer, Berlin, 1987.

[22] E. M. Stein, Interpolation of linear operators, *Trans. Amer. math. Soc.* **83** (1956), 482-492.

[23] E.M. Stein, *Singular Integrals and Differentiability Properties of Functions*, Princeton Mathematical Series, No. **30**, Princeton University Press, Princeton, N.J., 1970.

[24] M. Sugimoto, L^p-boundedness of pseudodifferential operators satisfying Besov estimates I, *J. Math. Soc. Japan* **40** (1) (1988), 105–122.

[25] M. Sugimoto, L^p-boundedness of pseudo-differential operators satisfying Besov estimates II, *J. Fac. Sci. Univ. Tokyo Sect. IA, Math.* **35** (1988), 149–162.

[26] M. E. Taylor, *Pseudodifferential Operators*, Princeton, University Press, 1981.

[27] M. E. Taylor, *Pseudodifferential Operators and Nonlinear PDE*, Progress in Mathematics **100**, Birkhuser, Boston, 1991.

[28] H. Triebel, General function spaces, III. Spaces $B^{g(x)}_{p,q}$ and $F^{g(x)}_{p,q}$, $1 < p < \infty$; basic properties, *Anal. Math.* **3** (3) (1977), 221–249.

[29] H. Triebel, *Theory of Function Spaces*, Birkhäuser Verlag, Basel, Boston, Stuttgart, 1983.

[30] L.R. Volevich and S.G. Gindikin, On a class of hypoelliptic polynomials, *Math. USSR Sbornik* **75** (1968), 131–174.

[31] S. Wainger, Special trigonometric series in k dimensions, *Mem. Amer. Math. Soc.* **59** (1965), 1-102.

[32] M. W. Wong, *An Introduction to Pseudo-Differential Operators*, Second Edition, World Scientific Publishing Co., Inc., River Edge, NJ, 1999.

Gianluca Garello and Alessandro Morando
Dipartimento di Matematica
Università di Torino
Via Carlo Alberto 10
I-10123 Torino
Italy
e-mail: gianluca.garello@unito.it
morando@dm.unito.it

Operator Theory:
Advances and Applications, Vol. 164, 95–114

Fredholm Property of Pseudo-Differential Operators on Weighted Hölder-Zygmund Spaces

V.S. Rabinovich

Abstract. We consider pseudo-differential operators in the L.Hörmander class $OPS^m_{1,0}$ acting on Hölder-Zygmund spaces with exponential weights. The necessary and sufficient conditions for operators under consideration to be Fredholm and a description of their essential spectra have been obtained. We also prove the Fragmen-Lindelöf principle for exponential decreasing of solutions of pseudo-differential equations.

Mathematics Subject Classification (2000). Primary 35S05, 35S30; Secondary 47G30, 58J40.

Keywords. Pseudo-differential operators, Hölder-Zygmund Spaces, essential spectrum.

1. Introduction

We study pseudo-differential operators with symbols $a(x,\xi)$ in the L. Hörmander class $S^m_{1,0}$ acting on the Hölder-Zygmund spaces $\Lambda^s(\mathbb{R}^N)$. We suppose that the symbols $a(x,\xi)$ are slowly oscillating with respect to the variable x and have an analytic extension with respect to the ξ variable in a tube domain $\mathbb{R}^N + iB$, where $B\left(\subset \mathbb{R}^N\right)$ is a bounded open set. The necessary and sufficient conditions for operators under consideration to be Fredholm in weighted Hölder-Zygmund spaces are obtained. Moreover, we study a location of their essential spectra. We also consider the Fragmen-Lindelöf principle for exponential decreasing of solutions of pseudo-differential equations. As example, we describe a location of essential spectrum of Schrödinger operators acting on Hölder-Zygmund spaces, and a behavior at infinity of solutions of Schrödinger equations.

Supported by the CONACYT project No. 43432.

The paper is organized as follows. In Chap. 1 we consider pseudo-differential operators in the class $OPS^m_{1,0}$ acting on the Hölder-Zygmund spaces $\Lambda^s(\mathbb{R}^N)$ without weights. We prove the boundedness of pseudo-differential operators $A\left(\in OPS^m_{1,0}\right)$ from $\Lambda^s(\mathbb{R}^N)$ to $\Lambda^{s-m}(\mathbb{R}^N)$ and establish the estimate of the norm of A via semi-norms of symbol. Notice that boundedness of operators in the class $OPS^m_{1,0}(\mathbb{R}^N)$ from $\Lambda^s(\mathbb{R}^N)$ to $\Lambda^{s-m}(\mathbb{R}^N)$ have been proved in [9], p. 253-257, for exponents $s > m$. An outline of the proof of boundedness for all $s \in \mathbb{R}$ can be find in [12], p.37-38. Also, there are many works, where boundedness of pseudo-differential operators on Hölder-Zygmund type spaces proved under minimal conditions of smoothness for symbols $a(x,\xi)$ with respect to x. (See for instance [13] and references given there). But these works do not give explicit estimates of the norms of pseudo-differential operators via the semi-norms of its symbols. However, such estimates are very important in our considerations. For the completeness of presentation we give a proof of such estimates. The main result of this Chapter is a criterion of Fredholmness, and a description of essential spectra of pseudo-differential operators on the Hölder-Zygmund spaces. It follows from results obtained here that the essential spectra (as sets) for operators acting from $H^s(\mathbb{R}^N)$ to $H^{s-m}(\mathbb{R}^N)$, and from $\Lambda^s(\mathbb{R}^N)$ to $\Lambda^{s-m}(\mathbb{R}^N)$ coincide.

Notice that the Fredholm property and the location of essential spectrum of pseudo-differential operators in the class $OPS^m_{0,0}(\mathbb{R}^N)$ acting from $H^s(\mathbb{R}^N)$ to $H^{s-m}(\mathbb{R}^N)$ have been studied in [4], Chap.4 by the limit operators method. (See also references given in [4]).

In Chap. 2 we study pseudo-differential operators with analytical in a tube domain $\mathbb{R}^N + iB$ symbols. We prove boundedness of pseudo-differential operators with analytical symbols on the weighted Hölder-Zygmund spaces. The main results here are: a criterion of Fredholmness, a description of essential spectra of pseudo-differential operators on weighted Hölder-Zygmund spaces, and the Fragmen-Lindelöf principle on a behavior of solutions of pseudo-differential equations at infinity. Notice that the similar problems for general classes of pseudo-differential operators with rapidly increasing and discontinuous symbols acting on the Sobolev type spaces were studied in [6], [7].

2. Pseudo-Differential Operators on Hölder-Zygmund Spaces

2.1. Calculus of Pseudo-Differential Operators on $\mathbb{R}^N$

The goal of this subsection is to set up some notations and summarize (without proof) some basic facts on pseudo-differential operators. Standard references are [2], [3], [8], [10], [11].

We say that a function a belongs to the L. Hörmander class $S^m_{1,0}(\mathbb{R}^N)$ if $a \in C^\infty\left(\mathbb{R}^N_x \times \mathbb{R}^N_\xi\right)$, and

$$|a|^m_{r,t} = \sum_{|\alpha|\le r,|\beta|\le t} \sup_{\mathbb{R}^N\times\mathbb{R}^N} \left|\partial^\alpha_\xi \partial^\beta_x a(x,\xi)\right| \langle\xi\rangle^{-m+|\alpha|} < \infty \tag{1}$$

for all multi-indices α, β. As usual we associate with a symbol a the pseudo-differential operator defined on the space $S(\mathbb{R}^N)$ by the formula

$$Op(a)u(x) = a(x,D)u(x) = (2\pi)^{-N} \int_{\mathbb{R}^N} d\xi \int_{\mathbb{R}^N} a(x,\xi)u(y)e^{i(x-y,\xi)}dy. \quad (2)$$

We denote by $S^m_{1,0,0}(\mathbb{R}^N)$ the class of double symbols $a \in C^\infty(\mathbb{R}^N_x \times \mathbb{R}^N_y \times \mathbb{R}^N_\xi)$ satisfying the estimates

$$|a|^m_{r,t,l} = \sum_{|\alpha|\le r, |\beta|\le t, |\gamma|\le l} \sup_{\mathbb{R}^N\times\mathbb{R}^N\times\mathbb{R}^N} \left|\partial^\alpha_\xi \partial^\beta_x \partial^\gamma_y a(x,y,\xi)\right| \langle\xi\rangle^{-m+|\alpha|} < \infty. \quad (3)$$

We associate with $a \in S^m_{1,0,0}(\mathbb{R}^N)$ a pseudo-differential operator with double symbol

$$Au(x) = Op_d(a)u(x) = (2\pi)^{-N} \int_{\mathbb{R}^N} d\xi \int_{\mathbb{R}^N} a(x,y,\xi)u(y)e^{i(x-y,\xi)}dy, \quad (4)$$

and we denote the class of such operators by $OPS^m_{1,0,0}(\mathbb{R}^N)$.

Proposition 1. (i) *Let $A = Op(a) \in OPS^{m_1}_{1,0}(\mathbb{R}^N)$, $B = Op(b) \in OPS^{m_2}_{1,0}(\mathbb{R}^N)$. Then $AB \in OPS^{m_1+m_2}_{1,0}(\mathbb{R}^N)$, and $AB = Op(c)$, where*

$$c(x,\xi) = (2\pi)^{-N} \int\int_{\mathbb{R}^N} a(x,\xi+\eta)b(x+y,\xi)e^{-i(y,\eta)}dyd\eta. \quad (5)$$

Moreover,

$$|c(x,\xi)|^{m_1+m_2}_{l_1,l_2} \le C(l_1,l_2)\,|a|^{m_1}_{2k_1+l_1+m_1,l_2}\,|b|^{m_2}_{l_1,l_2+2k_2}, \quad (6)$$

where $2k_1 > N + m_1, 2k_2 > N$.

(ii) *Let $A = Op_d(a) \in OPS^m_{1,0,0}(\mathbb{R}^N)$. Then $A = Op(c) \in OPS^m_{1,0}(\mathbb{R}^N)$, where*

$$c(x,\xi) = (2\pi)^{-N} \int\int_{\mathbb{R}^N} a(x,x+y,\xi+\eta)e^{-i(y,\eta)}dyd\eta \quad (7)$$

Moreover,

$$|c(x,\xi)|^m_{l_1,l_2} \le C(l_1,l_2)\,|a|^m_{2k_1+l_1,l_2,l_2+2k_2},$$

$2k_1 > N + m, 2k_2 > N$.

(iii) *Let A^t be a formal adjoint operator for $A = Op(a) \in OPS^m_{1,0}(\mathbb{R}^N)$ defined by the formula*

$$(Au,v) = \left(u, A^t v\right), u,v \in S\left(\mathbb{R}^N\right), \left((u,v) = \int_{\mathbb{R}^N} u(x)\bar{v}(x)dx\right). \quad (8)$$

Then $A^t = Op(a^t) \in OPS^m_{1,0}(\mathbb{R}^N)$, and

$$a^t(x,\xi) = (2\pi)^{-N} \int\int_{\mathbb{R}^N} \bar{a}(x+y,\xi+\eta)e^{-i(y,\eta)}dyd\eta. \quad (9)$$

The integrals in (5), (7), (9) are understood in the oscillatory sense.

Notice, that formula (8) allows us to extend the pseudo-differential operators on the space of distributions $S'\left(\mathbb{R}^N\right)$.

We denote by $C_b^\infty(\mathbb{R}^N)$ the class of functions a defined on $\mathbb{R}^N$ bounded with all their derivatives. The topology in $C_b^\infty(\mathbb{R}^N)$ is defined by the semi-norms

$$|a|_l = \sup_{\mathbb{R}^N} \sum_{|\beta|\leq l} \left|\partial^\beta u(x)\right|.$$

Corollary 2. *Let* $\varphi \in C_b^\infty(\mathbb{R}^N)$, $\varphi_R(x) = \varphi(x/R)$, $A = Op(a) \in OPS_{1,0}^m(\mathbb{R}^N)$. *Then* $[\varphi_R, A] = \varphi_R A - A\varphi_R \in OPS^{m-1}(\mathbb{R}^N)$, *and*

$$\|[\varphi_R, A]\|_{l_1,l_2}^{m-1} \leq CR^{-1} |\varphi|_{2k_2+l_2} |a|_{2k_1+l_1,l_2+2k_2}^m,$$

where $2k_1 > N+m, 2k_2 > N$.

Corollary 2 follows from formula (5) and estimate (6), where $2k_1 > N+m$, $2k_2 > N$.

Proposition 3. *Let* $A = Op(a) \in OPS_{1,0}^m(\mathbb{R}^N)$. *Then*

$$Au(x) = \int_{\mathbb{R}^N} k_a(x,z)u(x-z)dz, u \in S(\mathbb{R}^N),$$

where

$$k_a(x,z) = F_{\xi\to z}^{-1} a(x,\xi).$$

($F_{\xi\to z}^{-1}$ is the inverse Fourier transform in the sense of distributions.)

The kernel $k_a(x,z) \in C^\infty(\mathbb{R}^N \times \mathbb{R}^N \backslash 0)$, *and for every multi-indices* α, β, *and* $2k > N + m + |\alpha|$,

$$\left|\partial_x^\beta \partial_z^\alpha k_a(x,z)\right| \leq C\,|a|_{|\alpha|+2k,|\beta|}^m |z|^{-2k}, z \neq 0, \tag{10}$$

where the constant C *is independent of* a.

2.2. Operators with Slowly Oscillating Symbols

We need some facts on calculus of pseudo-differential operators with slowly oscillating symbols [5], [4], Chap. 4.

Definition 4. *A symbol* a *is called slowly oscillating if* $a \in S_{1,0}^m(\mathbb{R}^N)$, *and*

$$\left|\partial_\xi^\alpha \partial_x^\beta a(x,\xi)\right| \leq C_{\alpha\beta}(x) \langle\xi\rangle^{m-|\alpha|}, \tag{11}$$

where $\lim_{x\to\infty} C_{\alpha\beta}(x) = 0$ *for every* α *and* $\beta \neq 0$. *We denote by* $SO^m(\mathbb{R}^N)$ *the class of slowly oscillating symbols, and by* $SO_0^m(\mathbb{R}^N)$ *the subclass in* $SO^m(\mathbb{R}^N)$ *of symbols such that the* $\lim_{x\to\infty} C_{\alpha\beta}(x) = 0$ *for every* α *and* β. *We use the notations* $OPSO^m(\mathbb{R}^N)$, $OPSO_0^m(\mathbb{R}^N)$ *for the classes of operators with symbols in* $SO^m(\mathbb{R}^N), SO_0^m(\mathbb{R}^N)$, *respectively.*

A double symbol $a \in S_{1,0,0}^m(\mathbb{R}^N)$ *is called slowly oscillating if for every compact set* $K \subset \mathbb{R}^N$

$$\sup_{y\in K} \left|\partial_\xi^\alpha \partial_x^\beta \partial_y^\gamma a(x,x+y,\xi)\right| \leq C_{\alpha\beta\gamma}^K(x) \langle\xi\rangle^{m+|\alpha|},$$

where

$$\lim_{x\to\infty} C^K_{\alpha\beta\gamma}(x) = 0$$

for every α *and* $|\beta+\gamma| \neq 0$. *We denote by* $SO^m_d(\mathbb{R}^N)$ *the class of slowly oscillating double symbols, and by* $OPSO^m_d(\mathbb{R}^N)$ *the corresponding class of pseudo-differential operators.*

Proposition 5. (i) *Let* $A = Op(a) \in OPSO^{m_1}(\mathbb{R}^N)$, $B = Op(b) \in OPSO^{m_2}(\mathbb{R}^N)$. *Then* $AB \in OPSO^{m_1+m_2}(\mathbb{R}^N)$, *and*

$$AB = Op(a)Op(b) + Op(t(x,\xi)),$$

where $t(x,\xi) \in SO_0^{m_1+m_2-1}(\mathbb{R}^N)$.

(ii) *Let* $A = Op_d(a) \in OPSO^m_d(\mathbb{R}^N)$. *Then*

$$A = Op(a(x,x,\xi)) + Op(t(x,\xi)),$$

where $t(x,\xi) \in SO_0^{m-1}(\mathbb{R}^N)$.

2.3. Hölder-Zygmund Spaces

If $0 < s < 1$, we define $C^s(\mathbb{R}^N)$ to be a subspace of $C(\mathbb{R}^N)$ consisting of those bounded functions u which satisfies in $\mathbb{R}^N$ a Hölder conditions of exponent s, that is there exists a constant c such that $|u(x) - u(y)| \le c\,|x-y|^s$ for all $x, y \in \mathbb{R}^N$.

This space is a Banach space with norm given

$$\|u\|_{C^s(\mathbb{R}^N)} = \sup_{\mathbb{R}^N} |u(x)| + \sup_{x,y\in\mathbb{R}^N,\, |x-y|>0} \frac{|u(x)-u(y)|}{|x-y|^s}. \tag{12}$$

For $m = 0, 1, 2, \ldots$ we take $C^m_b(\mathbb{R}^N)$ to consist of bounded and continuous functions u such that all derivatives $\partial^\beta u$, $|\beta| \le m$ are continuous and bounded. If $s = m + r$, $0 < r < 1$, we define $C^s(\mathbb{R}^N)$ to consist of functions $u \in C^m_b(\mathbb{R}^N)$ such that, for $|\beta| = m$, $\partial^\beta u \in C^r(\mathbb{R}^N)$. The norm in $C^s(\mathbb{R}^N)$ is introduced as

$$\|u\|_{C^s(\mathbb{R}^N))} = \sum_{|\beta|\le m} \sup_{\mathbb{R}^N} \left|\partial^\beta u(x)\right| + \sum_{|\beta|=m} \sup_{x,y\in\mathbb{R}^N,|x-y|>0} \frac{\left|\partial^\beta u(x) - \partial^\beta u(y)\right|}{|x-y|^r}. \tag{13}$$

For non-integer s the Hölder space $C^s(\mathbb{R}^N)$ have a characterization via the Littlewood-Paley partition of unity

$$\sum_{k=0}^{\infty} \psi_k(\xi) = 1$$

with $\psi_0(\xi) = \Psi_0(\xi)$, $\psi_k(\xi) = \Psi_k(\xi) - \Psi_{k-1}(\xi)$, $k \in \mathbb{N}$, where $\Psi_0 \in C_0^\infty(\mathbb{R}^N)$, so that $\Psi_0(\xi) = 1$ for $|\xi| \le 1$ and 0 for $|\xi| \ge 2$, and $\Psi_k(\xi) = \Psi_0(2^{-k}\xi)$.

We say that $u \left(\in S'(\mathbb{R}^N)\right)$ belongs to the Hölder-Zygmund space $\Lambda^s(\mathbb{R}^N)$ if

$$\|u\|_{\Lambda^s(\mathbb{R}^N)} = \sup_{k\in\mathbb{N}} 2^{ks} \|\psi_k(D)u\|_{L^\infty(\mathbb{R}^N)} < \infty. \tag{14}$$

It is well-known that $C^s(\mathbb{R}^N) = \Lambda^s(\mathbb{R}^N)$, if $s \in \mathbb{R}_+ \backslash \mathbb{N}$ and $C^s(\mathbb{R}^N) \subset \Lambda^s(\mathbb{R}^N)$, if $s \in \mathbb{N}_0 = \mathbb{N} \cup 0$ (see for instance, [12] p. 37-40, [9], p. 253-257).

Notice that if $u \in \Lambda^s(\mathbb{R}^N), s > 0$, then

$$\|u\|_{L^\infty(\mathbb{R}^N)} \leq \sum_{k=0}^{\infty} \|\psi_k(D)u\|_{L^\infty(\mathbb{R}^N)} \leq \left(\sum_{k=0}^{\infty} 2^{-ks}\right) \|u\|_{\Lambda^s(\mathbb{R}^N)} = C_s \|u\|_{\Lambda^s(\mathbb{R}^N)} . \tag{15}$$

Given an open set $\Omega \subset \mathbb{R}^N$ we denote by $\Lambda^s(\Omega)$ the space of all functions in $\Lambda^s(\mathbb{R}^N)$ with supports in Ω, and we denote by $\Lambda^s_0(\mathbb{R}^N)$ the closure of functions in $\Lambda^s(\mathbb{R}^N)$ with compact supports in norm (14).

In what follows $\mathcal{B}(X, Y)$ is the space of all bounded linear operators acting from a Banach space X into a Banach space Y. If $X = Y$ we will write $\mathcal{B}(X)$.

Proposition 6. (*see,* [12], *p.* 39) *Let $r, s \in \mathbb{R}$, and $r < s$. Then $\Lambda^{\theta s+(1-\theta)r}(\mathbb{R}^N)$, $\theta \in (0, 1)$ is the interpolation space between $\Lambda^r(\mathbb{R}^N)$ and $\Lambda^s(\mathbb{R}^N)$.*

If $A \in \mathcal{B}(\Lambda^r(\mathbb{R}^N)) \cap \mathcal{B}(\Lambda^s(\mathbb{R}^N))$, then $A \in \mathcal{B}(\Lambda^{\theta s+(1-\theta)r}(\mathbb{R}^N))$, $\theta \in (0, 1)$, and

$$\|A\|_{\Lambda^{\theta s+(1-\theta)r}(\mathbb{R}^N)} \leq C \|A\|^{\theta}_{\Lambda^s(\mathbb{R}^N)} \|A\|^{1-\theta}_{\Lambda^r(\mathbb{R}^N)} .$$

Proposition 7. *Let $\varphi \in C_0^\infty(\mathbb{R}^N)$, $\varphi(x) = 1$ if $|x| \leq 1$, and $\varphi(x) = 0$ if $|x| \geq 2$, $\varphi_R(x) = \varphi(x/R)$. Then for every $s \in (0, 1)$*

$$\lim_{R\to\infty} \|\varphi_R u\|_{\Lambda^s(\mathbb{R}^N)} = \|u\|_{\Lambda^s(\mathbb{R}^N)} . \tag{16}$$

Proof. Using norm (13) we obtain that

$$\|\varphi_R u\|_{\Lambda^s(\mathbb{R}^N)} \leq |\varphi_R|_1 \|u\|_{\Lambda^s(\mathbb{R}^N)} . \tag{17}$$

Since $\partial^\alpha \varphi_R = O(1/R)$, $|\alpha| = 1$ we obtain that given $\varepsilon > 0$ there exists $R_0 > 0$ such that for $R > R_0$

$$\|\varphi_R u\|_{\Lambda^s(\mathbb{R}^N)} \leq \|u\|_{\Lambda^s(\mathbb{R}^N)} + \varepsilon. \tag{18}$$

Then using norm (13) again, we obtain that for arbitrary $\varepsilon > 0$ there exists point $x_0, y_0 \in \mathbb{R}^N, x_0 \neq y_0$ such that

$$\begin{aligned} \|u\|_{\Lambda^s(\mathbb{R}^N)} &\leq |u(x_0)| + \frac{|u(x_0) - u(y_0)|}{|x_0 - y_0|^s} + \varepsilon \\ &= |(\varphi_R u)(x_0)| + \frac{|(\varphi_R u)(x_0) - (\varphi_R u)(y_0)|}{|x_0 - y_0|^s} + \varepsilon \end{aligned} \tag{19}$$

for $R > 0$ large enough. Thus, (19) implies that given $\varepsilon > 0$ there exists $R_0 > 0$ such that for $R > R_0$

$$\|u\|_{\Lambda^s(\mathbb{R}^N)} \leq \|\varphi_R u\|_{\Lambda^s(\mathbb{R}^N)} + \varepsilon. \tag{20}$$

Hence, inequalities (17), (20) yield 16. □

Proposition 8. (*see,* [13], *p.* 52) *Let $s \leq r$. Then $\Lambda^r(\mathbb{R}^N) \subset \Lambda^s(\mathbb{R}^N)$. If $s < r$, and Ω is an open set with a compact closure, then the imbedding: $\Lambda^r(\Omega) \subset \Lambda^s(\Omega)$ is a compact operator.*

2.4. Boundedness and Compactness of Pseudo-Differential Operators on H ölder-Zygmund Spaces

Theorem 9. *A pseudo-differential operator $A = Op(a) \in OPS^m_{1,0}(\mathbb{R}^N)$ is a bounded operator from $\Lambda^s(\mathbb{R}^N)$ into $\Lambda^{s-m}(\mathbb{R}^N)$ for every $s \in \mathbb{R}$. Moreover, there exist constants $L_1, L_2 \in \mathbb{N}$ and $C > 0$ depending on s, m, N only such that*

$$\|Op(a)u\|_{\Lambda^{s-m}(\mathbb{R}^N)} \le C\,|a|^m_{L_1,L_2}\,\|u\|_{\Lambda^s(\mathbb{R}^N)}\,. \tag{21}$$

We separate the proof of Theorem 9 on a series of propositions.
In the sequel

$$\langle D\rangle^m = Op(\langle\xi\rangle^m)\ (\langle\xi\rangle = (1+|\xi|^2)^{1/2}),\ \ m \in \mathbb{R}.$$

Proposition 10. *Let $A = a(D) \in OPS^m_{1,0}$ $(m \in \mathbb{R})$. Then $a(D) : \Lambda^s(\mathbb{R}^N) \to \Lambda^{s-m}(\mathbb{R}^N)$ is a bounded operator for every $s \in \mathbb{R}$, and*

$$\|a(D)u\|_{\Lambda^{s-m}(\mathbb{R}^N)} \le C\,|a|^m_{2k}\,\|u\|_{\Lambda^s(\mathbb{R}^N)}\,, \tag{22}$$

where $2k > N$, and

$$|a|^m_{2k} = \sup_{\xi}\left|\langle D\rangle^k a(\xi)\right|\langle\xi\rangle^{-m+2k}.$$

Proof. Pick $\tilde{\psi}_1(\xi)$ such that $\tilde{\psi}_1(\xi) = 1$ on the support of ψ_1 and $\tilde{\psi}_j(\xi) = \tilde{\psi}_1(2^{1-j}\xi)$. Let $a_j(D) = a(D)\tilde{\psi}_j(D)$, and

$$k_j(z) = (2\pi)^{-N}\int_{\mathbb{R}^N} a(\xi)\tilde{\psi}_1(2^{1-j}\xi)e^{i(z,\xi)}d\xi.$$

Then

$$|z|^{2k}\,|k_j(z)| \le (2\pi)^{-N}\int_{\mathbb{R}^N}\left|\Delta^k\left(a(\xi)\tilde{\psi}_1(2^{1-j}\xi)\right)\right|d\xi \le C\,|a|_{2k}\,2^{j[N+m-2k]}. \tag{23}$$

It implies that

$$\begin{aligned} \|k_j\|_{L^1(\mathbb{Z}^N)} &\le C\,|a|_0\,2^{j[N+m]}\int_{|z|\le 2^{-j}} dz + C2^{j[N+m-2k]}\,|a|_{2k}\int_{|z|\ge 2^{-j}}|z|^{-2k}\,dz \\ &\le C\,|a|_{2k}\,2^{jm}, \quad 2k > N. \end{aligned} \tag{24}$$

Estimate (24) yields that

$$\begin{aligned} \|a(D)u\|_{\Lambda^{s-m}(\mathbb{R}^N)} &= \sup_j 2^{j(s-m)}\,\|\psi_j(D)a(D)u\|_{L^\infty(\mathbb{R}^N)} \\ &= \sup_j 2^{j(s-m)}\left\|a(D)\tilde{\psi}_j(D)\psi_j(D)u\right\|_{L^\infty(\mathbb{R}^N)} \\ &\le \sup_j 2^{j(s-m)}\,\|k_j\|_{L^1(\mathbb{Z}^N)}\,\|\psi_j(D)u\|_{L^\infty(\mathbb{R}^N)} \\ &\le C\,|a|_{2k}\sup_j 2^{js}\,\|\psi_j(D)u\|_{L^\infty(\mathbb{R}^N)} \\ &= C\,|a|_{2k}\,\|u\|_{\Lambda^s(\mathbb{R}^N)}\,. \end{aligned}$$

□

Corollary 11. *The operator* $\langle D\rangle^m$ $(m \in \mathbb{R})$ *is an isomorphism from* $\Lambda^s(\mathbb{R}^N)$ *on* $\Lambda^{s-m}(\mathbb{R}^N)$ *for every* $s \in \mathbb{R}$.

Proposition 12. *Let* $A = Op(a) \in OPS^m_{1,0}(\mathbb{R}^N)$, *and let* a *have a compact support in* x. *Then, for* $s > m$,

$$\|Op(a)u\|_{\Lambda^{s-m}(\mathbb{R}^N)} \leq CM\, |a|^m_{2k_1,2k_2} \|u\|_{\Lambda^s(\mathbb{R}^N)}\,, \tag{25}$$

where $2k_2 > N + s - m$, $2k_1 > N$, $M = mes(supp_x a(x,\xi))$, *and the constant* $C > 0$ *is independent of* a.

Proof. We can write

$$Op(a)u(x) = \int_{\mathbb{R}^N} e^{i(x,\eta)} \left(q_\eta(D)u\right)(x) d\eta, \tag{26}$$

where

$$q_\eta(\xi) = (2\pi)^{-N} \int_{\mathbb{R}^N} e^{-i(x,\eta)} a(x,\xi)dx. \tag{27}$$

One can see that

$$\langle\eta\rangle^{2k_2} q_\eta(\xi) = (2\pi)^{-N} \int_{\mathbb{R}^N} e^{-i(x,\eta)} \langle D_x\rangle^{2k_2} a(x,\xi)dx. \tag{28}$$

The inequality (28) implies the estimate

$$|q_\eta(\xi)|^m_{2k_1} \leq CM\, |a|^m_{2k_1,2k_2} \langle\eta\rangle^{-2k_2}. \tag{29}$$

It is easy to establish the estimate

$$\left\|e^{-i(x,\eta)}u\right\|_{\Lambda^{s-m}(\mathbb{R}^N_x)} \leq C_s \langle\eta\rangle^{s-m} \|u\|_{\Lambda^{s-m}(\mathbb{R}^N_x)}\ (s > m), \tag{30}$$

first, for $s-m \notin \mathbb{N}$ by using norm (13), then for general $s-m > 0$ by interpolation (see Proposition 6). Then, applying (29) and (30) we obtain

$$\begin{aligned} \|Op(a)u\|_{\Lambda^{s-m}(\mathbb{R}^N_x)} &\leq \int_{\mathbb{R}^N} \left\|e^{-i(x,\eta)} q_\eta(D)u(x)\right\|_{\Lambda^{s-m}(\mathbb{R}^N_x)} d\eta \\ &\leq CM\, |a|^m_{2k_1,2k_2} \int \langle\eta\rangle^{s-m-2k_2} d\eta, 2k_1 > N. \end{aligned}$$

Setting in the last estimate $2k_2 > N + s - m$ we obtain estimate (25). □

Let $f(x) \left(\in C_0^\infty(\mathbb{R}^N)\right)$ be a non negative function such that $f(x) = 1$ if $|x_i| \leq 2/3, i = 1, \ldots, N$, and $f(x) = 0$ if $|x_i| \geq 3/4$. Let

$$\varphi_0(x) = \frac{f(x)}{\sum_{\beta\in\mathbb{Z}^N} f(x-\beta)},\ \varphi_\alpha(x) = \varphi_0(x-\alpha),\ \alpha \in \mathbb{Z}^N.$$

Then we obtain the partition of unity on $\mathbb{R}^N$

$$\sum_{\alpha\in\mathbb{Z}^N} \varphi_\alpha(x) = 1,\ x \in \mathbb{R}^N. \tag{31}$$

Let $\phi_0 \in C_0^\infty(\mathbb{R}^N), 0 \leq \phi_0(x) \leq 1$, *supp* $\phi_0 \subset \left\{x \in \mathbb{R}^N : |x_i| \leq 1\right\}$, and $\varphi_0\phi_0 = \varphi_0$, $\phi_\alpha(x) = \phi_0(x-\alpha)$, $\alpha \in \mathbb{Z}^N$.

Proposition 13. *Let $A = Op(a) \in OPS^0_{1,0}(\mathbb{R}^N)$, and $s > 0$. Then*

$$\|\phi_\alpha Op(a)\phi_\beta\|_{\mathcal{B}(\Lambda^s(\mathbb{R}^N))} \leq C\,|a|_{4k_1,2k_2}\langle\alpha-\beta\rangle^{-2k_1}, \langle\alpha-\beta\rangle = (1+|\alpha-\beta|^2)^{1/2}, \tag{32}$$

where $2k_1 > N, 2k_2 > N+s$, and the constant $C > 0$ is independent of A.

Proof. Let

$$V_h f(x) = f(x-h), (x \in \mathbb{R}^N, h \in \mathbb{R}^N)$$

be the shift operator acting isometrically on $\Lambda^s(\mathbb{R}^N)$. Then

$$\|\phi_\alpha A\phi_\beta\|_{\mathcal{B}(\Lambda^s(\mathbb{R}^N))} = \|V_{-\alpha}\phi_\alpha A\phi_\beta V_\alpha\|_{\mathcal{B}(\Lambda^s(\mathbb{R}^N))} = \|\phi_0 V_{-\alpha}AV_\alpha\phi_{\beta-\alpha}\|_{\mathcal{B}(\Lambda^s(\mathbb{R}^N))}. \tag{33}$$

One can see that $A_\alpha = V_{-\alpha}AV_\alpha = Op(a(x+\alpha,\xi))$.

Let $\sigma_{\phi_0 A_\alpha\phi_{\beta-\alpha}}$ be the symbol of the pseudo-differential operator $\phi_0\, A_\alpha\phi_{\beta-\alpha}$. In light of Proposition 1 (i)

$$\sigma_{\phi_0 A_\alpha\phi_{\beta-\alpha}}(x,\xi) = \frac{\phi_0(x)}{(2\pi)^N}\iint_{\mathbb{R}^N} a(x+\alpha,\xi+\eta)\phi_{\beta-\alpha}(x+y)e^{-i(y,\eta)}dyd\eta$$

$$= \frac{\phi_0(x)}{(2\pi)^N}\iint_{\mathbb{R}^N}\langle y\rangle^{-2k_1}\langle D_\xi\rangle^{2k_1}\langle\xi\rangle^{-2k_2}\langle D_y\rangle^{2k_2}$$

$$\cdot\, a(x+\alpha,\xi+\eta)\phi_{\beta-\alpha}(x+y)e^{-i(y,\eta)}dyd\eta.$$

Hence,

$$\left|\sigma_{\phi_0 A_\alpha\phi_{\beta-\alpha}}(x,\xi)\right| \leq C\,|a|_{2k_1,0}\int_{\mathbb{R}^N}\langle y\rangle^{-2k_1}\left|\langle D_y\rangle^{2k_2}\phi_0(x+y+\alpha-\beta)\right|dy$$

$$= C\,|a|_{2k_1,0}\,\phi_0(x)\int_{\mathbb{R}^N}\langle z+(\beta-\alpha)\rangle^{-2k_1}\left|\langle D_z\rangle^{2k_2}\phi_0(x+z)\right|dz$$

$$\leq C'\,|a|_{2k_1,0}\langle\beta-\alpha\rangle^{-2k_1}\phi_0(x)\int_{\mathbb{R}^N}\langle z\rangle^{2k_1}\left|\langle D_z\rangle^{2k_2}\phi_0(x+z)\right|dz.$$

Because *supp* $\phi_0(x) \subset I_0 = [-1,1]^N$ the function

$$\phi_0(x)\int_{\mathbb{R}^N}\langle z\rangle^{2k_1}\left|\langle D_z\rangle^{2k_2}\phi_0(x+z)\right|dz$$

is bounded. Thus, we obtain the estimate

$$\left|\sigma_{\phi_0 A_\alpha\phi_{\beta-\alpha}}\right| \leq C'\,|a|_{2k_1,0}\langle\beta-\alpha\rangle^{-2k_1}.$$

In the same way we obtain:

$$\left|\sigma_{\phi_0 A_\alpha\phi_{\beta-\alpha}}(x,\xi)\right|_{2k_1,2k_2} \leq C\,|a|_{4k_1,2k_2.}\langle\beta-\alpha\rangle^{-2k_1}, \tag{34}$$

where the constant $C > 0$ is independent of A . Setting in estimate (34) $2k_1 > N$, $2k_2 > N+s$, and applying Proposition 6, and equality (33) we obtain estimate (33). □

Proposition 14. *An operator $A = Op(a) \in OPS^0_{1,0}(\mathbb{R}^N)$ is bounded from $\Lambda^s(\mathbb{R}^N)$ into $\Lambda^s(\mathbb{R}^N)$ for all $s \in (0,1)$. Moreover,*

$$\|Op(a)u\|_{\Lambda^s(\mathbb{R}^N)} \leq C\,|a|_{4k_1,2k_2}\,\|u\|_{\Lambda^s(\mathbb{R}^N)}\,, \tag{35}$$

where $2k_1 > N+s$, $2k_2 > N$, and the constant $C > 0$ is independent of A.

Proof. Applying the partition of unity (31) we obtain

$$Au = \sum_{\alpha\in\mathbb{Z}^N} \varphi_\alpha \sum_{\beta\in\mathbb{Z}^N} \phi_\alpha A \phi_\beta \varphi_\beta u. \tag{36}$$

Because $0 < s < 1$, we can use the norm (12). The elementary estimate gives

$$\|u\|_{\Lambda^s(\mathbb{R}^N)} = \left\| \sum_{\alpha\in\mathbb{Z}^N} \varphi_\alpha u \right\|_{\Lambda^s(\mathbb{R}^N)} \leq 2^N \sup_{\alpha\in\mathbb{Z}^N} \|\varphi_\alpha u\|_{\Lambda^s(\mathbb{R}^N)}\,. \tag{37}$$

Then applying (37) and Proposition 12 we obtain

$$\begin{aligned}
\|Au\|_{\Lambda^s(\mathbb{R}^N)} &\leq 2^N \sup_{\alpha\in\mathbb{Z}^N} \left\| \varphi_\alpha \sum_{\beta\in\mathbb{Z}^N} \phi_\alpha A \phi_\beta \varphi_\beta u. \right\|_{\Lambda^s(\mathbb{R}^N)} \\
&\leq 2^N \sup_{\alpha} \sum_{\beta\in\mathbb{Z}^N} \|\phi_\alpha A \phi_\beta\|_{\mathcal{B}(\Lambda^s(\mathbb{R}^N))} \|\varphi_\beta u\|_{\Lambda^s(\mathbb{R}^N)} \\
&\leq C \sup_{\alpha\in\mathbb{Z}^N} \sum_{\beta\in\mathbb{Z}^N} \langle \alpha - \beta\rangle^{-2k_1} \|\varphi_\beta u\|_{\Lambda^s(\mathbb{R}^N)} \\
&\leq C \left(\sum_{\beta\in\mathbb{Z}^N} \langle\beta\rangle^{-2k_1} \right) |a|_{4k_1,2k_2.} \sup_{\beta\in\mathbb{Z}^N} \|\varphi_\beta u\|_{\Lambda^s(\mathbb{R}^N)} \\
&\leq C\,|a|_{4k_1,2k_2.} \|u\|_{\Lambda^s(\mathbb{R}^N)}\,,
\end{aligned}$$

where $C > 0$ is independent of A, and $2k_1 > N+s, 2k_2 > N$. □

Proof of Theorem 9. Let $s = r + \delta$, where $\delta \in (0,1), A = Op(a) \in OPS^0_{1,0}(\mathbb{R}^N)$. Then $B = Op(b) = \langle D\rangle^{-r} A \langle D\rangle^{r} \in OPS^0(\mathbb{R}^N)$, and, by Proposition 14, $B = Op(b)$ is bounded on $\Lambda^\delta(\mathbb{R}^N)$. In light of Proposition 1 (i) and Corollary 11

$$\|A\|_{\mathcal{B}(\Lambda^s(\mathbb{R}^N))} \leq C\,\|B\|_{\mathcal{B}(\Lambda^\delta(\mathbb{R}^N))} \leq C\,|b|_{4k_1,2k_2} \leq C\,|a|_{L_1,L_2}\,. \tag{38}$$

If $A = Op(a) \in OPS^m_{1,0}(\mathbb{R}^N), m \neq 0$ then estimate (21) follows from the estimates

$$\|A\|_{\mathcal{B}(\Lambda^s(\mathbb{R}^N),\Lambda^{s-m}(\mathbb{R}^N))} \leq C \left\| \langle D\rangle^{-m} A \right\|_{\mathcal{B}(\Lambda^s(\mathbb{R}^N))} \leq C\,|a|^m_{L_1,L_2}\,,$$

where L_1, L_2 depend on s, m and N, only. □

Corollary 15. *The operator $A = Op(a) \in OPS^m_{1,0}(\mathbb{R}^N)$ is bounded from $\Lambda^s_0(\mathbb{R}^N)$ into $\Lambda^{s-m}_0(\mathbb{R}^N)$ for all $s \in \mathbb{R}$.*

Proof. Let $u(\in \Lambda^s(\mathbb{R}^N))$ have a compact support. Then there exist R_0 such that for $R > R_0$ $\varphi_R u = u$. Hence,

$$Au = A\varphi_R u = \varphi_R Au + [A, \varphi_R]\, u.$$

Then by Corollary 2

$$\|Au - \varphi_R Au\|_{\Lambda^{s-m}(\mathbb{R}^N)} \le \|[A, \varphi_R]\| \, \|u\|_{\Lambda^s(\mathbb{R}^N)} \le \frac{C}{R} \|u\|_{\Lambda^s(\mathbb{R}^N)} \to 0$$

if $R \to \infty$. Thus, for every $u \in \Lambda^s(\mathbb{R}^N)$ with a compact support $Au \in \Lambda_0^{s-m}(\mathbb{R}^N)$. Let $v \in \Lambda_0^s(\mathbb{R}^N)$. Given $\varepsilon > 0$ there exists u with a compact support such that $\|v - u\|_{\Lambda^s(\mathbb{R}^N)} < \varepsilon$. Then

$$\begin{aligned} \|Av - \varphi_R Av\|_{\Lambda^{s-m}(\mathbb{R}^N)} &\le \|Av - Au\|_{\Lambda^{s-m}(\mathbb{R}^N)} + \|Au - \varphi_R Au\|_{\Lambda^{s-m}(\mathbb{R}^N)} \\ &\quad + \|\varphi_R (Au - Av)\|_{\Lambda^{s-m}(\mathbb{R}^N)} \\ &\le C\, \|u - v\|_{\Lambda^s(\mathbb{R}^N)} + \|Au - \varphi_R Au\|_{\Lambda^{s-m}(\mathbb{R}^N)} \\ &\le (C+1)\, \varepsilon \end{aligned}$$

for $R > R_0$ large enough. □

Proposition 16. *Let $T = Op(t) \in SO_0^{m-\varepsilon}, \varepsilon > 0$. Then*

$$T : \Lambda^s(\mathbb{R}^N) \to \Lambda^{s-m}\ (\mathbb{R}^N), (\Lambda_0^s(\mathbb{R}^N) \to \Lambda_0^{s-m}\ (\mathbb{R}^N))$$

is a compact operator.

Proof. Let $\varphi \in C_0^\infty(\mathbb{R}^N), \varphi(x) = 1$ if $|x| \le 1$, and $\varphi(x) = 0$ if $|x| \ge 2, \varphi_R(x) = \varphi(x/R)$, $\psi_R = 1 - \varphi_R$. Then

$$\begin{aligned} &\|T - \varphi_R T\|_{\mathcal{B}(\Lambda^s(\mathbb{R}^N), \Lambda^{s-m}\ (\mathbb{R}^N))} \\ \le\ &\|\psi_R T\|_{\mathcal{B}(\Lambda^s(\mathbb{R}^N), \Lambda^{s-m}\ (\mathbb{R}^N))} \le C\, |\psi_R t|_{L_1, L_2}^{m-\varepsilon} \end{aligned}$$

where $C\,(> 0)$, and $L_1, L_2 \in \mathbb{N}$ are independent of t and R. The definition of the class $SO_0^{m-\varepsilon}$ implies that

$$\lim_{R\to\infty} |\psi_R t|_{L_1, L_2}^{m-\varepsilon} = 0.$$

Let us prove that $\varphi_R T : \Lambda^s(\mathbb{R}^N) \to \Lambda^{s-m}(\mathbb{R}^N)$ is a compact operator. Indeed, *supp* $\varphi_R T u \subset B_{2R} = \{x \in \mathbb{R}^N : |x| < 2R\}$ for every function $u \in \Lambda^s(\mathbb{R}^N)$. Hence, $\varphi_R T$ maps bounded sets in $\Lambda^s(\mathbb{R}^N)$ in bounded sets in $\Lambda^{s+\varepsilon-m}(B_{2R})$. By Proposition 8 the space $\Lambda^{s+\varepsilon-m}(B_{2R})$ is imbedded compactly in $\Lambda^{s-m}(B_{2R})$. It yields compactness of $\varphi_R T$. □

Corollary 17. *Let $A = Op(a) \in OPSO^{m_1}, B = Op(b) \in OPSO^{m_2}$. Then the commutator*

$$[A, B] = AB - BA$$

is a compact operator from $\Lambda^s(\mathbb{R}^N)$ in $\Lambda^{s-m_1-m_2}(\mathbb{R}^N)$.

Proof. We have by Proposition 5

$$AB = Op(ab) + Op(t_1), BA = Op(ab) + Op(t_2),$$

where $t_1, t_2 \in SO_0^{m_1+m_2-1}(\mathbb{R}^N)$, and by Proposition 16 $Op(t_1), Op(t_2)$ are compact operators from $\Lambda^s(\mathbb{R}^N)$ in $\Lambda^{s-m_1-m_2}(\mathbb{R}^N)$. □

3. Fredholmness and Essential spectra of Pseudo-Differential Operators on Weighted Hölder-Zygmund Spaces

3.1. Fredholmness and Essential spectra on Hölder-Zygmund Spaces without Weights

Definition 18. *We say that $Op(a) \in OPS_{1,0}^m(\mathbb{R}^N)$ is an uniformly elliptic operator on $\mathbb{R}^N$ if*

$$\lim_{r\to\infty} \inf_{|\xi|\geq r, x\in\mathbb{R}^N} |a(x,\xi)| \langle\xi\rangle^{-m} > 0.$$

It is well-known that if $A = Op(a)$ is uniformly elliptic then there exists an operator $B = Op(b) \in OPS_{1,0}^{-m}(\mathbb{R}^N)$ such that

$$BA = I + T_1, AB = I + T_2, \tag{39}$$

where $T_1, T_2 \in OPS_{1,0}^{-1}(\mathbb{R}^N)$.

We denote by $\widetilde{\mathbb{R}^N}$ the compactification of $\mathbb{R}^N$ homeomorphic to the unit ball $\mathbb{B} = \{x \in \mathbb{R}^N : |x| \leq 1\}$, and by $S_{1,0}^m(\widetilde{\mathbb{R}^N})$ the class of symbols $a \in S_{1,0}^m(\mathbb{R}^N)$ such that $a\langle\xi\rangle^{-m}$ is extended to a continuous function on $\mathbb{R}^N \times \widetilde{\mathbb{R}^N}$. If $a \in S_{1,0}^m(\widetilde{\mathbb{R}^N})$ then there exists the limit

$$a_0(x,\theta) = \lim_{r\to\infty} r^{-m} a(x, r\theta)$$

for every point $\theta \in S^{N-1}$, and the condition of ellipticity for operators in $OPS_{1,0}^m(\widetilde{\mathbb{R}^N})$ can be written as

$$\inf_{(x,\theta)\in\mathbb{R}^N\times S^{N-1}} |a_0(x,\theta)| > 0. \tag{40}$$

Proposition 19. *Let $A = Op(a) \in OPS_{1,0}^m(\widetilde{\mathbb{R}^N})$, and a sequence $h_n \to \infty$. Then there exists a subsequence h_{n_k} and a symbol $a_h \in S_{1,0}^m$ such that for every function $\varphi \in C_0^\infty(\mathbb{R}^N)$*

$$\lim_{k\to\infty} \left\| \left(V_{-h_{n_k}} A V_{h_{n_k}} - Op(a_h)\right) \varphi I \right\|_{\mathcal{B}(\Lambda^s(\mathbb{R}^N),\Lambda^{s-m}(\mathbb{R}^N))} = 0, \tag{41}$$

where $(V_h u)(x) = u(x-h), h \in \mathbb{R}^N$ is the shift operator, φI is the operator of multiplication by the function φ.

Proof. Let $A = Op(a) \in OPS_{1,0}^m(\widetilde{\mathbb{R}^N})$, and a sequence $h_n \to \infty$. Then $V_{-h_n} A V_{h_n} = Op(a(x+h_n,\xi))$.The functional sequence $a(x+h_n,\xi)\langle\xi\rangle^{-m}$ is uniformly bounded and equicontinuous on compact sets $K \times \widetilde{\mathbb{R}^N}$, where K is a compact set in $\mathbb{R}^N$.

Applying Arcella-Ascoli's Theorem we obtain that there exists a subsequence h_{n_k} such that

$$a(x+h_{n_k},\xi)\langle\xi\rangle^{-m} \to a_h(x,\xi)\langle\xi\rangle^{-m}$$

uniformly on every compact set $K\times\widetilde{\mathbb{R}^N}$, that is

$$\lim_{k\to\infty}\sup_{K\times\mathbb{R}^N}|a(x+h_{n_k},\xi)-a_h(x,\xi)|\langle\xi\rangle^{-m}=0. \tag{42}$$

By the well-known inequality

$$\sup_X\left|\frac{\partial u(x)}{\partial x_j}\right|\le C\sqrt{\sup_X|u(x)|}\sqrt{\sup_X\left|\frac{\partial^2 u(x)}{\partial x_j^2}\right|},$$

where X is a set in $\mathbb{R}$, we obtain that

$$\lim_{k\to\infty}\sup_{K\times\mathbb{R}^N}\left|\partial_\xi^\alpha\partial_x^\beta a(x+h_{n_k},\xi)-\partial_\xi^\alpha\partial_x^\beta a_h(x,\xi)\right|\langle\xi\rangle^{-m+|\alpha|}=0. \tag{43}$$

Note that (43) implies that the limit symbol $a_h(x,\xi)\in S^m_{1,0}(\mathbb{R}^N)$. Moreover, estimate (43) and Theorem 9 yield (41). □

Corollary 20. *For every function $u\in\Lambda_0^s(\mathbb{R}^{\mathrm{N}})$*

$$\lim_{k\to\infty}V_{-h_{n_k}}AV_{h_{n_k}}u=Op(a_h)u \tag{44}$$

in the sense of convergence in $\Lambda_0^{s-m}(\mathbb{R}^{\mathrm{N}})$

Proof. Let u be a function in $\Lambda_0^s(\mathbb{R}^N)$ with compact support, and R be such that $\varphi_R u=u$. Then

$$\begin{aligned}&\lim_{k\to\infty}\left\|V_{-h_{n_k}}AV_{h_{n_k}}u-Op(a_h)u\right\|_{\Lambda^{s-m}(\mathbb{R}^N)}\\ \le\ &\lim_{k\to\infty}\left\|\left(V_{-h_{n_k}}AV_{h_{n_k}}-Op(a_h)\right)\varphi_R I\right\|_{\mathcal{B}(\Lambda^s(\mathbb{R}^N),\Lambda^{s-m}(\mathbb{R}^N))}\|u\|_{\Lambda^s(\mathbb{R}^N)}=0.\end{aligned}$$

□

We set $OPSO^m(\widetilde{\mathbb{R}^N})=OPS^m_{1,0}(\widetilde{\mathbb{R}^N})\bigcap OPSO^m(\mathbb{R}^N)$.

Theorem 21. *Let $A=Op(a)\in OPSO^m(\widetilde{\mathbb{R}^N})$ be an uniformly elliptic pseudo-differential operator. Then $A:\Lambda^s(\mathbb{R}^N)\to\Lambda^{s-m}(\mathbb{R}^N),(s\in\mathbb{R})$ is a Fredholm operator if and only if*

$$\lim_{r\to\infty}\inf_{|x|\ge r,\xi\in\mathbb{R}^N}|a(x,\xi)|\langle\xi\rangle^{-m}>0. \tag{45}$$

Proof. Sufficiency. Let $\phi\in C_0^\infty(\mathbb{R}^N),\phi(x)=1$ if $|x|\le 1$, and $\phi(x)=0$ if $|x|\ge 2$. We set $\phi_R(x)=\phi(x/R),\psi_R=1-\phi_R$. Condition (45) yields that the symbol $\psi_R(x)a^{-1}(x,\xi)\in SO^{-m}(\mathbb{R}^N)$ for enough large $R>0$. Let $C_R=Op(\psi_R a^{-1})$.Then applying Proposition 5 (i) we obtain that

$$C_RA=\psi_R I+Op(t_1),$$

where $Op(t_1) \in SO_0^{-1}(\mathbb{R}^N)$. Hence,

$$C_R A = I - \phi_R I + Op(t_1),$$

where $\phi_R \in C_0^\infty(\mathbb{R}^N)$. Thus $-\phi_R I + Op(t_1) = T_1' \in OPSO_0^0$. In the same way we show that

$$AC_R = I + T_2',$$

where $T_2' \in OPSO_0^0$. Uniform ellipticity of A implies that there exists operator $B \in OPS_{1,0}^{-m}(\mathbb{R}^N)$ such that equalities (39) hold.

Set

$$\mathcal{R} = B + C_R - BAC_R \in OPS_{1,0}^{-m}(\mathbb{R}^N).$$

Then

$$\mathcal{R}A - I = -T_1'T_1, A\mathcal{R} - I = -T_2'T_2,$$

where $T_1'T_1, T_2'T_2$ belong $OPSO_0^{-1}(\mathbb{R}^N)$ and, consequently, by Proposition 16, $T_1'T_1$, $T_2'T_2$ are compact operators on $\Lambda^s(\mathbb{R}^N), \Lambda^{s-m}(\mathbb{R}^N)$, respectively.

Necessity. First, we consider the case $m = 0$ and $s \in (0,1)$. It follows from the definition of $\Lambda_0^s(\mathbb{R}^N)$ that the strong limit

$$s - \lim_{R\to\infty} \left(\psi_R I : \Lambda_0^s(\mathbb{R}^N) \to \Lambda_0^s(\mathbb{R}^N)\right) = 0.$$

Let $A : \Lambda^s(\mathbb{R}^N) \to \Lambda^s(\mathbb{R}^N)$ be a Fredholm operator. Then the following a priori estimate holds

$$\|Au\|_{\Lambda^s(\mathbb{R}^N)} \geq C(\|u\|_{\Lambda^s(\mathbb{R}^N)} - \|Tu\|_{\Lambda^s(\mathbb{R}^N)}), \tag{46}$$

where T is a compact operator. We can consider T as a compact operator from $\Lambda_0^s(\mathbb{R}^N)$ in $\Lambda^s(\mathbb{R}^N)$. Hence,

$$\lim_{R\to\infty} \|T\psi_R\|_{\Lambda_0^s(\mathbb{R}^N)\to\Lambda^s(\mathbb{R}^N)} = 0. \tag{47}$$

Formulas (46), (47) yield that there exist R_0 such that for $R > R_0$

$$\|A\psi_R u\|_{\Lambda^s(\mathbb{R}^N)} \geq C/2 \, \|\psi_R u\|_{\Lambda^s(\mathbb{R}^N)} \tag{48}$$

for every function $u \in \Lambda_0^s(\mathbb{R}^N)$. Let a sequence $h_m \in \mathbb{R}^N$ tend to infinity, and a function u have a compact support. Then for fixed $R > 0$ there exists $m \geq m_0$ such that $\psi_R V_{h_m} u = V_{h_m} u$. Thus, for $m \geq m_0$

$$\|V_{-h_m} A V_{h_m} u\|_{\Lambda^s(\mathbb{R}^N)} = \|V_{-h_m} A \psi_R V_{h_m} u\|_{\Lambda^s(\mathbb{R}^N)} \geq C/2 \, \|u\|_{\Lambda^s(\mathbb{R}^N)}. \tag{49}$$

Passing to the limit in (49) and applying Proposition 19 we obtain that for a compactly supported function $u \left(\in \Lambda^s(\mathbb{R}^N)\right)$

$$\|Op(a_h)u\|_{\Lambda^s(\mathbb{R}^N)} \geq C/2 \, \|u\|_{\Lambda^s(\mathbb{R}^N)}. \tag{50}$$

Let $u \in \Lambda^s(\mathbb{R}^N)$ be an arbitrary function. Then (50), Corollary (2), and Theorem 9 imply that

$$\begin{aligned} \|\varphi_R Op(a_h)u\|_{\Lambda^s(\mathbb{R}^N)} &\geq \|Op(a_h)\varphi_R u\|_{\Lambda^s(\mathbb{R}^N)} + O(1/R) \\ &\geq C/2 \, \|\varphi_R u\|_{\Lambda^s(\mathbb{R}^N)} + O(1/R). \end{aligned} \tag{51}$$

In light of Proposition 6

$$\lim_{R\to\infty}\|\varphi_R u\|_{\Lambda^s(\mathbb{R}^N)} = \|u\|_{\Lambda^s(\mathbb{R}^N)}\,. \tag{52}$$

Passing to the limit in (51), and applying (52) we obtain the estimate

$$\|Op(a_h)u\|_{\Lambda^s(\mathbb{R}^N)} \geq C/2\,\|u\|_{\Lambda^s(\mathbb{R}^N)} \tag{53}$$

for every function $u \in \Lambda^s(\mathbb{R}^N)$, $s \in (0,1]$.

Note that $a_h(x,\xi) = a_h(\xi)$ since $a(x,\xi) \in SO^0(\mathbb{R}^N)$. Thus, (53) implies that

$$\|a_h(D)u\|_{\Lambda^s(\mathbb{R}^N)} \geq C/2\,\|u\|_{\Lambda^s(\mathbb{R}^N)}\,. \tag{54}$$

Set in (54) $u = e_\xi = e^{i(x,\xi)}$. It is evident that $e_\xi \in \Lambda^s(\mathbb{R}^N)$ for every $s \geq 0$ and $a_h(D)e_\xi(x) = a_h(\xi)e_\xi(x)$. Thus, (54) implies that

$$\inf_\xi |a_h(\xi)| \geq C/2 > 0 \tag{55}$$

for every sequence $h_n \to \infty$ such that

$$\lim_{n\to\infty}\sup_{K\times\mathbb{R}^N}|a(x+h_n,\xi) - a_h(\xi)| = 0$$

for every compact set $K \subset \mathbb{R}^N$. Because $a \in SO^0(\mathbb{R}^N)$

$$\lim_{n\to\infty} a(x+h_n,\xi) = \lim_{n\to\infty} a(h_n,\xi) = a_h(\xi). \tag{56}$$

Let us show that (55) implies (45). Suppose that (55) holds, but (45) does not hold. Then there exists a sequence (h_n, p_n), $h_n \to \infty$ such that

$$\lim_{n\to\infty} a(h_n,p_n) = 0. \tag{57}$$

Let the sequence h_n be such that limit (56) exists. Then (56), (55) imply that there exists $N \in \mathbb{N}$ such that for all $n > N$

$$|a(h_n,p_n)| \geq C/4 > 0, \tag{58}$$

Inequality (58) contradicts to (57).

Let now $m \neq 0$, and $s \in \mathbb{R}$. Then $s = r + \delta$, where $\delta \in (0,1)$. We set $B = \langle D\rangle^{-m-r} A \langle D\rangle^r : \Lambda^\delta(\mathbb{R}^N) \to \Lambda^\delta(\mathbb{R}^N), \delta \in (0,1)$. It is clear that $A : \Lambda^s(\mathbb{R}^N) \to \Lambda^{s-m}(\mathbb{R}^N)$ is a Fredholm operator if and only if $B : \Lambda^\delta(\mathbb{R}^N) \to \Lambda^\delta(\mathbb{R}^N)$ is a Fredholm operator. In light of Proposition 5 (i) $B = Op(a(x,\xi)\langle\xi\rangle^{-m} + t(x,\xi))$, where

$$\lim_{x\to\infty}\sup_\xi |t(x,\xi)| = 0. \tag{59}$$

Thus we reduce the proof to the case $m = 0$ and $\delta \in (0,1)$. □

Let $A : X \to Y$, where $Y \supset X$. We say that $\lambda\,(\in \mathbb{C})$ is a point of essential spectrum of A, if $A - \lambda I : X \to Y$ is not Fredholm operator. We denote the essential spectrum of A by $sp_{ess}(A : X \to Y)$.

Theorem 21 has the following corollary.

Theorem 22. *Let* $A = Op(a) \in OPSO^m(\widetilde{\mathbb{R}^N})$ *be an uniformly elliptic pseudo-differential operator. Then*

$$\begin{aligned} sp_{ess}(A &: \Lambda^s(\mathbb{R}^N) \to \Lambda^{s-m}(\mathbb{R}^N)) \\ &= \bigcup_{h\in\Omega(a)} \overline{\{\lambda \in \mathbb{C} : \lambda = a_h(\xi), \xi \in \mathbb{R}^N\}}, \end{aligned}$$

where $\Omega(a)$ *is a set of all sequences* $h_n \to \infty$ *such that the limits* $\lim_{n\to\infty} a(h_n, \xi) = a_h(\xi)$ *exist.*

Remark 23. *It has been proved in* [4], *chapter 4, that uniformly elliptic operator* $A = Op(a) \in OPSO^m(\widetilde{\mathbb{R}^N})$ *acting from* $H^s(\mathbb{R}^N)$ *in* $H^{s-m}(\mathbb{R}^N), m \geq 0$ *has the essential spectrum defined by the right side part of the previous formula. Hence, the essential spectrum of a uniformly elliptic pseudo-differential operator acting on the Hölder-Zygmund spaces coincides with it essential spectra as the operator acting on the Sobolev spaces.*

Example 24. Let

$$A = -\Delta + aI$$

be a Schrödinger operator with slowly oscillating potential a. That is $a \in C_b^\infty(\mathbb{R}^N)$, and

$$\lim_{x\to\infty} \partial_{x_j} a(x) = 0, j = 1, \ldots, N.$$

Then A satisfies all conditions of Theorem 21. Hence,

$$\begin{aligned} sp_{ess}(A &: \Lambda^s(\mathbb{R}^N) \to \Lambda^{s-m}(\mathbb{R}^N)) \\ &= sp_{ess}(A : H^s(\mathbb{R}^N) \to H^{s-m}(\mathbb{R}^N)) \\ &= \bigcup_{h\in\Omega(a)} \overline{\left\{\lambda \in \mathbb{C} : \lambda = |\xi|^2 + a_h, \xi \in \mathbb{R}^N\right\}}, \end{aligned}$$

where $\Omega(a)$ is a closed set of all partial limits of a at infinity. Thus, the essential spectrum of Schrödinger operator with a slowly oscillating potential is a semi-strip in the complex plane. It is described in the following way. Let $L_r = \{z \in \mathbb{C} : \Im z = r\}$, and $M(a) = \{r : L_r \cap \Omega(a) \neq \emptyset\}$. Let m_r be the left extreme point of the straight line L_r for $r \in M(a)$. Then

$$sp_{ess}A = \bigcup_{r\in M(a)} \{z \in \mathbb{C} : z = m_r + t, t \in [0, +\infty)\}.$$

Thus, the essential spectrum of A is a massive set in the complex plane, in a distinction from the case of potentials with one limit at infinity.

If a is a real-valued potential, then Theorem 22 yields the following result for essential spectrum of A

$$\begin{aligned} sp_{ess}(A &: \Lambda^s(\mathbb{R}^N) \to \Lambda^{s-m}(\mathbb{R}^N)) \\ &= sp_{ess}(A : H^s(\mathbb{R}^N) \to H^{s-m}(\mathbb{R}^N)) = [\liminf_{x\to\infty} a(x), +\infty). \end{aligned}$$

3.2. Pseudo-Differential Operators with Analytical Symbols in Hölder-Zygmund Weighted Spaces

Definition 25. *Let B be an open convex domain in $\mathbb{R}^N$ containing the origin. We denote by $S^m_{1,0}(\mathbb{R}^N, B)$ a subclass of $S^m_{1,0}(\mathbb{R}^N)$ consisting of symbols $a(x,\xi)$ which have an analytic extension with respect to the variable ξ in a tube domain $\mathbb{R}^N_\xi + iB$, and such that for all $l_1, l_2 \in \mathbb{N}$*

$$|a|^m_{l_1,l_2,B} = \sup_{x\in\mathbb{R}^N, \xi\in\mathbf{R}^N_\xi, \eta\in B} \langle\xi\rangle^{-m} \sum_{|\alpha|\le l_1, |\beta|\le l_2} \left|\partial^\beta_x \partial^\alpha_\xi a(x, \xi + i\eta)\right| < \infty.$$

As above we correspond to a symbol $a \in S^m_{1,0}(\mathbb{R}^N, B)$ a pseudo-differential operator. The class of such pseudo-differential operators is denoted by $OPS^m_{1,0}(\mathbb{R}^N, B)$.

We denote by $\mathcal{R}(B)$ a class of positive weights w such that:

- $\log w \in C^\infty(\mathbb{R}^N)$, and

$$N_l(\log w) = \sup_x \sum_{|\beta|\le l} \left|\partial^\beta \nabla (\log w(x))\right| < \infty$$

  for all l.
- $\nabla(\log w(x)) \in B$ for every $x \in \mathbb{R}^N$.
- A weight $w(x) \in \mathcal{R}(B)$ is called slowly oscillating, if $\lim_{x\to\infty} \frac{\partial\nabla(\log w(x))}{\partial x_j} = 0$, $j = 1, \ldots, N$.

  We denote the class of slowly oscillating weights by $\mathcal{R}_{sl}(B)$.

  Let

$$g_w(x,y) = \int_0^1 (\nabla \log w)(x - t(x-y))dt.$$

It is easy to check that $g_w(x,y) \in C^\infty_b\left(\mathbb{R}^N \times \mathbb{R}^N, \mathcal{B}(\mathbb{R}^N)\right)$, and

$$\sup_{x,y} \sum_{|\alpha|\le l_1, |\beta|\le l_2} \left|\partial^\alpha_x \partial^\beta_y g_w(x,y)\right| \le C \sup_{x\in\mathbb{R}^N, 1\le|\beta|\le l_1+l_2} \left|\partial^\beta \log w(x)\right|.$$

Moreover, $g_w(x,y) \in B$ for every $(x,y) \in \mathbb{R}^N \times \mathbb{R}^N$.

The following Proposition is a key result for the study of pseudo-differential operators in exponential weighted spaces.

Proposition 26. *(see, [4], p. 243–247) Let*

$$A = Op(a(x,\xi)) \in OPS^m_{1,0}(\mathbb{R}^N, B);\ \ w(x) \in \mathcal{R}(B).$$

Then the operator $wOp(p)w^{-1} \in OPS^m_{1,0,0}(\mathbb{R}^N)$, and

$$wOp(a)w^{-1} = Op_d(a(x, \xi + ig_w(x,y))).$$

Proposition 27. *([4], p. 243–247) Let*

$$A = Op(a(x,\xi)) \in OPSO^m(\mathbb{R}^N, B) = OPSO^m(\mathbb{R}^N) \cap OPS^m_{1,0}(\mathbb{R}^N, B),$$

and a weight $w \in \mathcal{R}_{sl}(B)$. *Then*

$$wAw^{-1} = Op(a(x, \xi + i\nabla \log w(x))) + Op(t(x, \xi)), \tag{60}$$

where $t(x, \xi) \in SO_0^{m-1}(\mathbb{R}^N)$.

We denote by $\Lambda^{s,w}(\mathbb{R}^N), (\Lambda_0^{s,w}(\mathbb{R}^N))$ the weighted Hölder-Zygmund spaces with norms

$$\|u\|_{\Lambda^{s,w}(\mathbb{R}^N)} = \|wu\|_{\Lambda^{s}(\mathbb{R}^N)}, (\|u\|_{\Lambda_0^{s,w}(\mathbb{R}^N)} = \|wu\|_{\Lambda_0^{s}(\mathbb{R}^N)}).$$

Proposition 28. *Let* $A = Op(a) \in OPS^m(\mathbb{R}^N, B)$; $w(x) \in \mathcal{R}(B)$. *Then*

$$A : \Lambda^{s,w}(\mathbb{R}^N) \to \Lambda^{s-m,w}(\mathbb{R}^N), (A : \Lambda_0^{s,w}(\mathbb{R}^N) \to \Lambda_0^{s-m,w}(\mathbb{R}^N))$$

is a bounded operator.

Proof follows from Proposition 26 , Theorem 9, and Corollary 15.

Theorem 29. *Let* $A = Op(a) \in OPSO^m(\widetilde{\mathbb{R}^N}) \cap OPS^m(\mathbb{R}^N, B)$ *be an uniformly elliptic pseudo-differential operator, and* $w \in \mathcal{R}_{sl}(B)$. *Then*

$$A : \Lambda^{s,w}(\mathbb{R}^N) \to \Lambda^{s-m,w}(\mathbb{R}^N),\ (A : \Lambda_0^{s,w}(\mathbb{R}^N) \to \Lambda_0^{s-m,w}(\mathbb{R}^N))$$

is a Fredholm operator if and only if

$$\lim_{R\to\infty} \inf_{|x|\geq R, \xi\in\mathbb{R}^N} |a(x, \xi + i\nabla \log w(x))| \langle\xi\rangle^{-m} > 0. \tag{61}$$

Theorem 30. *Let* $A = Op(a) \in OPSO^m(\widetilde{\mathbb{R}^N}) \cap OPS^m_{1,0}(\mathbb{R}^N, B)$ *be an uniformly elliptic pseudo-differential operator,* $w \in \mathcal{R}_{sl}(B)$. *Then*

$$\begin{aligned} sp_{ess}(A &: \Lambda^{s,w}(\mathbb{R}^N) \to \Lambda^{s-m,w}(\mathbb{R}^N)) = sp_{ess}(A : \Lambda_0^{s,w}(\mathbb{R}^N) \to \Lambda_0^{s-m,w}(\mathbb{R}^N)) \\ &= \bigcup_{h\in\Omega(a,w)} \overline{\{\lambda \in \mathbb{C} : \lambda = a_h(\xi + i\,(\nabla \log w)_h), \xi \in \mathbb{R}^N\}}, \end{aligned}$$

where $\Omega(a, w)$ *is the set of all sequences* $h_m \to \infty$ *such that the limit*

$$a_h(\xi + i\,(\nabla \log w)_h) = \lim_{h_m\to\infty} a(h_m, \xi + i\,(\nabla \log w)\,(h_m)) \tag{62}$$

exists.

It follows from Theorem 30 that the spectrum of pseudo-differential operator is a massive set in the complex plane $\mathbb{C}$, depending on oscillations of symbol with respect to x, and oscillations of the characteristic $\nabla(\log w)$ of the weight w.

Theorem 31. *(Fragmen-Lindelöf principle) Let* $A = Op(a) \in OPSO^m(\widetilde{\mathbb{R}^N}) \cap OPS^m(\mathbb{R}^N, B)$ *be an uniformly elliptic pseudo-differential operator,* $w \in \mathcal{R}_{sl}(B)$, $\lim_{x\to\infty} w(x) = \infty$, *and the domain* B *be symmetric with respect to the origin. Let*

$$\lim_{R\to\infty} \inf_{|x|>R, \xi+i\eta\in\mathbb{R}^N+IB} |a(x, \xi + i\eta)| \langle\xi\rangle^{-m} > 0. \tag{63}$$

Then

$$u \in \Lambda_0^{s,w^{-1}}(\mathbb{R}^N), Au \in \Lambda_0^{s-m,w}(\mathbb{R}^N) \Longrightarrow u \in \Lambda_0^{s,w}(\mathbb{R}^N).$$

Proof. The operator $w^\theta Op(a)w^{-\theta}, \theta \in [-1,1]$ can be written as

$$w^\theta Op(a)w^{-\theta} = Op(a(x,\xi + i\theta\nabla \log w(x)) + Op(t_\theta(x,\xi)),$$

where $t_\theta(x,\xi)$ belongs to $SO_0^{m-1}(\mathbb{R}^N)$ uniformly with respect to $\theta \in [-1,1]$. In light of Theorem 29 and condition (63) $w^\theta Op(a)w^{-\theta} : \Lambda_0^s(\mathbb{R}^N) \to \Lambda_0^{s-m}(\mathbb{R}^N)$ is a Fredholm operator for all $\theta \in [-1,1]$. By Theorem 9 the mapping

$$[-1,1] \to \mathcal{B}(\Lambda_0^s(\mathbb{R}^N), \Lambda_0^{s-m}(\mathbb{R}^N)) : \theta \to w^\theta Op(a)w^{-\theta}$$

is continuous. Hence, $Index\; w^\theta Op(a)w^{-\theta} : \Lambda_0^s(\mathbb{R}^N) \to \Lambda_0^{s-m}(\mathbb{R}^N)$ does not depend on $\theta \in [-1,1]$. It yields that

$$\begin{aligned} Index\; Op(a) \;&:\; \Lambda_0^{s,w}(\mathbb{R}^N) \to \Lambda_0^{s-m,w}(\mathbb{R}^N) = \\ Index\; Op(a) \;&:\; \Lambda_0^{s,w^{-1}}(\mathbb{R}^N) \to \Lambda_0^{s-m,w^{-1}}(\mathbb{R}^N). \end{aligned}$$

Moreover, the conditions $\lim_{x\to\infty} w(x) = \infty$ implies that $\Lambda_0^{s,w}(\mathbb{R}^N) \subset \Lambda_0^{s,w^{-1}}(\mathbb{R}^N)$, and the last imbedding is dense.

Then

$$\ker Op(a) : \Lambda_0^{s,w^{-1}}(\mathbb{R}^N) \to \Lambda_0^{s-m,w^{-1}}(\mathbb{R}^N) \subset \Lambda_0^{s,w}(\mathbb{R}^N),$$

and it coincides with $\ker Op(a) : \Lambda_0^{s,w}(\mathbb{R}^N) \to \Lambda_0^{s-m,w}(\mathbb{R}^N)$ (see for instance [1], p. 308). Moreover, if the equation $Op(a)u = f,\;\; f \in \Lambda_0^{s-m,w}(\mathbb{R}^N))$ is solvable in $\Lambda_0^{s,w^{-1}}(\mathbb{R}^N)$, then $u \in \Lambda_0^{s,w}(\mathbb{R}^N)$. □

Example 32. Let

$$A = -\Delta + a$$

be a Schrödinger operator with real-valued slowly oscillating potential a, and let $\liminf_{x\to\infty} a(x) = a_- > 0$, $w(x) = e^{(a_- -\varepsilon)\langle x\rangle}$, where $\varepsilon \in (0, a_-)$. Then

$$u \in \Lambda^{s,w^{-1}}(\mathbb{R}^N), Au \in \Lambda^{s-2,w}(\mathbb{R}^N) \Longrightarrow u \in \Lambda^{s,w}(\mathbb{R}^N).$$

References

[1] I. Gohberg and I. Feldman, *Convolutions Equations and Projection Methods for Their Solutions,* Nauka, Moskva 1971 (Russian) (English Translation: Amer. Math. Soc. Transl. of Math. Monographs **41**, Providence, R.I., 1974).

[2] L. Hörmander, *The Analysis of Linear Partial Differential Operators, Vol. 1–4,* Springer-Verlag, Berlin, Heidelberg, New York, Tokyo, 1983–1985.

[3] H. Kumano-go, *Pseudo-Differential Operators,* MIT Presss, Cambrige, Mass., 1981.

[4] V. Rabinovich, S. Roch and B. Silbermann, *Limit Operators and Their Applications in Operator Theory,* Operator Theory: Advances and Applications, Vol. **150**, Birkhäuser-Verlag, Basel-Boston-Berlin, 2004.

[5] V. S. Rabinovich, *An Introductory Course on Pseudodifferential Operators,* -Textos de Matemática **1,** Instituto Superior Técnico, Lisboa 1998.

[6] V. S. Rabinovich, Pseudodifferential operators with analytic symbols and estimates for eigenfunctions of Schrödinger operators, *Zeit. Anal. und Anwend.* **21** (2) (2002), 351–370.

[7] V. S. Rabinovich, Exponential estimates for eigenfunctions of Schrödinger operators with rapidly increasing and discontinuous potentials, *Contemporary Mathematics* **364** (2004), 225–236.

[8] M. Shubin, *Pseudodifferential Operators and Spectral Theory,* Second Edition, Springer-Verlag, Berlin, Heidelberg, New York, Tokyo, 2001.

[9] E. M. Stein, *Harmonic Analysis, Real-Variable Methods, Orthogonality, and Oscillatory Integrals,* Princeton University Press, Princeton, New Jersey, 1993.

[10] M. E. Taylor, *Pseudodifferential Operators,* Princeton University Press, Princeton, New Jersay, 1981.

[11] M. E. Taylor, *Partial Differential Equations II, Qualitative Studies of Linear Equations,* Applied Math. Sciences **116**, Springer, New York, Berlin, Heidelberg, Tokyo, 1996.

[12] M. E. Taylor, *Partial Differential Equations III, Nonlinear Equations,* Applied Math. Sciences **117**, Springer, New York, Berlin, Heidelberg, Tokyo, 1996.

[13] M. E. Taylor, *Tools for PDE, Pseudodifferential Operators, Paradifferential Operators, and Layer Potentials,* Math. Surveys and Monographs, Vol. **81**, AMS, 2000.

V.S. Rabinovich
Instituto Politecnico Nacional
ESIME-Zacatenco
Av. IPN, Edif. 1, 2do piso
D.F. 07738, Mexico
e-mail: `vladimir.rabinovich@gmail.com`

Operator Theory:
Advances and Applications, Vol. 164, 115–120

Weyl Transforms and Convolution Operators on the Heisenberg Group

M.W. Wong

Abstract. The Fourier transform on the Heisenberg group, the Fourier transform along the center of the Heisenberg group and the Euclidean Fourier transform are used to prove that Weyl transforms and convolution operators on the Heisenberg group are, respectively, classical Weyl transforms and pseudo-differential operators.

Mathematics Subject Classification (2000). Primary 43A80, 47G30.

Keywords. Heisenberg group, left-invariant vector fields, Hamiltonians, representations, group Fourier transforms, Weyl transforms, twisted convolutions, group convolutions, pseudo-differential operators.

1. The Heisenberg Group

Let $\mathbb{H}^1 = \mathbb{R}^2 \times \mathbb{R}$. Then $\mathbb{H}^1$ is a non-commutative group when equipped with the multiplication $\cdot$ given by

$$(x, y, t) \cdot (u, v, s) = (x + u, y + v, t + s - 2(xv - yu))$$

for all (x, y, t) and (u, v, s) in $\mathbb{H}^1$. It is known as the Heisenberg group.

If we identify $\mathbb{R}^2$ with $\mathbb{C}$ by $z = x + iy$, then the group law becomes

$$(z, t) \cdot (w, s) = (z + w, t + s + [z, w])$$

for all (z, t) and (w, s) in $\mathbb{C} \times \mathbb{R}$, where $[z, w]$ is the symplectic form of z and w given by

$$[z, w] = 2\,\mathrm{Im}(z\overline{w}).$$

Let $\mathfrak{h}_1$ be the Lie algebra of left-invariant vector fields on $\mathbb{H}^1$. In order to determine a basis for $\mathfrak{h}_1$, we let $\gamma_1 : \mathbb{R} \to \mathbb{H}^1$, $\gamma_2 : \mathbb{R} \to \mathbb{H}^1$ and $\gamma_3 : \mathbb{R} \to \mathbb{H}^1$ be

This research has been partially supported by the Natural Sciences and Engineering Research Council of Canada.

homomorphic curves in $\mathbb{H}^1$ such that

$$\gamma_1(s) = (s, 0, 0), \quad s \in \mathbb{R},$$

$$\gamma_2(s) = (0, s, 0), \quad s \in \mathbb{R},$$

and

$$\gamma_3(s) = (0, 0, s), \quad s \in \mathbb{R}.$$

Then we define the left-invariant vector fields X, Y and T on $\mathbb{H}^1$ by

$$(Xf)(x,y,t) = \left.\frac{d}{ds}\right|_{s=0} f((x,y,t)\cdot\gamma_1(s)), \quad (x,y,t) \in \mathbb{H}^1,$$

$$(Yf)(x,y,t) = \left.\frac{d}{ds}\right|_{s=0} f((x,y,t)\cdot\gamma_2(s)), \quad (x,y,t) \in \mathbb{H}^1,$$

and

$$(Tf)(x,y,t) = \left.\frac{d}{ds}\right|_{s=0} f((x,y,t)\cdot\gamma_3(s)), \quad (x,y,t) \in \mathbb{H}^1,$$

for all $f \in C^\infty(\mathbb{H}^1)$. Thus,

$$X = \frac{\partial}{\partial x} + 2y\frac{\partial}{\partial t}, \qquad Y = \frac{\partial}{\partial y} - 2x\frac{\partial}{\partial t} \qquad \text{and} \qquad T = \frac{\partial}{\partial t}.$$

It can be checked easily that

$$[X, Y] = -4T,$$

and all other commutators are zero. If we replace $\frac{\partial}{\partial x}$ by ξ, $\frac{\partial}{\partial y}$ by η and $\frac{\partial}{\partial t}$ by λ, then the Hamiltonians for X, Y and T are, respectively, given by $\xi + 2y\lambda$, $\eta - 2x\lambda$ and λ.

Weyl transforms on the Heisenberg group $\mathbb{H}^1$ are introduced as group Fourier transforms in Section 2. They are then shown to be classical Weyl transforms studied extensively in the book [13] by Wong. While the formulas are not new and can be found in the books [11, 12] by Thangavelu, the viewpoint of looking at them as Weyl transforms with symbols on $\mathbb{C}$ is new and provides new insight in understanding the genesis of pseudo-differential operators. In Section 3, two types of convolutions intimately related to the Heisenberg group $\mathbb{H}^1$ are introduced and explicated by means of the Fourier transform along the center of $\mathbb{H}^1$ encountered in Section 2. These results are then used in Section 4 to prove that convolution operators on $\mathbb{H}^1$ are in fact classical pseudo-differential operators studied in, *e.g.*, the books [8] by Hörmander and [14] by Wong.

The Heisenberg group alluded to in this paper is different from the Weyl-Heisenberg group studied in the context of wavelet transforms and localization operators in, say, [1] by Boggiatto and Wong and [15, 16] by Wong. The significant difference is attributed to the fact that the center of the Heisenberg group is the real line $\mathbb{R}$, whereas the center of the Weyl-Heisenberg group is the unit circle centered at the origin. The compact center of the Weyl-Heisenberg group is an advantage in the sense that all its infinite-dimensional irreducible and unitary representations are square-integrable. The first hint of the role of the Heisenberg group in the study of harmonic analysis and partial differential equations can be traced to the paper [10] by Stein. The first important papers devoted to the analysis on the Heisenberg group are [4] by Folland and [6] by Folland and Stein. Recent references on the Heisenberg group include the books [5] by Folland, [9] by Stein and [11, 12] by Thangavelu. The papers [2, 3] by Dynin and [7] by Greiner also contain results related to this paper.

2. Weyl Transforms on $\mathbb{H}^1$

For each real number λ, we define the mapping R^λ from $\mathbb{H}^1$ into the group G of all unitary operators on $L^2(\mathbb{R})$ by

$$(R^\lambda(q,p,t)f)(x) = e^{i\lambda(qx+\frac{1}{2}qp+\frac{1}{4}t)}f(x+p), \quad x \in \mathbb{R},$$

for all (q,p,t) in $\mathbb{H}^1$ and f in $L^2(\mathbb{R})$ for all (q,p,t) in $\mathbb{H}^1$. Then it is proved in the book [13] that R^λ is an irreducible and unitary representation of $\mathbb{H}^1$ on $L^2(\mathbb{R})$. It is obvious that

$$R^\lambda(q,p,t) = e^{i\lambda\frac{1}{4}t}\rho^\lambda(q,p), \quad (q,p,t) \in \mathbb{H}^1,$$

where

$$(\rho^\lambda(q,p)f)(x) = e^{i\lambda(qx+\frac{1}{2}qp)}f(x+p), \quad x \in \mathbb{R},$$

for all measurable functions f on $\mathbb{R}$.

Let $F \in L^1(\mathbb{H}^1)$. Then we follow the work [16] by Wong and define the Weyl transform $W_{F,R^\lambda} : L^2(\mathbb{R}) \to L^2(\mathbb{R})$ corresponding to the symbol F and the representation R^λ by

$$(W_{F,R^\lambda}f,g)_{L^2(\mathbb{R})} = \int_{-\infty}^{\infty}\int_{-\infty}^{\infty}\int_{-\infty}^{\infty} F(q,p,t)(f,R^\lambda(q,p,t)g)_{L^2(\mathbb{R})}dq\,dp\,dt$$

for all f and g in $L^2(\mathbb{R})$. A more transparent formula is given by

$$W_{F,R^\lambda} = \int_{-\infty}^{\infty}\int_{-\infty}^{\infty}\int_{-\infty}^{\infty} F(q,p,t)R^\lambda((q,p,t)^{-1})dq\,dp\,dt.$$

A seasoned harmonic analyst can recognize immediately that W_{F,R^λ} is precisely the Fourier transform of the symbol F on the Heisenberg group $\mathbb{H}^1$ evaluated at R^λ. We note that

$$\begin{aligned} W_{F,R^\lambda} &= \int_{-\infty}^{\infty}\int_{-\infty}^{\infty}\left(\int_{-\infty}^{\infty} e^{-i\lambda\frac{1}{4}t}F(q,p,t)\,dt\right)\rho^\lambda(-q,-p)\,dq\,dp \\ &= (2\pi)^{1/2}\int_{-\infty}^{\infty}\int_{-\infty}^{\infty} F_{\lambda/4}(q,p)\rho^\lambda(-q,-p)\,dq\,dp, \end{aligned}$$

where $F_{\lambda/4}(q,p)$ is the Fourier transform along the center of $\mathbb{H}^1$, *i.e.*, the partial Fourier transform of F with respect to t evaluated at $\left(q,p,\frac{\lambda}{4}\right)$.

In the case when $\lambda = 1$, we see that W_{F,R^λ} is the classical Weyl transform with symbol $(2\pi)^{3/2}\mathcal{F}_1F_{1/4}$, where $\mathcal{F}_1$ denotes the spatial Euclidean Fourier transform on $\mathbb{C} \cong \mathbb{R}^2$. See the formula (9.3) in the book [13] by Wong.

3. Twisted Convolutions and Convolutions on $\mathbb{H}^1$

There are two types of convolutions, which are useful for the study of analysis on $\mathbb{H}^1$.

Definition 3.1. Let λ be a nonzero number. Let f and g be measurable functions on $\mathbb{C}$. Then we define the twisted convolution $f *_\lambda g$ of f and g by

$$(f *_\lambda g)(z) = \int_{\mathbb{C}} f(z-w)g(w)e^{i\lambda[z,w]}dw, \quad z \in \mathbb{C},$$

provided that the integral exists.

Definition 3.2. Let f and g be measurable functions on $\mathbb{H}^1$. Then we define the convolution $f *_{\mathbb{H}^1} g$ of f and g by

$$(f *_{\mathbb{H}^1} g)(z,t) = \int_{\mathbb{H}^1} f((z,t)\cdot(w,s)^{-1})g(w,s)\,dw\,ds$$

for all (z,t) in $\mathbb{H}^1$, provided that the integral exists.

Theorem 3.3. *Let f and g be functions in $L^1(\mathbb{H}^1)$. Then for all $\lambda \in \mathbb{R}\setminus\{0\}$,*

$$(f *_{\mathbb{H}^1} g)_\lambda = (2\pi)^{1/2}(f_\lambda *_\lambda g_\lambda).$$

Theorem 3.3 is an interesting result intertwining the twisted convolution and the convolution on $\mathbb{H}^1$ by the Fourier transform along the center of $\mathbb{H}^1$. This result is fairly well-known. It can be found in the books [9] by Stein and [11, 12] by Thangavelu. For the sake of being self-contained in this paper, we give a completely straightforward proof.

Proof of Theorem 3.3. For all z in $\mathbb{C}$,

$$\begin{aligned}
&(f *_{\mathbb{H}^1} g)_\lambda(z)\\
=\;& (2\pi)^{-1/2}\int_{-\infty}^{\infty} e^{-it\lambda}(f *_{\mathbb{H}^1} g)(z,t)\,dt\\
=\;& (2\pi)^{-1/2}\int_{-\infty}^{\infty} e^{-it\lambda}\left(\int_{-\infty}^{\infty}\int_{\mathbb{C}} f(z-w,t-s+[z,w])g(w,s)\,dw\,ds\right)dt.
\end{aligned}$$

Let $t' = t+[z,w]$. Then

$$(f *_{\mathbb{H}^1} g)_\lambda(z) = (2\pi)^{-1/2}\int_{-\infty}^{\infty}\int_{-\infty}^{\infty}\int_{\mathbb{C}} e^{-it'\lambda}f(z-w,t'-s)g(w,s)e^{i\lambda[z,w]}dw\,ds\,dt'.$$

On the other hand, for all z in $\mathbb{C}$, we get

$$\begin{aligned}
&(f_\lambda *_\lambda g_\lambda)(z)\\
=\;& \int_{\mathbb{C}} f_\lambda(z-w)g_\lambda(w)e^{i\lambda[z,w]}dw\\
=\;& (2\pi)^{-1/2}\int_{\mathbb{C}}\left(\int_{-\infty}^{\infty} f(z-w,\cdot-s)g(w,s)\,ds\right)^{\wedge}(\lambda)e^{i\lambda[z,w]}dw\\
=\;& (2\pi)^{-1}\int_{\mathbb{C}}\int_{-\infty}^{\infty} e^{-it\lambda}\left(\int_{-\infty}^{\infty} f(z-w,t-s)g(w,s)\,ds\right)e^{i\lambda[z,w]}dt\,dw
\end{aligned}$$

and the proof is complete. □

4. Convolution Operators on $\mathbb{H}^1$

It is shown in this section that convolution operators on $\mathbb{H}^1$ are precisely the classical pseudo-differential operators studied in, say, the book [8] by Hörmander and the book [14] by Wong.

Let F and φ be nice functions on $\mathbb{H}^1$. Then for all $\lambda \in \mathbb{R} \setminus \{0\}$,

$$(F *_{\mathbb{H}^1} \varphi)_\lambda = (2\pi)^{1/2}(F_\lambda *_\lambda \varphi_\lambda).$$

So, for all (z,t) in $\mathbb{H}^1$,

$$(F *_{\mathbb{H}^1} \varphi)(z,t) = \int_{-\infty}^{\infty} e^{it\lambda}(F_\lambda *_\lambda \varphi_\lambda)(z)\, d\lambda.$$

If we denote by $\wedge$ the Euclidean Fourier transform on $\mathbb{H}^1 \cong \mathbb{R}^3$, then for all z in $\mathbb{C}$,

$$\begin{aligned} &(F_\lambda *_\lambda \varphi_\lambda)(z) = (\varphi_\lambda *_{-\lambda} F_\lambda)(z) \\ = \;& \int_{\mathbb{C}} \varphi_\lambda(z-w)F_\lambda(w)e^{-i\lambda[z,w]}dw \\ = \;& \int_{\mathbb{C}} (\mathcal{F}_1^{-1}\hat{\varphi})(z-w,\lambda)(\mathcal{F}_1^{-1}\hat{F})(w,\lambda)e^{-i\lambda[z,w]}dw \\ = \;& \int_{-\infty}^{\infty}\int_{-\infty}^{\infty} (\mathcal{F}_1^{-1}\hat{\varphi})(x-\xi,y-\eta,\lambda)(\mathcal{F}_1^{-1}\hat{F})(\xi,\eta,\lambda)e^{-i2\lambda(\xi y-\eta x)}d\xi\, d\eta. \end{aligned}$$

If we let $M_{(x,y)}$ be the modulation by (x,y) and let $T_{(-2\lambda y, 2\lambda x)}$ be the translation by $(-2\lambda y, 2\lambda x)$, then for all z in $\mathbb{C}$,

$$\begin{aligned} &(F_\lambda *_\lambda \varphi_\lambda)(z) \\ = \;& \int_{-\infty}^{\infty}\int_{-\infty}^{\infty} (\mathcal{F}_1 M_{(x,y)}\hat{\varphi})(\xi,\eta,\lambda)(\mathcal{F}_1^{-1}T_{(-2\lambda y,2\lambda x)}\hat{F})(\xi,\eta,\lambda)\, d\xi\, d\eta \\ = \;& \int_{-\infty}^{\infty}\int_{-\infty}^{\infty} (M_{(x,y)}\hat{\varphi})(\xi,\eta,\lambda)(T_{(-2\lambda y,2\lambda x)}\hat{F})(\xi,\eta,\lambda)\, d\xi\, d\eta \\ = \;& \int_{-\infty}^{\infty}\int_{-\infty}^{\infty} e^{ix\xi+iy\eta}\hat{F}(\xi+2y\lambda,\eta-2x\lambda,\lambda,\lambda)\hat{\varphi}(\xi,\eta,\lambda)\, d\xi\, d\eta. \end{aligned}$$

So, for all (z,t) in $\mathbb{H}^1$,

$$\begin{aligned} &(F *_{\mathbb{H}^1} \varphi)(z,t) \\ = \;& \int_{-\infty}^{\infty}\int_{-\infty}^{\infty}\int_{-\infty}^{\infty} e^{i(x\xi+y\eta+t\lambda)}\hat{F}(\xi+2y\lambda,\eta-2x\lambda,\lambda)\hat{\varphi}(\xi,\eta,\lambda)\, d\xi\, d\eta\, d\lambda. \end{aligned}$$

The conclusion then is that the convolution operator $F*_{\mathbb{H}^1}$ on $\mathbb{H}^1$ with kernel F is in fact a classical pseudo-differential operator with symbol σ on $\mathbb{R}^3$ given by

$$\sigma(\xi,\eta,\lambda) = (2\pi)^{3/2}\hat{F}(\xi+2y\lambda,\eta-2x\lambda,\lambda)$$

for all (ξ, η, λ) in $\mathbb{R}^3$. It is worth noting that the symbol is a function of the Hamiltonians of the vector fields X, Y and T given in Section 1.

References

[1] P. Boggiatto and M. W. Wong, Two-wavelet localization operators on $L^p(\mathbb{R}^n)$ for the Weyl-Heisenberg group, *Integral Equations Operator Theory* **49** (2004), 1–10.

[2] A. Dynin, An algebra of pseudo-differential operators on the Heisenberg group: symbolic calculus, *Sov. Math. Doklady* **17** (1976), 508–512.

[3] A. Dynin, Pseudo-differential operators on Heisenberg groups, in *Pseudo-Differential Operators with Applications, C.I.M.E. Bressanone,* 1977, 5–18.

[4] G. B. Folland, A fundamental solution for a subelliptic operator, *Bull. Amer. Math. Soc.* **79** (1973), 373–376.

[5] G. B. Folland, *Harmonic Analysis in Phase Space*, Princeton University Press, 1989.

[6] G. B. Folland and E. M. Stein, Estimates for the $\overline{\partial}_b$ complex and analysis on the Heisenberg group, *Comm. Pure Appl. Math.* **27** (1974), 429–522.

[7] P. C. Greiner, On the Laguerre calculus of left-invariant convolution (pseudo-differential) operators on the Heisenberg group, *Séminaire Goulaouic-Meyer-Schwartz XI* (1980–1981), 1–39.

[8] L. Hörmander, *The Analysis of Linear Partial Differential Operators III*, Springer-Verlag, 1985.

[9] E. M. Stein, *Harmonic Analysis: Real-Variable Methods, Orthogonality, and Oscillatory Integrals,* Princeton University Press, 1993.

[10] E. M. Stein, Some problems in harmonic analysis suggested by symmetric spaces and semi-simple groups, *Actes Congrès Intern. Math.* Tome **1** (1970), 173–189.

[11] S. Thangavelu, *Harmonic Analysis on the Heisenberg Group,* Birkhäuser, 1998.

[12] S. Thangavelu, *An Introduction to the Uncertainty Principle: Hardy's Theorem on Lie Groups,* Birkhäuser, 2004.

[13] M. W. Wong, *Weyl Transforms*, Springer-Verlag, 1998.

[14] M. W. Wong, *An Introduction to Pseudo-Differential Operators,* Second Edition, World Scientific, 1999.

[15] M. W. Wong, *Wavelet Transforms and Localization Operators,* Birkhäuser, 2002.

[16] M. W. Wong, A new look at pseudo-differential operators, in *Wavelets and Their Applications,* Editors: M. Krishna, R. Radha and S. Thangavelu, Allied Publishers, 2003, 283–290.

M.W. Wong
Department of Mathematics and Statistics
York University
4700 Keele Street
Toronto, Ontario M3J 1P3
Canada
e-mail: `mwwong@pascal.math.yorku.ca`

Operator Theory:
Advances and Applications, Vol. 164, 121–132

Uncertainty Principle, Phase Space Ellipsoids and Weyl Calculus

Maurice de Gosson

Abstract. We state a precise form of the uncertainty principle in terms of phase space ellipsoids, which we then express in terms of the symplectic capacity of phase space ellipsoids. We apply our approach to the study of the positivity of the Wigner transform of a pure quantum state, and of that of the Weyl operator associated to the average of a positive symbol over a phase space ellipsoid.

Mathematics Subject Classification (2000). Primary 53D05, 53D50; Secondary 35S05.

Keywords. Uncertainty principle, Weyl transform, Wigner function, symplectic capacity.

1. Introduction

The uncertainty principle from quantum mechanics is stated in most introductory texts in physics (and in practically *all* mathematical texts!) as the set of "Heisenberg inequalities"

$$\Delta p_j \Delta x_j \geq \tfrac{1}{2}\hbar \quad , \quad 1 \leq j \leq n \tag{1}$$

where Δx_j and Δp_j are the standard (RMS) deviations for the position and momentum quantum operators $X = (X_1, \ldots, X_n)$ and $P = (P_1, \ldots, P_n)$ in a given quantum state Ψ. (More generally X and P could be any pair of vector-valued quantum operators satisfying the commutation relations $[P_j, X_k] = i\hbar\delta_{jk}$). In physics the inequalities (1) are interpreted as follows: suppose we perform position and momentum measurements on the system represented by the state Ψ; then the measured values x_j and p_j will lie in some intervals $[\bar{x}_j - \Delta x_j/2, \bar{x}_j + \Delta x_j/2]$ and $[\bar{p}_j - \Delta p_j/2, \bar{p}_j + \Delta p_j/2]$ where $\Delta p_j \Delta x_j \geq \frac{1}{2}\hbar$. Observing that the Heisenberg inequalities (1) are equivalent to

$$\left(\tfrac{1}{\lambda}\right)^2 \Delta p_j^2 + \lambda^2 \Delta x_j^2 \leq \hbar \text{ for all } \lambda > 0 \tag{2}$$

a heuristic interpretation of the uncertainty principle is the that it does not make sense to distinguish two points (x,p) and (x',p') of phase space if their x_j, p_j coordinates lie inside an ellipse

$$\left(\tfrac{1}{\lambda}\right)^2 (x_j - \bar{x}_j)^2 + \lambda^2 (p_j - \bar{p}_j)^2 = \hbar \ , \ \lambda > 0 \tag{3}$$

in the x_j, p_j plane; note that the intersection of all these ellipses consists of the two hyperbolae $(x_j - \bar{x}_j)(p_j - \bar{p}_j) = \pm\frac{1}{2}\hbar$.

So far, so good. There are however two rubs with this geometric interpretation quantum uncertainty, the second of which being completely fatal when one wants to make changes of coordinates. The first is that one is entitled to ask why one makes particular coordinate planes play such a primordial role in the expression of the uncertainty principle; the second (which is related to the first) is that the ellipses (3) have very few symmetries and are invariant only under a very limited number of linear changes of coordinates; in particular they are certainly not invariant under arbitrary linear *symplectic* transformations. The difficulty does actually not come from the geometric picture itself; it is easy to check that the Heisenberg inequalities (1) themselves are not invariant under symplectic transformations of variables. This is due to the fact that any non-trivial linear transformation of x_j, p_j will make introduce non-zero covariances between these variables even if they were originally uncorrelated. As we will see, it turns out that the inequalities (1) only express the quantum uncertainty principle in a very crude form; it is precisely the consideration of the covariances that will make possible a global and symplectically invariant statement of that principle. We will deduce several interesting consequences from this reformulation of the uncertainty principle.

Notations. Our notations are standard; the natural symplectic form on $\mathbb{R}^{2n}_z \equiv \mathbb{R}^n_x \times \mathbb{R}^n_p$ is given by

$$\sigma(z,z') = \langle p, x'\rangle - \langle p', x\rangle$$

and the associated symplectic group is denoted by $Sp(n)$, we will without further comments identify $S \in Sp(n)$ with its matrix in the canonical basis; $U(n)$ is the image of the unitary group $U(n,\mathbb{C})$ in $Sp(n)$ by the usual embedding $A + iB \longmapsto \begin{bmatrix} A & -B \\ B & A \end{bmatrix}$.

Acknowledgements. *It is my duty –and immense pleasure!– to thank the organizers of the Växjö Conference for a very congenial and friendly environment; a special thanks to Professor Joachim Toft for his hospitality and kind invitation to give a talk.*

2. A Strong Version of the Uncertainty Principle

In classical (Hamiltonian) mechanics a *state* is a point of phase space $\mathbb{R}^{2n}_z \equiv \mathbb{R}^n_x \times \mathbb{R}^n_p$, or more generally a symplectic manifold M (for instance the cotangent bundle T^*X of the "configuration space" X of the system): the important thing is to have a symplectic form at one's disposal. A *classical observable* is a *real measurable* function a ("symbol") on this phase space. In quantum mechanics one assumes that the state of a physical system is represented by a vector $\Psi \in \mathcal{H}$, $\Psi \neq 0$, where $\mathcal{H}$ is a Hilbert space (such well-defined states are called *pure states* in quantum mechanics). In what follows we will identify that Hilbert space with $L^2(\mathbb{R}^n)$; its scalar product is denoted by $(\cdot,\cdot)_{L^2}$ and the associated norm by $||\cdot||_{L^2}$.

Suppose that we are given a classical observable a; to a one associates a *quantum observable*, which is a self-adjoint operator A obtained by some "quantization procedure" (for instance, by the Weyl correspondence). The mathematical *expectation* (or: *average*) *value* of A in the state Ψ is then the real number

$$\langle A\rangle_\Psi = \frac{(A\Psi,\Psi)_{L^2}}{(\Psi,\Psi)_{L^2}}$$

that is:

$$\langle A\rangle_\Psi = (A\Psi,\Psi)_{L^2} \text{ if } ||\Psi||_{L^2} = 1. \tag{4}$$

Assume that $\langle A^2\rangle_\Psi$ exists (and is finite); then the quantity

$$(\Delta A)^2_\Psi = \langle A^2\rangle_\Psi - \langle A\rangle^2_\Psi$$

is called the "variance of A in the state Ψ" ; its positive square root $(\Delta A)_\Psi$ is called "standard deviation". If B is a second observable with the same properties then (when defined) the quantity

$$\operatorname{Cov}(A,B)_\Psi = \tfrac{1}{2}\langle AB + BA\rangle_\Psi$$

is called the "covariance" of the pair (A,B) in the state Ψ.

Proposition 1. *Assume that the self-adjoint operators A and B admit variances and a covariance. Then*

$$(\Delta A)^2_\Psi(\Delta B)^2_\Psi \geq \tfrac{1}{4}\operatorname{Cov}(A,B)_\Psi - \tfrac{1}{4}[A,B]^2_\Psi \tag{5}$$

where $[A,B]_\Psi$ is the pure imaginary number $([A,B]\Psi,\Psi)_{L^2}$.

Proof. Replacing if necessary the operators A and B by $A - \langle A\rangle_\Psi$ and $B - \langle B\rangle_\Psi$ we may assume that $\langle A\rangle_\Psi = \langle B\rangle_\Psi = 0$; it is thus sufficient to prove the inequality

$$\langle A^2\rangle_\Psi\langle B^2\rangle_\Psi \geq \tfrac{1}{4}\operatorname{Cov}(A,B)_\Psi + \tfrac{1}{4}[A,B]^2_\Psi. \tag{6}$$

Since A and B are self-adjoint we have $\langle A^2\rangle_\Psi = ||A\Psi||^2_{L^2}$ and $\langle B^2\rangle_\Psi = ||B\Psi||^2_{L^2}$ hence, using Cauchy–Schwarz's inequality:

$$\langle A^2\rangle_\Psi\langle B^2\rangle_\Psi \geq |(A\Psi,B\Psi)_{L^2}|^2 = |(AB\Psi,\Psi)_{L^2}|^2.$$

Noting that $AB = \frac{1}{2}(AB+BA) + \frac{1}{2}[A,B]$ this inequality can be rewritten as

$$\langle A^2\rangle_\Psi\langle B^2\rangle_\Psi \geq |\tfrac{1}{2}((AB+BA)\Psi,\Psi)_{L^2} + \tfrac{1}{2}([A,B]_\Psi\Psi,\Psi)_{L^2}|^2.$$

The self-adjointness of A and B implies that $(AB+BA)\Psi,\Psi)_{L^2}$ is real and that $([A,B]\Psi,\Psi)_{L^2}$ is pure imaginary; the inequality (6) follows. □

The following consequence of this result is immediate:

Corollary 2. *Assume that* $A = X_j$, *the operator of multiplication by* x_j *and* $B = P_j = -i\hbar\partial/\partial x_j$. *Then*

$$(\Delta X_j)^2_\Psi(\Delta P_j)^2_\Psi \geq \tfrac{1}{4}\operatorname{Cov}(X_j, P_j)_\Psi + \tfrac{1}{4}\hbar^2. \tag{7}$$

From now on we will always assume that A is he *Weyl operator* $a^w = a^w(x, D)$ with symbol a, defined for $\Psi \in \mathcal{S}(\mathbb{R}^n)$ by

$$a^w\Psi(x) = \left(\tfrac{1}{2\pi\hbar}\right)^n \iint e^{\frac{i}{\hbar}\langle p, x-y\rangle} a(\tfrac{1}{2}(x+y), p)\Psi(y)d^n y d^n p$$

the integral being interpreted as an "oscillatory integral" (see *e.g.* [3] or [11]; we are assuming that a belongs to some reasonable symbol class, for instance the Hörmander classes $S^m_{\rho,\delta}$).

Remark 3. *One should be aware of the fact that the choice* $A = a^w$ *is neither the only possible, nor necessarily the most realistic "physical" choice: see* [13] *for other possible quantization schemes.*

3. Metaplectic Covariance

Let $\pi : Sp_2(n) \longrightarrow Sp(n)$ be the connected double covering of the symplectic group. $Sp_2(n)$ has a (faithful) representation by a group $Mp(n)$ of unitary operators $L^2(\mathbb{R}^n) \longrightarrow L^2(\mathbb{R}^n)$. That group, the *metaplectic group*, is generated by the quadratic Fourier transforms $\widehat{S}_{W,m}$ defined, for $\Psi \in \mathcal{S}(\mathbb{R}^n)$, by

$$\widehat{S}_{W,m}\Psi(x) = \left(\frac{1}{2\pi i\hbar}\right)^{n/2} i^m\sqrt{|\det L|}\int e^{\frac{i}{\hbar}W(x,x')}\Psi(x')d^n x';$$

in the formula above W is any real quadratic form on $\mathbb{R}^n \times \mathbb{R}^n$ of the type

$$W(x, x') = \frac{1}{2}\langle Px, x\rangle - \langle Lx, x'\rangle + \frac{1}{2}\langle Qx', x'\rangle$$

with $P = P^T$, $Q = Q^T$ and L invertible; $m\pi$ is a choice of $\arg\det L$ ($\widehat{S}_{W,m}$ thus only depends on that choice $\bmod 4\pi$); the integer m is the *Maslov index* of the operator $\widehat{S}_{W,m}$. Identifying $Mp(n)$ with $Sp_2(n)$ the projection π is then unambiguously determined by

$$\pi(\widehat{S}_{W,m}) = S_W = \begin{bmatrix} L^{-1}Q & L^{-1} \\ PL^{-1}Q - L^T & PL^{-1} \end{bmatrix}.$$

A fundamental property is that for every $S \in Sp(n)$ the "metaplectic covariance formula"

$$(a \circ S^{-1})^w = \widehat{S}a^w\widehat{S}^{-1} \tag{8}$$

holds; $\widehat{S}$ is here any of the two metaplectic operators $\pm\widehat{S} \in Mp(n)$ associated to $S = \pi(\widehat{S})$.

We claim that the "uncertainty principle" (5) is invariant under linear symplectic transformations. Let us glorify this statement by giving it the status of a Proposition:

Proposition 4. *Assume that the Weyl operators* $A = a^w$ *and* $B = b^w$ *satisfy the uncertainty relations* (5). *Let* $S \in Sp(n)$. *Then* $A_S = (a \circ S^{-1})^w$ *and* $A_S = (a \circ S^{-1})^w$ *satisfy*

$$(\Delta A_S)^2_{\widehat{S}\Psi}(\Delta B_S)^2_{\widehat{S}\Psi} \geq \tfrac{1}{4}\operatorname{Cov}(A,B)_\Psi - \tfrac{1}{4}[A,B]^2_\Psi \tag{9}$$

for any $\widehat{S} \in Mp(n)$ *with* $\pi(\widehat{S}) = S$. *In particular*

$$(\Delta X_j)^2_{\widehat{S}\Psi}(\Delta P_j)^2_{\widehat{S}\Psi} \geq \tfrac{1}{4}\operatorname{Cov}(X_j,P_j)_\Psi + \tfrac{1}{4}\hbar^2.$$

Proof. Using (8) we have

$$\langle A_S\rangle^2_{\widehat{S}\Psi} = \int |\widehat{S}A\Psi(x)|^2 d^n x = ||\widehat{S}A\Psi||^2_{L^2}$$

hence, since $\widehat{S}$ is a unitary operator:

$$\langle A_S\rangle_{\widehat{S}\Psi} = ||A\Psi||^2_{L^2} = \langle A\rangle^2_\Psi.$$

Writing similar relations for $\langle B_S\rangle_\Psi$, $\langle A_S^2\rangle_\Psi$ and $\langle B_S^2\rangle_\Psi$ formula (9) follows. □

4. Admissible Covariance Matrices

Assume that $X = (X_1,\ldots,X_n)$ and $P = (P_1,\ldots,P_n)$ are the quantum position and momentum operators: X_j is multiplication by x_j and $P_j = -i\hbar(\partial/\partial x_j)$ (as mentioned in the introduction all what we are going to do actually generalizes to any operators X and P such that $[P_j,X_k] = i\hbar\delta_{jk}$).

The covariance matrix is, by definition, the symmetric $2n\times 2n$ matrix written in block matrix form as

$$\Sigma = \begin{bmatrix} \Sigma_{XX} & \Sigma_{XP} \\ \Sigma_{PX} & \Sigma_{PP} \end{bmatrix}, \tag{10}$$

where $\Sigma_{PX} = \Sigma_{XP}^T$ and $\Sigma_{XX}, \Sigma_{PP}, \Sigma_{XP}$ are defined as follows:

- the entry of Σ_{XX} at the i-th row and j-th column is $\operatorname{Cov}(X_i,X_j)$;
- the entry of Σ_{PP} at the i-th row and j-th column is $\operatorname{Cov}(P_i,P_j)$;
- the entry of Σ_{XP} at the i-th row and j-th column is $\operatorname{Cov}(X_i,P_j)$.

We will always assume that Σ is a positive definite matrix: $\Sigma > 0$.

A simple, but essential, observation is the following: since $J^T = -J$ the matrix $\Sigma + i\frac{\hbar}{2}J$ is Hermitian:

$$(\Sigma + i\tfrac{\hbar}{2}J)^* = \Sigma - i\tfrac{\hbar}{2}J^T = \Sigma + i\tfrac{\hbar}{2}J.$$

It follows, in particular, that all the eigenvalues of $\Sigma + i\frac{\hbar}{2}J$ are *real*. It turns out that the uncertainty principle is equivalent to the property that all the eigenvalues are *non-negative*:

Proposition 5. *The uncertainty relations*

$$(\Delta X_j)^2_\Psi(\Delta P_j)^2_\Psi \geq \tfrac{1}{4}\operatorname{Cov}(X_j,P_j)_\Psi + \tfrac{1}{4}\hbar^2$$

are equivalent to either of the properties:

(i)

$$\Sigma + i\tfrac{\hbar}{2}J \ \text{ is positive semi-definite.} \tag{11}$$

(ii) *The moduli* $\lambda_1, \dots, \lambda_n$ *of the eigenvalues* $\pm i\lambda_j$, $1 \leq j \leq n$, *of* $J\Sigma$ *are all* $\geq \frac{1}{2}\hbar$.

Proof. (i) See [14, 4, 5]. (ii) Recall Williamson's diagonalization theorem [18]: for every real symmetric positive-definite $2n \times 2n$ matrix M there exists $S \in Sp(n)$ such that

$$M = S^T \Lambda S \ \ , \ \ \Lambda = \begin{bmatrix} D_\sigma & 0 \\ 0 & D_\sigma \end{bmatrix}$$

where D_σ is the diagonal $n \times n$ matrix whose non-zero entries are the moduli $\lambda_{\sigma,j}$ of the eigenvalues $\pm i\lambda_{\sigma,j}$ ($\lambda_{\sigma,j} > 0$) of JM. The diagonalizing symplectic matrix S is not unique; we have however proven in [5] that if $S, S' \in Sp(n)$ are such that $M = S^T \Lambda S = S'^T \Lambda S'$, then there exists $U \in U(n)$ such that $S' = US$. In view of the discussion above the condition $\Sigma + i\frac{\hbar}{2}J \geq 0$ is equivalent to $D + i\frac{\hbar}{2}J \geq 0$. The characteristic polynomial $\mathcal{P}(\lambda)$ of $D + i\frac{\hbar}{2}J \geq 0$ is $\mathcal{P}(\lambda) = \mathcal{P}_1(\lambda) \cdots \mathcal{P}_n(\lambda)$ where

$$\mathcal{P}_j(\lambda) = \lambda^2 - 2\lambda_{\sigma,j}\lambda + \lambda_{\sigma,j}^2 - \tfrac{1}{4}\hbar^2.$$

The eigenvalues of $D+i\frac{\hbar}{2}J$ are thus the numbers $\lambda_{\sigma,j} \pm \frac{\hbar}{2}$; the condition $\Sigma + i\frac{\hbar}{2}J \geq 0$ means that $\lambda_{\sigma,j} \pm \frac{\hbar}{2} \geq 0$ for $j = 1, \dots, n$ which is equivalent to $\lambda_{\sigma,j} \geq \frac{1}{2}\hbar$ for $j = 1, \dots, n$. □

For instance in the case $n = 1$ we have

$$\Sigma + \tfrac{i\hbar}{2}J = \begin{bmatrix} (\Delta X)^2 & \Delta(X,P) + \frac{i\hbar}{2} \\ \Delta(X,P) - \frac{i\hbar}{2} & (\Delta P)^2 \end{bmatrix}$$

and condition (11) is equivalent to $\det(\Sigma + \frac{i\hbar}{2}J) \geq 0$ that is to

$$(\Delta P)^2 (\Delta X)^2 \geq \Delta(X,P)^2 + \frac{\hbar^2}{4} \tag{12}$$

which is the uncertainty principle.

We will from now on say that a real symmetric $2n \times 2n$ matrix Σ is *quantum mechanically admissible* (or, for short: *admissible*) if it satisfies condition (11).

The property for a covariance matrix to be admissible is invariant under linear symplectic transformations: the equality $S^T J S = J$ implies that

$$S\Sigma S^T + i\tfrac{\hbar}{2}J = S(\Sigma + i\tfrac{\hbar}{2}J)S^T$$

and the Hermitian matrix $S\Sigma S^T + i\frac{\hbar}{2}J$ is thus semi-definite positive if and only $\Sigma + i\frac{\hbar}{2}J$ is.

We are going to restate Proposition 5 in a concise geometric way, using the notion of *symplectic capacity*, due to Ekeland and Hofer [2] (and whose existence is guaranteed by Gromov's non-squeezing theorem [6]). Recall that a symplectic capacity on $(\mathbb{R}^{2n}, \sigma)$ is the assignment to every subset Ω of $\mathbb{R}^{2n}$ of a number $c(\Omega) \geq 0$ or ∞ such that the following properties hold:

1. $c(\Omega) \leq c(\Omega')$ if $\Omega \subset \Omega'$;
2. $c(f(\Omega)) = c(\Omega)$ if $f : \Omega \longrightarrow \Omega'$ is a symplectomorphism;
3. $c(k\Omega) = k^2 c(\Omega)$ for every $k \in \mathbb{R}$;
4. $c(B(R)) = \pi R^2 = Z_j(R)$.

($B(R)$ is any phase space ball with radius R and $Z_j(R)$ is any cylinder with radius 1 based on a conjugate coordinate plane x_j, p_j).

We will need the following result (see [9] for proofs and details):

Lemma 6. *Let $B_M : \frac{1}{2}\langle Mz, z\rangle \leq 1$ ($M = M^T > 0$) be an ellipsoid in $\mathbb{R}^{2n}$. All symplectic capacities agree on B_M, and we have*

$$c(B_M) = c_{\text{lin}}(B_M) = 2\pi/\lambda$$

where c_{lin} is the linear symplectic capacity on $(\mathbb{R}^{2n}, \sigma)$, i.e.

$$c_{\text{lin}}(M) = \{\pi R^2 / \exists f \in ASp(n) : f(B(R)) \subset M\},$$

and $\lambda = \sup\{\lambda_1, \dots, \lambda_n\}$ where the numbers $\pm i\lambda_j$ ($\lambda_j > 0$, $1 \leq j \leq n$) are the eigenvalues of JM.

($ASp(n)$ is the affine symplectic group, semi-direct product of $Sp(n)$ and $\mathbb{R}^{2n}$.)

It follows that:

Proposition 7. *The covariance matrix Σ is admissible if and only if the ellipsoid $B_{\Sigma^{-1}} : \frac{1}{2} z^T \Sigma^{-1} z \leq 1$ has symplectic capacity $c(B_{\Sigma^{-1}}) \geq \frac{1}{2}h$.*

Proof. Let, as before, $\lambda_1, \dots, \lambda_n$ be the moduli of the eigenvalues of $J\Sigma$; then $\lambda_1^{-1}, \dots, \lambda_n^{-1}$ are the moduli of the eigenvalues of $J\Sigma^{-1}$. Let $\lambda_{\text{inf}} = \inf\{\lambda_j\}$. In view of Lemma 6 we have $c(B_{\Sigma^{-1}}) = 2\pi\lambda_{\text{inf}}$; using the condition $c(B_{\Sigma^{-1}}) \geq \frac{1}{2}h$ is thus equivalent to $\lambda_{\text{inf}} \geq \frac{1}{2}\hbar$ which concludes the proof. □

5. Gaussian Averaging

Recall [19] that the Wigner transform $W(\Psi, \Phi)$ (also called *radar ambiguity function* in engineering, and *Blokhintsev transform* by Soviet authors) of a pair of functions $\Psi, \Phi \in \mathcal{S}(\mathbb{R}^n)$ (or, more generally, $\Psi, \Phi \in L^2(\mathbb{R}^n)$) is the Weyl symbol of the operator with kernel $K_{\Psi,\Phi}(x, y) = \Psi(x)\overline{\Phi}(y)$; that is

$$W(\Psi, \Phi)(z) = \left(\tfrac{1}{2\pi\hbar}\right)^n \int e^{-\frac{i}{\hbar}\langle p,y\rangle} \Psi(x + \tfrac{1}{2}y)\overline{\Phi(x - \tfrac{1}{2}y)} d^n y.$$

When $\Psi = \Phi$ we write $W\Psi$, and one recovers the familiar expression

$$W\Psi(z) = \left(\tfrac{1}{2\pi\hbar}\right)^n \int e^{-\frac{i}{\hbar}\langle p,y\rangle} \Psi(x + \tfrac{1}{2}y)\overline{\Psi(x - \tfrac{1}{2}y)} d^n y. \tag{13}$$

A very important result, and which a posteriori justifies the introduction of Weyl operators in quantum mechanics, is the following: if $A = a^w(x, D)$ is a Weyl pseudo-differential operator then the mathematical expectation value of A corresponding to a state Ψ is just the average of the Weyl symbol with respect to the Wigner transform of Ψ:

$$\langle A\rangle_\Psi = (A\Psi, \Psi)_{L^2} = \int a(z)W\Psi(z)d^{2n}z. \tag{14}$$

It is thus as if $W\Psi$ played the role of a probability density over which the classical observable is averaged (behold: $W\Psi$ can however take negative values; see below).

Let Ψ be a normalized and centered Gaussian, that is

$$\Psi_{X,Y}(x) = c\left(\frac{\det X}{(\pi\hbar)^n}\right)^{1/4} \exp\left(-\tfrac{1}{2\hbar}\langle (X+iY)x, x\rangle\right) \tag{15}$$

where $|c| = 1$ and X and Y are real $n\times n$ symmetric matrices, $X > 0$. A tedious but straightforward calculation of Gaussian integrals shows that the Wigner transform of $\Psi_{X,Y}$ is positive; it is in fact the real Gaussian defined by

$$W\Psi_{X,Y}(z) = \left(\tfrac{1}{\pi\hbar}\right)^n e^{-\frac{1}{\hbar}\langle Gz,z\rangle} \quad , \quad G = \begin{bmatrix} X + YX^{-1}Y & YX^{-1} \\ X^{-1}Y & X^{-1}\end{bmatrix}. \tag{16}$$

One moreover immediately verifies that $G \in Sp(n)$ and that $G = G^T > 0$. On the other hand, it is known since Hudson [10] that the Wigner transform of a function $\Psi \in \mathcal{S}(\mathbb{R}^n)$ is non-negative if and only if Ψ is proportional to $\Psi_{X,Y}$ for some choice of X, Y (Toft gives an elegant proof of this property in his thesis [16]). It is also known since de Bruijn [1] that the "average" $W\Psi * \Phi_R$ over a Gaussian $\Phi_R(z) = e^{-|z|^2/R^2}$ satisfies

$$W\Psi * \Phi_R \geq 0 \text{ if } R^2 = \hbar \ , \ W\Psi * \Phi_R > 0 \text{ if } R^2 > \hbar; \tag{17}$$

this actually follows from the observation (see for instance [3], Proposition (1.99)) that the convolution of two Wigner functions is always non-negative:

$$W\Psi * W\Phi = |W(\Psi^\vee, \Phi)\circ J|^2 \geq 0 \quad , \quad \Psi^\vee(x) = \Psi(-x). \tag{18}$$

Let us extend de Bruijn's result to arbitrary Gaussians. To any symmetric $2n \times 2n$ matrix $\Sigma > 0$ we associate the normalized Gaussian ρ_Σ defined by

$$\rho_\Sigma(z) = \left(\tfrac{1}{2\pi}\right)^n (\det\Sigma)^{-1/2} e^{-\frac{1}{2}\langle\Sigma^{-1}z,z\rangle}.$$

A straightforward calculation shows that ρ_Σ has integral one (and is hence a probability density); computing the characteristic function of $\rho_\Sigma * \rho_{\Sigma'}$ (see e.g. [15]) one moreover easily verifies the semigroup property

$$\rho_\Sigma * \rho_{\Sigma'} = \rho_{\Sigma+\Sigma'}. \tag{19}$$

Proposition 8. *Let Σ be a covariance matrix and B_Σ the associated ellipsoid.*

(i) *If there exists $S \in Sp(n)$ such that $B_\Sigma = S(B(\sqrt{\hbar}))$ (and hence $c(B_\Sigma) = \frac{1}{2}h$) then $W\Psi * \rho_\Sigma \geq 0$.*

(ii) *If $c(B_\Sigma) > \frac{1}{2}h$ then $W\Psi * \rho_\Sigma > 0$.*

Proof. (i) The condition $B_\Sigma = S(B(\sqrt{\hbar}))$ is equivalent to $\Sigma = \frac{1}{\hbar}S^T S$ and hence there exists a Gaussian $\Psi_{X,Y} \in L^2(\mathbb{R}^n)$ such that $\rho_\Sigma = W\Psi_{X,Y}$. It follows from (18) that we have

$$W\Psi * \rho_\Sigma = W\Psi * W\Psi_{X,Y} \geq 0$$

which proves the assertion.

(ii) If $c(B_\Sigma) > \frac{1}{2}h$ then there exists $S \in Sp(n)$ such that B_Σ contains $B_{\Sigma_0} = S(B(\sqrt{\hbar}))$ as a proper subset; it follows that $\Sigma - \Sigma_0 > 0$ and hence

$$W\Psi * \rho_\Sigma = (W\Psi * \rho_{\Sigma_0}) * \rho_{\Sigma - \Sigma_0} > 0.$$ □

Using Proposition 8 we can prove a positivity result for the average of a Weyl operators with positive symbol. This result is actually no more than a predictable generalization of calculations that can be found elsewhere (a good summary being [3]); it however very clearly shows that phase space "coarse graining" by ellipsoids with symplectic capacity $\geq \frac{1}{2}\hbar$ eliminates positivity difficulties appearing in Weyl calculus: we are going to see that the Gaussian average of a symbol $a \geq 0$ over an ellipsoid with symplectic capacity $\geq \frac{1}{2}\hbar$ *always* leads to a positive operator. This is interesting, since the condition $a \geq 0$ does not guarantee the positivity of the associated Weyl operator $a^w = a^w(x, D)$; in fact we have, for a non-negative $a \in S^m_{\rho,\delta}$, the so-called "sharp Gårding inequality"

$$(a^w\Psi, \Psi)_{L^2} \geq -C||\Psi||^2_{(m-\rho+\delta)/2} \quad \text{for all} \quad \Psi \in \mathcal{S}(\mathbb{R}^n)$$

($C = C_{m,\rho,\delta}$ a constant > 0; $||\cdot||_s$ is the norm of the usual Sobolev space H^s).

Corollary 9. *Assume that $a \geq 0$. If $c(B_\Sigma) \geq \frac{1}{2}h$ then the operator $A_\Sigma = (a * \rho_\Sigma)^w$ satisfies $\langle A_\Sigma\Psi, \Psi\rangle \geq 0$ for all $\Psi \in \mathcal{S}(\mathbb{R}^n)$.*

Proof. (*cf.* Lemma (2.85) in Folland [3]). We have, in view of (14),

$$(A_\Sigma\Psi, \Psi)_{L^2} = \int (a * \rho_\Sigma)(z) W\Psi(z) d^{2n}z$$

that is, taking into account the fact that ρ_Σ is an even function,

$$\begin{aligned}(A_\Sigma\Psi, \Psi)_{L^2} &= \iint a(u)\rho_\Sigma(z-u)W\Psi(z)d^n u d^n z \\ &= \int a(u)\left(\int \rho_\Sigma(u-z)W\Psi(z)d^n z\right) d^n u \\ &= \int a(u)(\rho_\Sigma * W\Psi)(u) d^n u.\end{aligned}$$

In view of Proposition 8 we have $W\Psi * \rho_\Sigma \geq 0$ hence the result. □

Let us illustrate this result on a simple and – hopefully – suggestive example. Consider the harmonic oscillator Hamiltonian

$$H(z) = \frac{1}{2m}(p^2 + m^2\omega^2 x^2) \ , \ m > 0, \ \omega > 0$$

(with $n = 1$) and choose for ρ_Σ the Gaussian

$$\rho_\sigma(z) = \tfrac{1}{2\pi\sigma^2} e^{-(x^2+p^2)/2\sigma^2}.$$

Set $H_\sigma = H * \rho_\sigma$; a straightforward calculation yields the sharp lower bound

$$H_\sigma(z) = H(z) + \frac{\sigma^2}{2m}(1 + m\omega^2) \geq H(z) + \sigma^2\omega.$$

Assume that the ellipsoid $(x^2+p^2)/2\sigma^2 \leq 1$ is admissible; this is equivalent to the condition $\sigma^2 \geq \frac{1}{2}\hbar$ and hence $H_\sigma(z) \geq \frac{1}{2}\hbar\omega$: we have thus recovered the ground state energy of the one-dimensional harmonic oscillator. We leave it to the reader to check that we would still get the same result if we had used a more general Gaussian ρ_Σ instead of ρ_σ. This example actually generalizes to quadratic Hamiltonian without difficulty using Williamson's symplectic diagonalization theorem [18] (or its variants [9]) which asserts that any semi-positive definite matrix can be diagonalized using linear symplectic transformations.

6. Concluding Remarks

We hope that we have succeeded in convincing the reader of the fact that the uncertainty principle of quantum mechanics strongly indicates the existence of a "quantum phase space" whose points are subsets of the classical phase space. Of course, since the notion of symplectic capacity itself depends on the choice of symplectic structure, but this fact should only be moderately annoying: as already noted and commented in [7], Chapter 1, it is after all the fields which determine the symplectic structure. It should therefore not be so surprising if every (Hamiltonian) physical system could be treated by the methods outlined above by assigning to it its own private quantum phase space, as suggested and discussed in de Gosson [4, 5].

In the last section we recovered the ground energy level for the harmonic oscillator by averaging the Hamiltonian function over admissible ellipsoids. It would be extremely interesting to extend this kind of results to the case of arbitrary Hamiltonians. This might be not too difficult in the case of "physical" Hamiltonians of the type "kinetic energy + potential". One should probably have to replace the averaging over ellipsoids by an averaging over arbitrary subsets of phase space having symplectic capacity $\geq \frac{1}{2}\hbar$. This might actually not be too difficult since general symplectic capacities are invariant under symplectomorphisms. It would actually not be too surprising if this averaging process would lead to semiclassical estimates for the calculation of the spectrum of the Hamiltonian operator, using Gutzwiller-type trace formulae. We leave these fascinating perspectives open for further research.

References

[1] N. G. de Bruijn, A theory of generalized functions, with applications to Wigner distribution and Weyl correspondence, *Nieuw Archiev voor Wiskunde,* **21** (1973), 205–280.

[2] I. Ekeland and H. Hofer, Symplectic topology and Hamiltonian dynamics, I and II, *Math. Zeit.* **200** (1990), 355–378 and **203** (1990), 553–567.

[3] G. B. Folland, *Harmonic Analysis in Phase space*, Annals of Mathematics Studies, Princeton University Press, Princeton, N. J., 1989.

[4] M. de Gosson, The optimal pure Gaussian state canonically associated to a Gaussian quantum state, *Phys. Lett. A*, **330** no. 3–4 (2004), 161–167.

[5] M. de Gosson, Cellules quantiques symplectiques et fonctions de Husimi–Wigner. *Bull. Sci. Math.*, to appear.

[6] M. Gromov, Pseudoholomorphic curves in symplectic manifolds, *Invent. Math.* **82** (1985), 307–347.

[7] V. Guillemin and S. Sternberg, *Symplectic Techniques in Physics*, Cambridge University Press, Cambridge, Mass., 1984.

[8] M. J. W. Hall and M. Reginatto, Schrödinger equation from an exact uncertainty principle, *J. Phys. A: Math. Gen.* **35** (2002), 3289–3303.

[9] H. Hofer and E. Zehnder, *Symplectic Invariants and Hamiltonian Dynamics*, Birkhäuser Advanced texts (Basler Lehrbücher, Birkhäuser Verlag, 1994.

[10] R. L. Hudson, When is the Wigner quasi-probability density non-negative? *Rep. Math. Phys.* **6** (1974), 249–252.

[11] L. Hörmander, *The Analysis of Linear Partial differential Operators III.* Springer-Verlag, 1985.

[12] K. Husimi, Some formal properties of the density matrix. *Proc. Physico-Math. Soc. Japan* **22** (1940), 264.

[13] V. Nazaikiinskii, B.-W. Schulze and B. Sternin, *Quantization Methods in Differential Equations*, Differential and Integral Equations and Their Applications, Taylor & Francis, 2002.

[14] R. Simon, E. C. G. Sudarshan and N. Mukunda, Gaussian–Wigner distributions in quantum mechanics and optics, *Phys. Rev. A* **36** (8) (1987), 3868–3880.

[15] Y. G. Sinai, *Probability Theory: an Introductory Course*, Springer-Verlag, 1992.

[16] J. Toft, *Continuity and Positivity Problems in Pseudo-Differential Calculus*, Thesis, University of Lund, 1996.

[17] J. Toft, Continuity properties in non-commutative convolution algebras, with applications in pseudo-differential calculus, *Bull. Sci. Math.* **126** (2002), 115–142.

[18] J. Williamson, On the algebraic problem concerning the normal forms of linear dynamical systems, *Amer. J. of Math.* **58** (1963), 141–163.

[19] M. W. Wong, *Weyl Transforms*, Universitext, Springer, 1998.

Maurice de Gosson
Universität Potsdam
Institut für Mathematik
Am Neuen Palais 10
D-14415 Potsdam
Germany

and

Universidade de São Paulo
Departamento de Matemãtica
Rua do Matão 1010
CEP 05508-900 São Paulo
Brazil
e-mail: maurice.degosson@gmail.com

Operator Theory:
Advances and Applications, Vol. 164, 133–151

Pseudo-Differential Operator and Reproducing Kernels Arising in Geometric Quantization

Kenro Furutani

Abstract. We show that an operator defined on the quaternion projective space is a zeroth order selfadjoint pseudo-differential operator of Hörmander class $L^0_{1,0}$. This operator arises when we compare two quantization operators of the geodesic flow on the quaternion projective space. Such quantization operators are defined on a Hilbert space consisting of holomorphic functions and the Hilbert space has reproducing kernel. We describe the reproducing kernels in the cases of sphere and quaternion projective space in terms of hypergeometric functions, and discuss their relation through fiber integration with respect to the complexified Hopf fibration.

Mathematics Subject Classification (2000). Primary 53D12; Secondary 58J30, 58B15.

Keywords. Quaternion projective space, Hopf fibration, polarization, geometric quantization, geodesic flow, Kähler form, reproducing kernel, fiber integration, pseudo-differential operator.

1. Introduction

This is a continuation of our previous paper [3], where we constructed a Kähler structure on the punctured cotangent bundle $T_0^*(P^n\mathbb{H})$ of the quaternion projective space $P^n\mathbb{H}$ whose Kähler form coincides with the natural symplectic form on the cotangent bundle, and based on this structure we constructed two quantization operators T and $\widetilde{T}$ of the geodesic flow. The later one is obtained from an operator in the sphere case through the complexified Hopf fibration. These operators are defined on a same Hilbert space consisting of holomorphic functions on the punctured cotangent bundle with respect to the Kähler structure. The method we employed there is called "Pairing of Polarizations" ([9], [14], [6]). Here two polarizations are the vertical polarization, that is, a real polarization (= Lagrangian foliation) whose leaves are the fibers of the projection map from cotangent bundle to the base space, and the Kähler polarization, that is, a positive complex polarization which is given by embedding $T_0^*(P^n\mathbb{H})$ into a matrix space $M(2n+2,\mathbb{C})$. Also we showed that this Hilbert space has the reproducing kernel ([8], [3]). In this note we prove that the operator $\widetilde{T}\circ T^{-1}$ is a zeroth order pseudo-differential operator in the Hörmander class $L^0_{1,0}$ and give an expression of the reproducing kernel in a form of a fiber integration of the reproducing kernel corresponding to the sphere case through the complexified Hopf fibration.

Qantization procedure is not unique. Even our method "pairing polarizations" gives us two kinds of correspondences from classical observables (= functions on phase space) to quantum states (= functions on configuration space). These are not the quantization in the sense of N. Bohr, but in our cases each eigenfunction of the Laplacian is expressed in a form of fiber integration of a polynomial. Our basic interests in this topics are to find (compact) manifolds (= compact configuration spaces) whose punctured cotangent bundle (= phase space) has such a Kähler structure that the natural symplectic form coincides with the Kähler form and the explicit determination of the pairing of polarizations. This pairing will give us an interesting operator governed by a class of special functions. In all the known compact cases it is the Gamma function ([9], [4], [3]). At the moment we know that all rank one compact symmetric spaces have such structures ([8], [10], [2], [5], [7], [11], [12]). Note that we can not construct such a Kähler structure on the whole space of these cotangent bundles but on the subspace being zero section deleted always. Among them it remains only the case of the Cayley projective plane for which the quantization operator is not constructed yet.

To fix the notations, in §2 we recall basic facts on the Riemannian structure of the quaternion projective space and a Kähler structure on its punctured cotangent bundle. Then we explain the complexified Hopf fibration. In §3 we give the relation of Liouville volume forms on the cotangent bundle of sphere and quaternion projective space. In §4 we recall briefly the operators constructed in [3] (see also [8] and [9]) and we prove our main theorem 4.2. In the final section we express the reproducing kernels of Hilbert spaces consisting of holomorphic functions on punctured cotangent bundles in terms of hypergeometric functions and discuss their relation.

2. Complexification of the Hopf Fibration

We describe the quaternion projective space and recall the complexified Hopf fibration on the punctured (co)tangent bundle of the quaternion projective space ([3]).

Let $\mathbb{H}$ be the quaternion number field with the basis $\{\mathbf{e}_0,\ \mathbf{e}_1,\ \mathbf{e}_2,\ \mathbf{e}_3\}$ satisfying the relations:

$$\mathbf{e}_0 = \text{ identity},\ \mathbf{e}_i^2 = -\mathbf{e}_0\ (i=1,2,3),\ \mathbf{e}_1\mathbf{e}_2 = \mathbf{e}_3 = -\mathbf{e}_3\mathbf{e}_2,$$
$$\mathbf{e}_2\mathbf{e}_3 = \mathbf{e}_1 = -\mathbf{e}_3\mathbf{e}_2,\ \mathbf{e}_1\mathbf{e}_3 = -\mathbf{e}_3\mathbf{e}_1 = -\mathbf{e}_2.$$

We denote the elements in $\mathbb{H}^{n+1}$ by $p = (p_0, p_1, \ldots, p_n)$ with

$$p_i = \sum_{j=0}^{3} x_{4i+j}\mathbf{e}_j,\ x_{4i+j} \in \mathbb{R},\ 0 \le 4i+j \le 4n+3,\ 0 \le i \le n,\ 0 \le j \le 3,$$

and regard it as a right $\mathbb{H}$-vector space equipped with the $\mathbb{H}$-inner product

$$(p,q)_{\mathbb{H}} = \sum_{i=0}^{n} \theta(p_i)q_i, \tag{2.1}$$

where $p = (p_0, \ldots, p_n),\ q = (q_0, \ldots, q_n) \in \mathbb{H}^{n+1}$ and

$$\theta(x) = x_0\mathbf{e}_0 - \sum_{i=1}^{3} x_i\mathbf{e}_i \tag{2.2}$$

for $x = \sum_{i=0}^{3} x_i\mathbf{e}_i \in \mathbb{H}$ (we call θ quaternion conjugation). Then the $\mathbb{R}$-bilinear form $< p,q >= \frac{1}{2}\{(p,q)_{\mathbb{H}} + (q,p)_{\mathbb{H}}\}$ defines the usual Euclidean inner product on $\mathbb{H}^{n+1} \cong \mathbb{R}^{4n+4}$ as a real vector space. We will denote its complex bilinear extension to the complexification $\mathbb{H}^{n+1} \otimes \mathbb{C} \cong (\mathbb{H} \otimes \mathbb{C})^{n+1}$ with the same notation $< \bullet, \bullet >$. Then the Hermitian inner product on $\mathbb{H}^{n+1} \otimes \mathbb{C}$ is given by $< h, \overline{k} >$, where $\overline{k} = (\overline{k}_0, \overline{k}_1, \ldots, \overline{k}_n)$ and $\overline{k}_i = \overline{p_i \otimes 1 + q_i \otimes \sqrt{-1}} = p_i \otimes 1 - q_i \otimes \sqrt{-1},\ p_i,\ q_i \in \mathbb{H}$.

Let $M(n,\mathbb{H})$ be the space of $n \times n$ matrices with the elements in $\mathbb{H}$, and for $X = (x_{ij}) \in M(n,\mathbb{H})$ define respectively

$$\theta(X) = \Big(\theta(x_{ij})\Big), \tag{2.3}$$

$$({}^tX)_{ij} = x_{ji}, \tag{2.4}$$

$$\operatorname{tr} X = \sum x_{ii}, \tag{2.5}$$

$${}^TX = \theta({}^tX). \tag{2.6}$$

Each matrix $X \in M(n,\mathbb{H})$ defines a right $\mathbb{H}$-linear map $X : \mathbb{H}^n \to \mathbb{H}^n$, and we have

$$(X\,h, k)_{\mathbb{H}} = (h, {}^TX\,k)_{\mathbb{H}}. \tag{2.7}$$

The Symplectic group $\mathrm{Sp}(n)$ is a group consisting of those matrices $g \in M(n,\mathbb{H})$ which preserves the $\mathbb{H}$-inner product $(\bullet,\bullet)_{\mathbb{H}}$.

The space $\mathcal{H}(n,\mathbb{H}) = \Big\{X \in M(n,\mathbb{H}) \Big| X = {}^TX\Big\}$ is called a Jordan algebra with the Jordan product

$$X \circ Y = \frac{1}{2}(XY + YX). \tag{2.8}$$

Note that for $X \in \mathcal{H}(n,\mathbb{H})$, $\operatorname{tr} X \in \mathbb{R}\mathbf{e}_0 \cong \mathbb{R}$ and $\mathcal{H}(n,\mathbb{H})$ is equipped with the Euclidean inner product $< \bullet, \bullet >_{\mathbb{R}}$ given by

$$< X, Y >_{\mathbb{R}} = \operatorname{tr}(X \circ Y). \tag{2.9}$$

This inner product has the property:

$$< X \circ Y, Z >_{\mathbb{R}} = < X, Y \circ Z >_{\mathbb{R}} . \tag{2.10}$$

We also denote by $< \bullet, \bullet >_{\mathbb{C}}$ its complex linear extension to the complexified Jordan algebra $\mathcal{H}(n, \mathbb{H}) \otimes \mathbb{C}$ and its realization in the complex matrix space $\rho(\mathcal{H}(n, \mathbb{H}) \otimes \mathbb{C})$ in $M(2n, \mathbb{C})$.

Let $S^{4n+3} = \left\{ p = (p_0, \ldots, p_n) \in \mathbb{H}^{n+1} \;\middle|\; < p,\ p >= 1,\ p_i = \sum_{j=0}^{3} x_{4i+j} \mathbf{e}_j \right\}$ be the unit sphere with the standard metric, then its volume form is expressed as

$$\mathbf{v}_S = \sum_{i=0}^{4n+3} (-1)^i x_i dx_0 \wedge \cdots \wedge \widehat{dx_i} \wedge \cdots \wedge dx_{4n+3}. \tag{2.11}$$

The quaternion projective space $P^n\mathbb{H}$ is the set of all (right)$\mathbb{H}$-one-dimensional subspaces in $\mathbb{H}^{n+1}$. We identify $P^n\mathbb{H}$ with the subset in $\mathcal{H}(n+1, \mathbb{H})$:

$$P^n(\mathbb{H}) = \left\{ P \in \mathcal{H}(n+1, \mathbb{H}) \;\middle|\; P = \left(p_i \theta(p_j)\right),\ p = (p_0, \ldots, p_n) \in \mathbb{H}^{n+1},\ < p,\ p >= 1 \right\}.$$

The tangent bundle $T(P^n\mathbb{H})$ is identified with the space

$$T(P^n\mathbb{H}) = \left\{ (P,\ Q) \in \mathcal{H}(n+1, \mathbb{H}) \times \mathcal{H}(n+1, \mathbb{H}) \;\middle|\; P \in P^n(\mathbb{H}),\ 2P \circ Q = Q \right\}. \tag{2.12}$$

Let $\pi_{\mathbb{H}} \ :\ S^{4n+3} \to P^n\mathbb{H}$ be the Hopf fibration:

$$\pi_{\mathbb{H}} \ :\ S^{4n+3} \to P^n\mathbb{H},\ \pi_{\mathbb{H}}(p) = \left(p_i \theta(p_j)\right) \in \mathcal{H}(n+1, \mathbb{H}). \tag{2.13}$$

The differential

$$d\pi_{\mathbb{H}} \ :\ T(S^{4n+3}) = \left\{ (p, q) \in \mathbb{H}^{n+1} \times \mathbb{H}^{n+1} \;\middle|\; <p, p>= 1,\ <p, q>= 0 \right\} \to T(P^n\mathbb{H}), \tag{2.14}$$

is expressed as

$$d\pi_{\mathbb{H}}(p, q) = (P,\ Q),\ P = \left(p_i \theta(p_j)\right),\ Q = \left(p_i \theta(q_j)\right) + \left(q_i \theta(p_j)\right). \tag{2.15}$$

We introduce the Riemann metric on $P^n\mathbb{H}$ through the Hopf fibration in an obvious way and identify each of the tangent bundles and the cotangent bundles of the sphere and the quaternion projective space through the Riemann metrics respectively.

Let V_i $(i = 1, 2, 3)$ be the fundamental vector fields corresponding to the basis $\mathbf{e}_i$ $(i = 1, 2, 3)$ of the Lie algebra $\mathfrak{sp}(1)$, the Lie algebra of the structure group Sp(1) of the Hopf fibration $\pi_{\mathbb{H}} : S^{4n+3} \to P^n\mathbb{H}$. Then at each point $p \in S^{4n+3}$,

$$< V_i,\ V_j >= \delta_{i\,j}.$$

So we fix the orientation of $P^n\mathbb{H}$ by fixing the Riemann volume form $\mathbf{v}_{\mathbb{H}}$ on $P^n\mathbb{H}$ through the relation

$$\pi_{\mathbb{H}}^*(\mathbf{v}_{\mathbb{H}}) = \mathbf{i}_{V_3} \circ \mathbf{i}_{V_2} \circ \mathbf{i}_{V_1}(\mathbf{v}_S), \tag{2.16}$$

where $\mathbf{i}_{V_j}$ is the interior product by the vector field V_j.

Let

$$\tau_S : T(S^{4n+3}) \to \mathbb{H}^{n+1} \otimes \mathbb{C} \cong (\mathbb{H} \otimes \mathbb{C})^{n+1}$$

be a map defined by

$$\tau_S(p,q) = \|q\| p \otimes 1 + q \otimes \sqrt{-1} = (W_0, W_1, \dots, W_n),$$

where

$$p = (p_0, \dots, p_n),\ p_i = \sum_{j=0}^{3} x_{4i+j}\mathbf{e}_j,\ q = (q_0, \dots, q_n),\ q_i = \sum_{j=0}^{3} y_{4i+j}\mathbf{e}_j,$$

$$W_i = \sum_{j=0}^{3} \left(u_{4i+j} + v_{4i+j}\sqrt{-1}\right)\mathbf{e}_j = \sum_{j=0}^{3} \left(\|q\| x_{4i+j} + y_{4i+j}\sqrt{-1}\right)\mathbf{e}_j,$$

$$\|q\| = \sqrt{\sum_{\ell=0}^{4n+3} y_\ell^2}.$$

The quaternion number field $\mathbb{H}$ is mapped into the space of 2×2 complex matrices $M(2,\mathbb{C})$ by the map $\rho : \mathbb{H} \to M(2,\mathbb{C})$,

$$\rho : \mathbb{H} \ni h = \sum_{i=0}^{3} x_i \mathbf{e}_i \longmapsto \begin{pmatrix} x_0 + \sqrt{-1}x_1 & x_2 + \sqrt{-1}x_3 \\ -x_2 + \sqrt{-1}x_3 & x_0 - \sqrt{-1}x_1 \end{pmatrix} \in M(2,\mathbb{C}),$$

and its complexification is the isomorphism:

$$\mathbb{H} \otimes \mathbb{C} \overset{\rho}{\underset{\sim}{\to}} M(2,\mathbb{C}). \tag{2.17}$$

Also we express the isomorphisms

$$(\mathbb{H} \otimes \mathbb{C})^{n+1} \tilde{\to} M(2,\mathbb{C})^{n+1}$$

and

$$M(n+1,\mathbb{H}) \otimes \mathbb{C} \tilde{\to} M(2n+2,\mathbb{C})$$

by the same ρ.

Let $J = \begin{pmatrix} 0 & 1 \\ -1 & 0 \end{pmatrix}$. Then $\rho \circ \theta \circ \rho^{-1}(B) = -J^t B J$ for $B \in M(2,\mathbb{C})$, where θ denotes the complex linear extension of quaternion conjugation (2.2) to $\mathbb{H} \otimes \mathbb{C}$.

We denote by

$$T_0 S^{4n+3} = \left\{ (p,q) \in T(S^{4n+3}) \subset \mathbb{H}^{n+1} \times \mathbb{H}^{n+1} \;\middle|\; q \neq 0 \right\}$$

the punctured tangent bundle, then

Proposition 2.1. ([8], [10])

$$\tau_S : T_0 S^{4n+3} \to Z_S = \left\{ (W_0, \dots, W_n) \in (\mathbb{H} \otimes \mathbb{C})^{n+1} \setminus \{0\} \;\middle|\; \sum_{\ell=0}^{4n+3} \left(u_\ell + v_\ell\sqrt{-1}\right)^2 = 0 \right\} \tag{2.18}$$

is an isomorphism and through this map the punctured (co)tangent bundle $T_0 S^{4n+3}$ *has a Kähler structure, whose Kähler form coincides with the natural symplectic form. Here* $W_i = \sum_{j=0}^{3} \big(u_{4i+j} \otimes 1 + v_{4i+j} \otimes \sqrt{-1}\big)\mathbf{e}_j$, u_{4i+j}, $v_{4i+j} \in \mathbb{R}$.

Let

$$\tilde{\tau}_S \ : \ T(S^{4n+3}) \to M(2,\mathbb{C})^{n+1},$$
$$\tilde{\tau}_S(p,q) = \rho \circ \tau_S(p,q) \ = \big(\rho(W_0), \dots, \rho(W_n)\big).$$

For $(p,q) \in T_0 S^{4n+3}$, we have

$$\begin{aligned}\tilde{\tau}_S(p,q) &= (B_0, \dots, B_n)\\ &= \left(\begin{pmatrix} z_0 & w_0 \\ z_1 & w_1 \end{pmatrix}, \dots, \begin{pmatrix} z_{2i} & w_{2i} \\ z_{2i+1} & w_{2i+1} \end{pmatrix}, \dots, \begin{pmatrix} z_{2n} & w_{2n} \\ z_{2n+1} & w_{2n+1} \end{pmatrix}\right),\end{aligned}$$

where

$$\begin{aligned} z_{2i} &= \big(\|q\| x_{4i} + y_{4i}\sqrt{-1}\big) + \sqrt{-1}\big(\|q\| x_{4i+1} + y_{4i+1}\sqrt{-1}\big),\\ z_{2i+1} &= -\big(\|q\| x_{4i+2} + y_{4i+2}\sqrt{-1}\big) + \sqrt{-1}\big(\|q\| x_{4i+3} + y_{4i+3}\sqrt{-1}\big),\\ w_{2i} &= \big(\|q\| x_{4i+2} + y_{4i+2}\sqrt{-1}\big) + \sqrt{-1}\big(\|q\| x_{4i+3} + y_{4i+3}\sqrt{-1}\big)\end{aligned}$$

and

$$w_{2i+1} = \big(\|q\| x_{4i} + y_{4i}\sqrt{-1}\big) - \sqrt{-1}\big(\|q\| x_{4i+1} + y_{4i+1}\sqrt{-1}\big).$$

Then

Proposition 2.2.

$$\tilde{\tau}_S(T_0 S^{4n+3}) = \Big\{(B_0, \dots, B_n) \in M(2,\mathbb{C})^{n+1} \backslash \{0\} \ \Big| \ \sum_{i=0}^{n} \det B_i = 0\Big\}. \tag{2.19}$$

Let $\mathbf{E}_S$ be a subspace in $T_0 S^{4n+3}$ such that

$$\mathbf{E}_S = \Big\{(p,q) \in \mathbb{H}^{n+1} \times \mathbb{H}^{n+1} \ \Big| \ < p,\, p >= 1, < p,\, q >= 0, q + p(q,p)_{\mathbb{H}} \neq 0\Big\}. \tag{2.20}$$

The space $\mathbf{E}_S$ consists of those elements that $d\pi_{\mathbb{H}}(p,q) \neq 0$ and is (left) $\mathrm{Sp}(n+1)$-invariant.

Proposition 2.3.

$$\tau_S(\mathbf{E}_S) = \Big\{(W_0, \dots, W_n) \in Z_S \ \Big| \ \exists i, \exists j,\ W_i \theta(W_j) \neq 0\Big\},$$
$$\tilde{\tau}_S(\mathbf{E}_S) = \Big\{(B_0, \dots, B_n) \in M(2n+2,\mathbb{C})^{n+1} \ \Big| \ \sum_{i=0}^{n} \det B_i = 0,\ z \wedge w \neq 0\Big\},$$

where $B_i = \begin{pmatrix} z_{2i} & w_{2i} \\ z_{2i+1} & w_{2i+1} \end{pmatrix}$ *and* $z = (z_0, \dots, z_{2n+1})$, $w = (w_0, \dots, w_{2n+1})$.

We put $\tau_S(\mathbf{E}_S) = \mathbf{X}_S$.

Let $G_1 = \{h \in \mathbb{H} \otimes \mathbb{C} \mid \theta(h)h = h\theta(h) = 1\}$, then $\rho(G_1) = SL(2, \mathbb{C})$.

Put $\mathbf{E}_{\mathbb{H}} = T_0(P^n\mathbb{H})$, the punctured (co)tangent bundle of the quaternion projective space, and also put

$$\widetilde{\mathbf{E}}_{\mathbb{H}} = \left\{A \in M(2n+2, \mathbb{C}) \;\middle|\; \mathbb{J}\, A = {}^tA\, \mathbb{J},\ \operatorname{rank} A = 2,\ A^2 = 0\right\}. \tag{2.21}$$

Here we denote by $\mathbb{J}$

$$\mathbb{J} = \begin{pmatrix} J & & & O \\ & J & & \\ & & \ddots & \\ O & & & J \end{pmatrix} \in M(2n+2, \mathbb{C}),$$

and $J = \begin{pmatrix} 0 & 1 \\ -1 & 0 \end{pmatrix}$.

Let $\tau_{\mathbb{H}}$ be a map

$$\tau_{\mathbb{H}} : \mathbf{E}_{\mathbb{H}} \to \mathcal{H}(n+1, \mathbb{H}) \otimes \mathbb{C},$$

defined by

$$\tau_{\mathbb{H}}(P, Q) = \left(\|Q\|^2 P - Q^2\right) \otimes 1 + \frac{\|Q\|}{\sqrt{2}} Q \otimes \sqrt{-1},$$

where $\|Q\| = \sqrt{\operatorname{tr} Q^2}$ and

$$\begin{aligned} &\tilde{\tau}_{\mathbb{H}} : \mathbf{E}_{\mathbb{H}} \to M(2n+2, \mathbb{C}) \\ &\tilde{\tau}_{\mathbb{H}}(P, Q) = \tilde{\tau}_{\mathbb{H}}\Big((p_{i\,j}), (q_{i\,j})\Big) \\ &= \|Q\|^2 \Big(\rho(p_{i\,j})\Big) - \Big(\rho(q_{i\,j})\Big)^2 + \frac{\|Q\|}{\sqrt{2}} \Big(\rho(q_{i\,j})\Big) \sqrt{-1}. \end{aligned}$$

Proposition 2.4. ([3]) *The maps $\tau_{\mathbb{H}}$ and $\tilde{\tau}_{\mathbb{H}}$ are isomorphisms:*

$$\tau_{\mathbb{H}} : \mathbf{E}_{\mathbb{H}} \xrightarrow{\sim} \left\{W \in \mathcal{H}(n+1, \mathbb{H} \otimes \mathbb{C}) \backslash \{0\} \,\middle|\, W = \Big(W_i \theta(W_j)\Big),\ \sum_{i=0}^{n} \theta(W_i) W_i = 0\right\}$$

$$\tilde{\tau}_{\mathbb{H}} : \mathbf{E}_{\mathbb{H}} \xrightarrow{\sim} \widetilde{\mathbf{E}}_{\mathbb{H}}.$$

We put $\tau_{\mathbb{H}}(\mathbf{E}_{\mathbb{H}}) = \mathbf{X}_{\mathbb{H}}$ and define maps α and β as follows:

$$\begin{aligned} &\alpha : \tau_S(\mathbf{E}_S) = \mathbf{X}_S \to \mathcal{H}(n+1, \mathbb{H}) \otimes \mathbb{C} \\ &\quad \alpha(W_0, \dots W_n) = \Big(W_i \theta(W_j)\Big) \\ &\beta : \tilde{\tau}_S(\mathbf{E}_S) = \widetilde{\mathbf{E}}_S \to M(2n+2, \mathbb{C}) \\ &\quad \beta(B_0, \dots, B_n) = \Big(A_{i\,j}\Big),\ A_{i\,j} = -B_i J^t B_j J. \end{aligned}$$

Proposition 2.5. ([3]) *The triple $\beta : \tilde{\tau}_S(\mathbf{E}_S) = \widetilde{\mathbf{E}}_S \to \widetilde{\mathbf{E}}_{\mathbb{H}}$ is a holomorphic principal bundle with the structure group $SL(2, \mathbb{C})$.*

This can be seen as a complexification of the Hopf fibration. Of course this is isomorphic with the principal bundle

$$\alpha : \mathbf{X}_S \to \mathbf{X}_{\mathbb{H}}$$

with the structure group G_1 through the map ρ.

Remark 2.6. The diagram

$$\begin{array}{ccc} \mathbf{E}_S & \xrightarrow{\tau_S} & \mathbf{X}_S \\ {\scriptstyle d\pi_H}\big\downarrow & & \big\downarrow{\scriptstyle \alpha} \\ \mathbf{E}_{\mathbb{H}} & \xrightarrow[\tau_{\mathbb{H}}]{} & \mathbf{X}_{\mathbb{H}} \end{array} \tag{2.22}$$

is not commutative. When we restrict the map τ_S to a subspace

$$\mathbf{E}_S^0 = \left\{(p,q) \in T_0 S^{4n+3} \middle| (p,q)_{\mathbb{H}} = 0\right\}$$

then it is commutative.

3. Liouville Volume Forms

In this section we describe the nowhere vanishing holomorphic global sections on the complex manifolds $\tilde{\tau}_S(\mathbf{E}_S) = \widetilde{\mathbf{E}}_S$ and $\tilde{\tau}_{\mathbb{H}}(\mathbf{E}_{\mathbb{H}}) = \widetilde{\mathbf{E}}_{\mathbb{H}}$ and their pairings ([3]). Also we describe the relation of the volume forms on $\widetilde{\mathbf{E}}_S$ and $\widetilde{\mathbf{E}}_{\mathbb{H}}$, both of which pull-back by the maps $\tilde{\tau}_S$ and $\tilde{\tau}_{\mathbb{H}}$ are the Liouville volume forms on $T_0^*(S^{4n+3})$ and $T_0^*(P^n\mathbb{H})$ respectively.

Let Z be a vector field on $M(2,\mathbb{C})^{n+1}$ such that

$$Z = \frac{1}{\|B\|}\left(\sum_{i=0}^{2n+1}\overline{\frac{\partial D}{\partial z_i}}\frac{\partial}{\partial z_i} + \sum_{i=0}^{2n+1}\overline{\frac{\partial D}{\partial w_i}}\frac{\partial}{\partial w_i}\right),$$

where $B = (B_0,\ldots,B_n)$, $B_i = \begin{pmatrix} z_{2i} & w_{2i} \\ z_{2i+1} & w_{2i+1}\end{pmatrix}$, $D = \sum_{i=0}^{n}\det B_i$ and $\|B\| = \sqrt{\sum_{i=0}^{n}\operatorname{tr} B_i\, B_i^{\,*}}$.

Let $(4n+3)$-form σ_S on $M(2,\mathbb{C})^{n+1}\backslash\{0\}$ be

$$\sigma_S = \frac{1}{(2\sqrt{-1})^{2n+2}}\mathbf{i}_Z(dz_0 \wedge \cdots dz_{2n+1} \wedge dw_0 \wedge \cdots \wedge dw_{2n+1}).$$

This satisfies

$$(2\sqrt{-1})^{2n+2} \cdot dD \wedge \sigma_S = dz_0 \wedge \cdots \wedge dz_{2n+1} \wedge dw_0 \wedge \cdots \wedge dw_{2n+1},$$

and σ_S is holomorphic, nowhere vanishing $(4n+3)$-form on the subspace $\widetilde{\mathbf{E}}_S$.

Let $\rho(\mathbf{e}_1) = \begin{pmatrix}\sqrt{-1} & 0 \\ 0 & -\sqrt{-1}\end{pmatrix}$, $\rho(\mathbf{e}_2) = \begin{pmatrix} 0 & 1 \\ -1 & 0\end{pmatrix}$ and $\rho(\mathbf{e}_3) = \begin{pmatrix} 0 & \sqrt{-1} \\ \sqrt{-1} & 0\end{pmatrix}$ be the basis of $\mathfrak{su}(2) \subset \mathfrak{sl}(2,\mathbb{C})$ and we denote by Y_1, Y_2 and Y_3 the vector fields on $\widetilde{\mathbf{E}}_S$ corresponding to these three elements defined by the right action of $SL(2,\mathbb{C})$.

Let σ be a $4n$-form defined by

$$\sigma = \mathbf{i}_{Y_3} \circ \mathbf{i}_{Y_2} \circ \mathbf{i}_{Y_1}(\sigma_S).$$

Then σ is $SL(2,\mathbb{C})$-invariant.

We decompose $Y_i = Y_i^{'} + Y_i^{''}$ into holomorphic and anti-holomorphic components in the complexified tangent bundle $T(\widetilde{\mathbf{E}}_S) \otimes \mathbb{C} = T^{'}(\widetilde{\mathbf{E}}_S) \oplus T^{''}(\widetilde{\mathbf{E}}_S)$:

$$\begin{aligned}
Y_1 &= Y_1^{'} + Y_1^{''} \\
&= \sum_{k=0}^{2n+1} \left(\sqrt{-1} z_k \frac{\partial}{\partial z_k} + \sqrt{-1} w_k \frac{\partial}{\partial w_k} - \sqrt{-1}\overline{z}_k \frac{\partial}{\partial \overline{z}_k} - \sqrt{-1}\overline{w}_k \frac{\partial}{\partial \overline{w}_k} \right), \\
Y_2 &= Y_2^{'} + Y_2^{''} \\
&= \sum_{k=0}^{2n+1} \left(-w_k \frac{\partial}{\partial z_k} + z_k \frac{\partial}{\partial w_k} - \overline{w}_k \frac{\partial}{\partial \overline{z}_k} + \overline{z}_k \frac{\partial}{\partial \overline{w}_k} \right), \\
Y_3 &= Y_3^{'} + Y_3^{''} \\
&= \sum_{k=0}^{2n+1} \left(\sqrt{-1} w_k \frac{\partial}{\partial z_k} + \sqrt{-1} z_k \frac{\partial}{\partial w_k} - \sqrt{-1}\overline{w}_k \frac{\partial}{\partial \overline{z}_k} - \sqrt{-1}\overline{z}_k \frac{\partial}{\partial \overline{w}_k} \right).
\end{aligned}$$

Since the coefficients of the holomorphic components $Y_i^{'}$ are holomorphic, the form σ on $\widetilde{\mathbf{E}}_S$ is a holomorphic form, and so we have a unique, nowhere vanishing holomorphic $4n$-form $\sigma_{\mathbb{H}}$ on $\widetilde{\mathbf{E}}_{\mathbb{H}}$ such that

$$\beta^*(\sigma_{\mathbb{H}}) = \sigma.$$

So

Proposition 3.1. *The nowhere vanishing global holomorphic $4n$-form $\sigma_{\mathbb{H}}$ gives a holomorphic trivialization of the canonical line bundle $K_{\mathbb{H}}^{\mathcal{G}} = \bigwedge^{4n} T^{*'}(\widetilde{\mathbf{E}}_{\mathbb{H}})$ of $\widetilde{\mathbf{E}}_{\mathbb{H}}$.*

Let η_i, $i = 1,2,3$, be smooth one-forms on S^{4n+3} defined by the conditions:

$$\begin{aligned}
&\eta_i(V_j) = \delta_{i\,j},\ i,\ j = 1,2,3, \\
&\eta_i(Y) = 0 \text{ for any } Y \text{ orthogonal to } V_i,\ i = 1,2,3.
\end{aligned}$$

Let π_S be the projection map $\pi_S : \mathbf{E}_S \to S^{4n+3}$ and decompose the one-forms $(\pi_S \circ (\tau_S)^{-1})^*(\eta_i)$ on $\mathbf{X}_S$ into holomorphic and anti-holomorphic components:

$$(\pi_S \circ (\tau_S)^{-1})^*(\eta_i) = \theta_i^{'} + \theta_i^{''} \in T^{*'}(\mathbf{E}_S) \oplus T^{*''}(\mathbf{E}_S) = T^*(\mathbf{E}_S) \otimes \mathbb{C}.$$

We denote the Liouville volume form on the cotangent bundle of the sphere and quaternion projective space by Ω_S and $\Omega_{\mathbb{H}}$ respectively, and denote $(\tau_S^{-1})^*(\Omega_S)$ and $(\tau_{\mathbb{H}}^{-1})^*(\Omega_{\mathbb{H}})$ with the same notations.

Proposition 3.2. ([3])

$$\Omega_S = \frac{C}{|\det(\theta_i^{'}(Y_j))|^2}\theta_1^{'} \wedge \theta_2^{'} \wedge \theta_3^{'} \wedge \theta_1^{''} \wedge \theta_2^{''} \wedge \theta_3^{''} \wedge \beta^*(\Omega_{\mathbb{H}}).$$

Remark 3.3. Since the constant C is proportional to the constant multiple of the Rieman metric on $P^n\mathbb{H}$ which we introduced through the Hopf fibration, we shall ignore it in this note.

Here we determine the coefficient $\det\left(\theta_i^{'}(Y_j)\right)$ explicitly.

For $X \in \mathfrak{sp}(1)$, let V_X be a vector field on $\mathbf{X}_S$ defined by

$$V_X(f)(W) = f(W \cdot \exp(tX))^{'}_{|t=0}.$$

V_X can be identified with the triple

$$\left(u \otimes 1 + v \otimes \sqrt{-1}, u \cdot X, v \cdot X\right) \in \mathbf{X}_S \times \mathbb{H}^{n+1} \times \mathbb{H}^{n+1},$$

where $W = (W_0, \ldots, W_n),\ W_i = u_i \otimes 1 + v_i \otimes \sqrt{-1}$. Let $\mathbf{J}$ be the almost complex structure on $\mathbf{X}_S$.

Then the vector field $\mathbf{J}(V_X)$ is identified with

$$\mathbf{J}(V_X) = \left(u \otimes 1 + v \otimes \sqrt{-1},\ -v \cdot X, u \cdot X\right).$$

We consider a curve $\left\{c(t)\right\}_{t\in\mathbb{R}}$ such that

$$c(t) = \left(u + v - v \cdot \exp(tX)\right) \otimes 1 + \left(v - u + u \cdot \exp(tX)\right) \otimes \sqrt{-1}.$$

By a direct calculation it turns out that $c(t) \in \mathbf{X}_S$ for any $t \in \mathbb{R}$, and we know by differentiating it at $t = 0$ the vector field

$$\left.\frac{d\,c(t)}{dt}\right|_{t=0} = \mathbf{J}(V_X). \tag{3.1}$$

Here we used the conditions $\|u\| = \|v\|$ and $< u, v >= 0$.

Next consider the curve on S^{4n+3}, $\left\{\pi_S \circ (\tau_S)^{-1}(c(t))\right\}$:

$$\pi_S \circ (\tau_S)^{-1}(c(t)) = \frac{u + v - v \cdot \exp(tX)}{\|u + v - v \cdot \exp(tX)\|}.$$

Again by differentiating this curve at $t = 0$ we have a tangent vector at a point $\frac{u}{\|u\|} \in S^{4n+3}$:

$$\left(\frac{u}{\|u\|}, -\frac{v \cdot X}{\|u\|} + \frac{< u, v \cdot X > u}{\|u\|^3}\right) \in T(S^{4n+3}) \subset S^{4n+3} \times \mathbb{H}^{n+1}.$$

Now we can express the values

$$\theta_i(Y_j^{'})\left(u \otimes 1 + v \otimes \sqrt{-1}\right) = \frac{1}{2}\delta_{ij} + \frac{< u \cdot \mathbf{e}_i,\ v \cdot \mathbf{e}_j >}{2\|u\|^2}\sqrt{-1}.$$

So we have

$$\left(\theta_i(Y_j')\right)$$

$$= \begin{pmatrix} \frac{1}{2} & \frac{\sqrt{-1}}{2\|u\|^2} < u\cdot \mathbf{e}_1, v\cdot \mathbf{e}_2 > & \frac{\sqrt{-1}}{2\|u\|^2} < u\cdot \mathbf{e}_1, v\cdot \mathbf{e}_3 > \\ \frac{\sqrt{-1}}{2\|u\|^2} < u\cdot \mathbf{e}_2, v\cdot \mathbf{e}_1 > & \frac{1}{2} & \frac{\sqrt{-1}}{2\|u\|^2} < u\cdot \mathbf{e}_2, v\cdot \mathbf{e}_3 > \\ \frac{\sqrt{-1}}{2\|u\|^2} < u\cdot \mathbf{e}_3, v\cdot \mathbf{e}_1 > & \frac{\sqrt{-1}}{2\|u\|^2} < u\cdot \mathbf{e}_3, v\cdot \mathbf{e}_2 > & \frac{1}{2} \end{pmatrix}$$

$$= \left(\frac{1}{2\|u\|^2}\right)^3 \begin{pmatrix} \|u\|^2 & \sqrt{-1} < u, v\cdot \mathbf{e}_3 > & -\sqrt{-1} < u, v\cdot \mathbf{e}_2 > \\ -\sqrt{-1} < u, v\cdot \mathbf{e}_3 > & \|u\|^2 & \sqrt{-1} < u, v\cdot \mathbf{e}_1 > \\ \sqrt{-1} < u, v\cdot \mathbf{e}_2 > & -\sqrt{-1} < u, v\cdot \mathbf{e}_1 > & \|u\|^2 \end{pmatrix}.$$

Hence

Proposition 3.4.

$$\det\left(\theta_i'(Y_j)\right) = \det\left(\theta_i(Y_j')\right) = \det\left(\theta_i'(Y_j')\right)$$

$$= \frac{1}{8\|u\|^4}\left(\|u\|^4 - \sum_{i=1}^{3} < u,\ v\cdot \mathbf{e}_i >^2\right) = \frac{1}{8\|u\|^4}\left(\|u\|^4 - \|(u,v)_{\mathbb{H}}\|^2\right).$$

4. Quantization Operators

In the paper [3] we constructed two operators $\widetilde{T}$ and T defined on a function space $\mathfrak{h}_{\mathbb{H}}^{\mathcal{G}}$ on the punctured cotangent bundle $T_0^*(P^n\mathbb{H})$ to $L_2(P^n\mathbb{H})$ by means of the method, pairing of the polarizations([8]). These operators give us exact quantizations of the geodesic flow. In this section we sum up these operators following [9] and [3] and prove our main theorem 4.2.

By the embedding

$$\tilde{\tau}_{\mathbb{H}} : T_0^*(P^n\mathbb{H}) \cong T_0(P^n\mathbb{H}) \to M(2n+2, \mathbb{C})$$

$$(P,Q) = \left((p_{ij}), (q_{ij})\right) \mapsto A = \|Q\|^2\left(\rho(p_{ij})\right) - \left(\rho(q_{ij})\right)^2 + \frac{\|Q\|}{2}\left(\rho(q_{ij})\right)\sqrt{-1}$$

we can introduce a Kähler structure, i.e. a positive complex polarization $\mathcal{G} \subset T^*(T_0^*(P^n\mathbb{H})) \otimes \mathbb{C}$, on $T_0^*(P^n\mathbb{H})$ whose Kähler form coincides with the natural symplectic form:

$$\omega_{\mathbb{H}} = d\,\theta_{P^n\mathbb{H}} = \tau_{\mathbb{H}}^*\left(2^{\frac{1}{4}}\sqrt{-1}\,\overline{\partial}\,\partial\sqrt{\|A\|}\right) = d\left(\tau_{\mathbb{H}}^*\left(2^{\frac{1}{4}}\sqrt{-1}\,\partial\sqrt{\|A\|}\right)\right), \tag{4.1}$$

and more precisely we have an expression for the canonical one-form $\theta_{P^n\mathbb{H}}$ on the punctured cotangent bundle $T_0^*(P^n\mathbb{H})$:

$$2^{\frac{3}{4}}\theta_{P^n\mathbb{H}} = \sqrt{-1}\tau_{\mathbb{H}}^*\left(\partial\sqrt{\|A\|} - \overline{\partial}\sqrt{\|A\|}\right). \tag{4.2}$$

Let $\mathbf{L}$ be the trivial line bundle on $T_0^*(P^n\mathbb{H})$ with the trivialization and the connection given by the relation (4.2). We can identify the space of parallel sections of the line bundle $\mathbf{L}\otimes\sqrt{K_{\mathbb{H}}^{\mathcal{G}}}$ and a space of holomorphic functions through a trivialization of this line bundle. In fact, the connection on the canonical line bundle $K_{\mathbb{H}}^{\mathcal{G}}$ along the polarization $\mathcal{G}$ (and that of the square root $\sqrt{K_{\mathbb{H}}^{\mathcal{G}}}$) is defined by making use of the homotopy formula of Lie derivative. The particular trivialization we use to identify the space of parallel sections of $\mathbf{L}\otimes\sqrt{K_{\mathbb{H}}^{\mathcal{G}}}$ is given by the nowhere vanishing global section which is the tensor product of a canonical section of $\mathbf{L}$ and the square root of the holomorphic global section $\sigma_{\mathbb{H}}$. Note that the canonical section of $\mathbf{L}$ incidentally determines a connection on $\mathbf{L}$ following the relation (4.2). Then we introduce the inner product on the function space on $T_0^*(P^n\mathbb{H})$. This is defined naturally by making use of the data obtained by determining the pairing $\sigma_{\mathbb{H}}\wedge\overline{\sigma_{\mathbb{H}}}$:

$$\sigma_{\mathbb{H}}\wedge\overline{\sigma_{\mathbb{H}}} = 2^{n-2}\|A\|^{2n+2}\Omega_{\mathbb{H}} \tag{4.3}$$

and is written as

$$(f,\ g)_{\widetilde{\mathbf{E}}_{\mathbb{H}}} = 2^{\frac{n-2}{2}}\int_{\widetilde{\mathbf{E}}_{\mathbb{H}}} f(A)\overline{g(A)}e^{-2\sqrt[4]{2}\pi\sqrt{\|A\|}}\|A\|^{n+1}|\Omega_{\mathbb{H}}|, \tag{4.4}$$

for functions $f,\ g$ on $T_0^*(P^n\mathbb{H})$.

The function space $\mathfrak{h}_{\mathbb{H}}^{\mathcal{G}}$ we are dealing with in the framework of geometric quantization theory is the completion of the space of polynomials on $M(2n+2,\mathbb{C})$ restricted to $\widetilde{\mathbf{E}}_{\mathbb{H}}$ with respect to this inner product (4.4).

Then, by calculating the pairing

$$(\pi_{\mathbb{H}}\circ\tilde{\tau}_{\mathbb{H}}^{-1})^*(\mathbf{v}_{\mathbb{H}})\wedge\sigma_{\mathbb{H}} = (\pi_{\mathbb{H}}\circ\tilde{\tau}_{\mathbb{H}}^{-1})^*(\mathbf{v}_{\mathbb{H}})\wedge\overline{\sigma_{\mathbb{H}}} = -\frac{1}{\sqrt{2\pi^2}}\|A\|\Omega_{\mathbb{H}}$$

we construct an operator T by taking a fiber integration of the map $\pi_{\mathbb{H}}$: $T_0^*(P^n\mathbb{H})\to P^n\mathbb{H}$:

$$\int_{P^n\mathbb{H}} f(P)T(g)(P)\mathbf{v}_{\mathbb{H}} - \frac{1}{\sqrt[4]{2}\pi}\int_{\widetilde{\mathbf{E}}_{\mathbb{H}}}(\pi_{\mathbb{H}}\circ\tilde{\tau}_{\mathbb{H}}^{-1})^*(f)(A)g(A)e^{-\sqrt[4]{2}\pi\sqrt{\|A\|}}\|A\|^{1/?}\Omega_{\mathbb{H}},$$

that is,

$$\begin{aligned} T &: \mathfrak{h}_{\mathbb{H}}^{\mathcal{G}}\to L_2(P^n\mathbb{H}) \\ T(f)\mathbf{v}_{\mathbb{H}} &= \frac{1}{\sqrt[4]{2}\pi}(\pi_{\mathbb{H}})_*(fe^{-\sqrt[4]{2}\pi\sqrt{\|A\|}}\|A\|^{1/2}\Omega_{\mathbb{H}}) \end{aligned}$$

and another one $\widetilde{T}:\mathfrak{h}_{\mathbb{H}}^{\mathcal{G}}\to L_2(P^n\mathbb{H})$ which is a composition of the pull-back by the map β and the quantization operator constructed for the sphere S^{4n+3} by the same method of pairing polarizations ([9],[3]).

Now we describe the operator

$$\widetilde{T}\circ T^{-1}: L_2(P^n\mathbb{H})\to L_2(P^n\mathbb{H})$$

on each eigenspace of the Laplacian $\Delta_{\mathbb{H}}$.

Let $\mathcal{S}_k$ $(k=0,1,\ldots)$ be the space of harmonic polynomials of degree k on $\mathbb{R}^{4n+4}\cong\mathbb{H}^{n+1}$, and denote by $\mathcal{S}_k^0$ a subspace of $\mathcal{S}_k$ consisting of polynomials which are invariant under the action of the group $\mathrm{Sp}(1)$ on $\mathbb{H}^{n+1}$ from the right. The

ℓ-th eigenspace H_ℓ of the Laplacian $\Delta_{\mathbb{H}}$ on $P^n\mathbb{H}$ is isomorphic to $\mathcal{S}^0_{2\ell}$ by the map $\pi_{\mathbb{H}} : S^{4n+3} \to P^n\mathbb{H}$,

$$\pi_{\mathbb{H}}^* : H_\ell \to \mathcal{S}^0_{2\ell},$$

and the corresponding eigenvalue λ_ℓ is $4\ell(2n+1+\ell)$. Note $\mathcal{S}^0_{2\ell+1} = \{0\}$ and

$$\dim H_\ell = \frac{2n(\ell+1)}{(2n+1)(\ell+2n)} \cdot (2\ell+2n+1) \cdot \left(\frac{\Gamma(\ell+2n+1)}{\Gamma(2n+1)\Gamma(\ell+2)}\right)^2 .$$

Each space H_ℓ is an invariant subspace of $\widetilde{T} \circ T$ and

Proposition 4.1. ([3]) *On the space H_ℓ the operator $\widetilde{T} \circ T^{-1}$ is equal to a constant:*

$$C \cdot \frac{(\ell+2n+1/4)(\ell+2n+3/4)}{(\ell+2n)(\ell+2n+1/2)} \frac{\Gamma(\ell+2n+1/4)\Gamma(\ell+2n+3/4)}{\Gamma(\ell+2n+1/2)^2}.$$

Here the constant C is common for all ℓ (see *Remark* 3.3), so that again we shall ignore it in the arguments below for the sake of simplicity.

Based on this expression we show

Theorem 4.2. *The operator $\widetilde{T} \circ T^{-1}$ is an elliptic pseudo-differential operator of order zero.*

To prove this, first we recall a criterion for an operator defined by functional calculus of a selfadjoint elliptic operator to be a pseudo-differential operator:

Proposition 4.3. ([13]) *Let A be a first order positive elliptic pseudo-differential operator of classical type defined on a closed manifold and let*

$$A = \int_0^{+\infty} \lambda dE_A(\lambda)$$

be the spectral decomposition of A. Let $f \in S^m_{1,0}(\mathbb{R})$ $(m \in \mathbb{R})$ be in the symbol class of Hörmander, i.e., $f \in C^\infty(\mathbb{R})$ satisfies the inequalities:

$$\left|\frac{d^k f(t)}{dt^k}\right| \le C_k(1+|t|)^{m-k},\ t \in \mathbb{R},$$

for each $k = 0,\ 1,\ 2, \ldots,$ with a suitable constant $C_k > 0$. Then

$$f(A) = \int_0^{+\infty} f(\lambda)dE_A(\lambda) = (2\pi)^{-1/2} \int_{-\infty}^{\infty} \hat{f}(t)e^{\sqrt{-1}tA}dt \tag{4.5}$$

is a pseudo-differential operator in $L^m_{1,0}$, where $\hat{f}(t) = (2\pi)^{-1/2} \int_{-\infty}^{\infty} e^{-\sqrt{-1}tx} f(x)dx$ is the Fourier transformation of f.

Let $\sigma_A : T_0^*M \to \mathbb{R}$ be the principal symbol of the operator A. Then the principal symbol $\sigma_{f(A)}$ of the operator $f(A)$ is given by

$$\sigma_{f(A)}(x,\xi) = f(\sigma_A(x,\xi)).$$

Let

$$\sqrt{\Delta_{\mathbb{H}}} = \int_0^\infty \lambda dE_{\mathbb{H}}(\lambda)$$

be the spectral decomposition of the square root of the Laplacian on $P^n\mathbb{H}$ and let

$$f_1(\lambda) = \frac{\left(\sqrt{\lambda^2 + (2n+1)^2} + 2n - 1/2\right)\left(\sqrt{\lambda^2 + (2n+1)^2} + 2n + 1/2\right)}{\left(\sqrt{\lambda^2 + (2n+1)^2} + 2n - 1\right)\left(\sqrt{\lambda^2 + (2n+1)^2} + 2n\right)}$$

and

$$f_2(\lambda) = \frac{\Gamma\left(\frac{\sqrt{\lambda^2+(2n+1)^2}+2n-1/2}{2}\right)\Gamma\left(\frac{\sqrt{\lambda^2+(2n+1)^2}+2n+1/2}{2}\right)}{\Gamma\left(\frac{\sqrt{\lambda^2+(2n+1)^2}+2n}{2}\right)^2},$$

then

$$\widetilde{T} \circ T^{-1} = C \int_0^\infty f_1(\lambda) f_2(\lambda) dE_{\mathbb{H}}(\lambda).$$

It will be easy to see that $f_1 \in S^0_{1,0}(\mathbb{R})$ and so we prove

Proposition 4.4. $f_2 \in S^0_{1,0}(\mathbb{R})$.

Proof. First note that

$$\lim_{\lambda \to \infty} f_2(\lambda) = 1.$$

The derivative of the function f_2 is of the form

$$\begin{aligned}\frac{df_2(\lambda)}{d\lambda} = f_2(\lambda) \cdot &\left(\frac{\lambda}{2\sqrt{\lambda^2 + (2n+1)^2}}\right) \\ \times &\left(\frac{\Gamma'\left(\frac{\sqrt{\lambda^2+(2n+1)^2}+2n-1/2}{2}\right)}{\Gamma\left(\frac{\sqrt{\lambda^2+(2n+1)^2}+2n-1/2}{2}\right)} - \frac{\Gamma'\left(\frac{\sqrt{\lambda^2+(2n+1)^2}+2n}{2}\right)}{\Gamma\left(\frac{\sqrt{\lambda^2+(2n+1)^2}+2n}{2}\right)}\right. \\ &\left. + \frac{\Gamma'\left(\frac{\sqrt{\lambda^2+(2n+1)^2}+2n+1/2}{2}\right)}{\Gamma\left(\frac{\sqrt{\lambda^2+(2n+1)^2}+2n+1/2}{2}\right)} - \frac{\Gamma'\left(\frac{\sqrt{\lambda^2+(2n+1)^2}+2n}{2}\right)}{\Gamma\left(\frac{\sqrt{\lambda^2+(2n+1)^2}+2n}{2}\right)}\right).\end{aligned} \tag{4.6}$$

Let

$$\phi(z) = \frac{d \log \Gamma(z)}{dz} = \frac{\Gamma'(z)}{\Gamma(z)},$$

then this has an expression:

$$\phi(z) = -C_{\mathbf{e}} + \sum_{n=0}^{\infty} \left(\frac{1}{n+1} - \frac{1}{n+z}\right),$$

where $C_{\mathbf{e}}$ is the Euler constant(see [1]).

By making use of this expression we have for $z >> 0$ and (locally fixed) $a > 0$

$$\phi(z+a) - \phi(z) = O(|z|^{-1}).$$

Applying this estimate to (4.6) we have

$$\frac{df_2(\lambda)}{d\lambda} = O(|\lambda|^{-1}).$$

Also with the expression ($k = 1, \ldots$):

$$\frac{d^k\phi(z)}{dz^k} = \sum_{n=0}^{\infty} \frac{(-1)^{k+1}k!}{(n+z)^{k+1}}$$

we have

$$\frac{d^k\phi}{dz^k}(z+a) - \frac{d^k\phi}{dz^k}(z) = O(|z|^{-k-1}).$$

Now by induction, for any $k \geq 0$

$$\frac{d^k f_2(\lambda)}{d\lambda^k} = O(|\lambda|^{-k})$$

which shows that $f_2 \in S^0_{1,0}$. □

Hence our operator $\widetilde{T} \circ T^{-1}$ is a pseudo-differential operator of order zero and the principal symbol $\sigma_{\widetilde{T}\circ T^{-1}}$ is given by

Proposition 4.5.

$$\sigma_{\widetilde{T}\circ T^{-1}}(P,Q) = C \cdot f_1\big(\|Q\|\big) f_2\big(\|Q\|\big), \ (P,Q) \in T_0^*(P^n\mathbb{H}).$$

5. Reproducing Kernels

We have also a Hilbert space $\mathfrak{h}_S^{\mathcal{G}}$ consisting of holomorphic functions on $\tilde{\tau}_S(T_0^*(S^{4n+3}))$, which is the completion of the space of polynomials on $M(2,\mathbb{C})^{n+1}$ restricted to $\tilde{\tau}_S(T_0^*(S^{4n+3}))$ ([9]). The inner product of this space is given by:

$$< f,\ g >_{\widetilde{\mathbf{E}}_S} = 2^2 \int_{\widetilde{\mathbf{E}}_S} f(B)\overline{g(B)} e^{-2\pi\|B\|} \|B\|^{2n+(1/2)} |\Omega_S|.$$

Note that the complement of $\tilde{\tau}(\mathbf{E}_S)$ in $\tilde{\tau}(T_0^*(S^{4n+3}))$ is measure zero with respect to the volume form $e^{-2\pi\|B\|}\|B\|^{2n+(1/2)}|\Omega_S|$.

This Hilbert space has the reproducing kernel $\mathfrak{K}_S(B,\ B')$:

$$\mathfrak{R}_S(B',\ B) = \sum_{\ell=0}^{\infty} \frac{s_\ell}{2^{2\ell}} \int_{S^{4n+3}} \Big(\sum_{i=0}^{n} \operatorname{tr} \rho(p_i) J^t \rho(z_i) J\Big)^\ell \Big(\sum_{i=0}^{n} \operatorname{tr} \rho(p_i) J^t \rho(\overline{w}_i) J\Big)^\ell \mathbf{v}_S,$$

where $B' = (\rho(z_0), \rho(z_1), \ldots, \rho(z_n)),\ B = (\rho(w_0), \rho(w_1), \ldots, \rho(w_n)) \in M(2,\mathbb{C})^{n+1}$ and the constants s_ℓ are given by

$$s_\ell = \frac{(2\pi)^\ell (\ell+n)\Gamma(\ell+2n)}{2\sqrt{\pi}\cdot \ell! \cdot \Gamma(\ell+2n+1/2)}.$$

Let ${}_pF_q\big(a_1, a_2, \dots, a_p; b_1, b_2, \dots, b_q; t\big)$ be a hypergeometric function:

$$ {}_pF_q\big(a_1, a_2, \dots, a_p; b_1, b_2, \dots, b_q; t\big) = \sum_{k=0}^{\infty} \frac{(a_1)_k \cdot (a_2)_k \cdots (a_p)_k}{(b_1)_k \cdot (b_2)_k \cdots (b_q)_k} \cdot \frac{t^k}{k!}, $$

where $(a)_k = a \cdot (a+1) \cdots (a+k-1)$ for non-negative a (and so on).

Let $G_S(t)$ be a function

$$ G_S(t) = \frac{\Gamma(2n)}{2\sqrt{\pi}\Gamma(2n+1/2)} \Big(t \cdot {}_1F_1(2n; 2n+1/2; t)' + n \cdot {}_1F_1(2n; 2n+1/2; t)\Big), $$

then the reproducing kernel $\mathfrak{R}_S$ is expressed as

$$ \mathfrak{R}_S(B',\ B) = \int_{S^{4n+3}} G_S\big(2\pi < p,\ z >< p,\ \overline{w} >\big) \mathbf{v}_S(p), \tag{5.1} $$

where $\rho(z) = B',\ \rho(w) = B \in M(2, \mathbb{C})^{n+1}$.

The Hilbert space $\mathfrak{h}^{\mathcal{G}}_{\mathbb{H}}$ has the reproducing kernel $\mathfrak{R}_{\mathbb{H}}(A,\ A')$ ([3]):

$$ \mathfrak{R}_{\mathbb{H}}(A',\ A) = \sum_{\ell=0}^{\infty} \frac{1}{b_\ell} \int_{P^n\mathbb{H}} < P,\ A' >_{\mathbb{C}}^{\ell} \cdot < P,\ \overline{A} >_{\mathbb{C}}^{\ell}\ \mathbf{v}_{\mathbb{H}},\ A,\ A' \in \widetilde{\mathbf{E}}_{\mathbb{H}} \tag{5.2} $$

where the coefficients b_ℓ are given by

$$ b_\ell = \pi^{-4\ell-3} \cdot \frac{1}{(2\ell+2n+1)^2} \cdot \frac{\Gamma(\ell+1)^2\Gamma(\ell+2)^2}{\Gamma(\ell+n+1/2)\Gamma(\ell+n+1)\Gamma(\ell+2n)\Gamma(\ell+2n+1)} \times \Gamma\left(\ell + \frac{6n+2}{4}\right) \Gamma\left(\ell + \frac{6n+3}{4}\right) \Gamma\left(\ell + \frac{6n+4}{4}\right) \Gamma\left(\ell + \frac{6n+5}{4}\right), $$

that is,

Proposition 5.1. *For $f \in \mathfrak{h}^{\mathcal{G}}_{\mathbb{H}}$,*

$$ f(A') - \int_{\widetilde{\mathbf{E}}_{\mathbb{H}}} \mathfrak{R}_{\mathbb{H}}(A',\ A) f(A) 2^{(2n-2)/2} e^{-2\sqrt[4]{2}\pi\sqrt{\|A\|}} \|A\|^{n+1} |\Omega_{\mathbb{H}}|. $$

Let $C_{\mathbb{H}}$ be the constant:

$$ C_{\mathbb{H}} = \pi^3 \frac{n! \cdot (2n-1)! \cdot (2n)! \cdot \Gamma(n+1/2)}{\Gamma\left(\frac{6n+2}{4}\right)\Gamma\left(\frac{6n+3}{4}\right)\Gamma\left(\frac{6n+4}{4}\right)\Gamma\left(\frac{6n+5}{4}\right)}, $$

and let the function $F_{\mathbb{H}}(t)$ be

$$ C_{\mathbb{H}} \cdot {}_4F_7\left(n+1/2, n+1, n+2, n+1; 1, 2, 2, \frac{6n+2}{4}, \frac{6n+3}{4}, \frac{6n+4}{4}, \frac{6n+5}{4}; t\right). $$

We denote by $G_{\mathbb{H}}(t)$ the function

$$ G_{\mathbb{H}}(t) = 4t^2 F''_{\mathbb{H}}(t) + 4(2n+1) F'_{\mathbb{H}}(t) + (2n+1)^2 F_{\mathbb{H}}(t), $$

then,

$$ \mathfrak{R}_{\mathbb{H}}(A',\ A) = \int_{P^n\mathbb{H}} G_{\mathbb{H}}\big(< P, A' >_{\mathbb{C}} \overline{< P, A >_{\mathbb{C}}}\big) \mathbf{v}_{\mathbb{H}}. \tag{5.3} $$

Now let $f \in \mathfrak{h}^{\mathcal{G}}_{\mathbb{H}}$, then we have a *week form* of a relation between these two reproducing kernels:

Theorem 5.2.

$$\beta^*(f)(B') \tag{5.4}$$
$$= \int_{\widetilde{\mathbf{E}}_S} \mathfrak{R}_S(B',\ B)\cdot\beta^*(f)(B)\cdot\left(e^{-2\pi\|B\|}\|B\|^{2n+(1/2)}\right)|\Omega_S|$$
$$= \int_{\widetilde{\mathbf{E}}_S} \mathfrak{R}_S(B',\ B)\cdot\beta^*(f)(B)\cdot\left(e^{-2\pi\|B\|}\|B\|^{2n+(1/2)}\left|\frac{8\|u\|^4}{(\|u\|^4-\|(u,v)_{\mathbb{H}}\|^2)}\right|^2\right)$$
$$\times|\theta_1'\wedge\theta_2'\wedge\theta_3'\wedge\theta_1''\wedge\theta_2''\wedge\theta_3''\wedge\beta^*(\Omega_{\mathbb{H}})|$$
$$= \int_{\widetilde{\mathbf{E}}_{\mathbb{H}}} \beta_*\left(\mathfrak{R}_S(B',\ \bullet)\cdot E(\bullet)\cdot\theta_1'\wedge\theta_2'\wedge\theta_3'\wedge\theta_1''\wedge\theta_2''\wedge\theta_3''\right)(A)\cdot f(A)\cdot|\Omega_{\mathbb{H}}|$$
$$= \int_{\widetilde{\mathbf{E}}_{\mathbb{H}}} \mathfrak{R}_{\mathbb{H}}(\beta(B'),\ A)\cdot\left(2^{(2n-2)/2}e^{-2\sqrt[4]{2}\pi\sqrt{\|A\|}}\|A\|^{n+1}\right)\cdot f(A)\cdot|\Omega_{\mathbb{H}}| = f(\beta(B')),$$

where in the expression above we put

$$E(B) = e^{-2\pi\|B\|}\|B\|^{2n+(1/2)}.\left|\frac{8\|u\|^4}{(\|u\|^4-\|(u,v)_{\mathbb{H}}\|^2)}\right|^2$$

and

$$B = (B_0,\dots,B_n) = \left(\rho\left(u_0\otimes 1+v_0\otimes\sqrt{-1}\right),\dots,\rho\left(u_n\otimes 1+v_n\otimes\sqrt{-1}\right)\right)$$

to express the fiber integration

$$\beta_*\left(\mathfrak{R}_S(B',\ \bullet)\cdot E(\bullet)\cdot\theta_1'\wedge\theta_2'\wedge\theta_3'\wedge\theta_1''\wedge\theta_2''\wedge\theta_3''\right)(A)$$
$$= \int_{\beta^{-1}(A)} \mathfrak{R}_S(B',\ B)\cdot e^{-2\pi\|B\|}\|B\|^{2n+(1/2)}.\left|\frac{8\|u\|^4}{(\|u\|^4-\|(u,v)_{\mathbb{H}}\|^2)}\right|^2$$
$$\times\,\theta_1'\wedge\theta_2'\wedge\theta_3'\wedge\theta_1''\wedge\theta_2''\wedge\theta_3''.$$

From the equality (5.4) we expect the strong equality might hold :

$$\mathfrak{R}_{\mathbb{H}}(\beta(B'),\ A) \tag{5.5}$$
$$= e^{2\sqrt[4]{2}\pi\sqrt{\|A\|}}\|A\|^{-n-1}\cdot\beta_*\left(\mathfrak{R}_S(B',\ \bullet)\cdot E(\bullet)\cdot\theta_1'\wedge\theta_2'\wedge\theta_3'\wedge\theta_1''\wedge\theta_2''\wedge\theta_3''\right)(A),$$

or further it might hold the equality

$$G_{\mathbb{H}}(<P,A'>_{\mathbb{C}}\overline{<P,A>_{\mathbb{C}}}) = e^{2\sqrt[4]{2}\pi\sqrt{\|A\|}}\|A\|^{-n-1}\cdot\beta_*\Big(\overline{G}(P,A',\ \bullet) \tag{5.6}$$
$$\times\,E(\bullet)\cdot\theta_1'\wedge\theta_2'\wedge\theta_3'\wedge\theta_1''\wedge\theta_2''\wedge\theta_3''\Big)(A),$$

where we express the fiber integration with respect to the map $\pi_{\mathbb{H}}$ as

$$(\pi_{\mathbb{H}})_*\left(G_S(2\pi<\bullet,\ z><\bullet,\ \overline{w}>)\cdot\eta_1\wedge\eta_2\wedge\eta_3\right)(P)$$
$$= \int_{\pi_{\mathbb{H}}(p)=P} G_S(2\pi<p,\ z><p,\ \overline{w}>)\cdot\eta_1(p)\wedge\eta_2(p)\wedge\eta_3(p)$$
$$= \overline{G}(P,\ A',\ A),$$

and $\rho(z) = B'$, $\rho(w) = B$, $\beta(B') = A'$, $\beta(B) = A$.

Note that $\eta_1 \wedge \eta_2 \wedge \eta_3 \wedge \pi_{\mathbb{H}}^*(\mathbf{v}_{\mathbb{H}}) = \mathbf{v}_S$.

For the moment it is not clear whether these (5.5) *and/or* (5.6) *hold or not. To prove it we need to know, for example, whether a L_2-holomorphic function on $\widetilde{\mathbf{E}}_{\mathbb{H}}$ is in $\mathfrak{h}_{\mathbb{H}}^{\mathcal{G}}$, i.e., such functions are always approximated by polynomials in $\mathfrak{h}_{\mathbb{H}}^{\mathcal{G}}$-norm or not? Or rather, relations among hypergeometric functions might give the equality* (5.6) *directly.*

Acknowledgment

The author would like to express his hearty thanks to Prof. Maurice and Mrs. Charlyne de Gosson and Prof. J. Toft (Växjö university) for their encouragement, kind hospitality and various advices during his stay at Blekinge institute of technology (Sweden) where a first version of this manuscript was prepared, and also to the referee for the valuable suggestions.

References

[1] G. E. Andrews, R. Askey and R. Roy, *Special functions*, Encyclopedia of Mathematics and Its Applications, Vol. **71**, Cambridge University Press, 1999.

[2] K. Furutani, A Kähler structure on the punctured cotangent bundle of the Cayley projective plane, *Math. Phy. Studies*, **24**, Kluwer Academic Publishers, 2003, 163–182.

[3] K. Furutani, Quantization of the Geodesic Flow on Quaternion Projective Spaces, *Ann. Global Anal. Geom.* **22**, No. 1 (2002), 1–27.

[4] K. Furutani and S. Yoshizawa, A Kähler structure on the punctured cotangent bundle of complex and quaternion projective spaces and its application to geometric quantization II, *Japan. J. Math.* **21** (1995), 355–392.

[5] K. Furutani and R. Tanaka, A Kähler structure on the punctured cotangent bundle of complex and quaternion projective spaces and its application to geometric quantization I, *J. Math. Kyoto Univ.* **34** (1994), 719–737.

[6] K. Ii, On a Bargmann-type transform and a Hilbert space of holomorphic functions, *Tôhoku Math. J.* **38** (1) (1986), 57–69.

[7] K. Ii and T. Morikawa, Kähler structures on the tangent bundle of Riemannian manifolds of constant positive curvature, *Bull. Yamagata Univ. Natur. Sci.* **14** (1999), 141–154.

[8] J. H. Rawnsley, Coherent states and Kähler manifolds, *Quart. J. Math. Oxford Ser.* **28** (1977), 403–415.

[9] ———, A non-unitary pairing of polarization for the Kepler problem, *Trans. Amer. Math. Soc.* **250** (1979), 167–180.

[10] J. M. Souriau, Sur la variété de Kepler, *Symposia Math.* **14** (1974), 343–360.

[11] R. Szőke, Adapted complex structures and geometric quantization, *Nagoya J. Math.* **154** (1999), 171–183.

[12] ———, Involutive structures on tangent bundles of symmetric spaces, *Math. Ann.* **319** (2001), 319-348.

[13] M. Taylor, *Pseudo-Differential Operators*, Princeton Mathematical Series **34**, Princeton University Press, 1981.

[14] N. M. J. Woodhouse, *Geometric Quantization*, 2nd Edition, Oxford Mathematical Monographs, Oxford, Clarendon Press, 1997.

Kenro Furutani
Department of Mathematics
Faculty of Science and Technology
Science University of Tokyo
2641 Noda, Chiba (278-8510)
Japan
e-mail: `furutani_kenro@ma.noda.tus.ac.jp`

Operator Theory:
Advances and Applications, Vol. 164, 153–159

Hudson's Theorem and Rank One Operators in Weyl Calculus

Joachim Toft

Abstract. A proof of Hudson's theorem in several dimension is presented. Some consequences for pseudo-differential operators of rank one are given.

Mathematics Subject Classification (2000). Primary 81R30, 35S05.

Keywords. Hudson's theorem, Wigner distributions, positivity, rank one operators.

0. Introduction and Preliminaries

In [5], Hudson proves that any Wigner distribution (which is closely related to Radar ambiguity functions and coherent state transformed functions) of two variables is nonnegative if and only if it is Gaussian. Some more general positivity results were thereafter obtained by Janssen in [8] and [9]. A direct proof of Hudson's theorem of arbitrary dimension was thereafter presented by Folland in [3], and Lieb remarked in [10] that Hudson's result was also a consequence of certain L^p inequalities for Wigner distributions and radar ambiguity functions. Later on, an alternative proof of Hudson's theorem was presented in [4] by Gröchenig.

In the present paper, we give, as in Section 2.5 in [15], an alternative proof of Hudson's theorem of arbitrary dimension. We also show some consequences for rank one operators in the Weyl calculus of pseudo-differential operators.

In order to describe the results more in details, we recall some facts. Assume that $f, g \in \mathscr{S}(\mathbf{R}^n)$. (We use the same notations for the usual function and distribution spaces as in [7].) Then the *Wigner distribution* $W_{f,g}$ *for* f *and* g is defined by the formula

$$W_{f,g}(x,\xi) \equiv (2\pi)^{-n/2} \int e^{i\langle y,\xi\rangle} f(x-y/2)\overline{g(x+y/2)}\, dy \tag{0.1}$$

Obviously, the map $(f,g) \mapsto W_{f,g}$ is continuous from $\mathscr{S}(\mathbf{R}^n) \times \mathscr{S}(\mathbf{R}^n)$ to $\mathscr{S}(\mathbf{R}^{2n})$. Moreover, by Fourier's inversion formula it follows that $(f,g) \mapsto W_{f,g}$ extends to a continuous map from $\mathscr{S}'(\mathbf{R}^n)\times\mathscr{S}'(\mathbf{R}^n)$ to $\mathscr{S}'(\mathbf{R}^{2n})$, and from $L^2(\mathbf{R}^n)\times L^2(\mathbf{R}^n)$ to $L^2(\mathbf{R}^{2n})$ and

$$\|W_{f,g}\|_{L^2}\le \|f\|_{L^2}\|g\|_{L^2}.$$

(See [3].)

It was proved by Hudson in [5] that if $f\in L^2(\mathbf{R})$, then $W_{f,f}$ is nonnegative if and only if f is Gaussian. Recall that f is a Gaussian means that $f(x)=Ce^{-Q(x)}$, where C is a complex number and Q is a polynomial of degree 2 such that $\operatorname{Re}Q(x)\to+\infty$ as $|x|\to\infty$. In [5] it also was remarked that if instead $f\in\mathscr{S}'(\mathbf{R})$ then $W_{f,f}$ is a non-negative measure if and only if f is Gaussian, $f=c\,\delta_y$ or $f=c$ for some (complex) constant c and $y\in\mathbf{R}$.

In the next section we present a generalization of this result. In particular we prove that if $f,g\in L^2(\mathbf{R}^n)$, then $W_{f,g}$ is nonnegative if and only if $g=c\,f$ is Gaussian for some constant $c\ge 0$.

Thereafter we present some applications to the Weyl calculus of pseudo-differential operators. These investigations are based on the fact that the symbol of a Weyl operator of rank one is a Wigner distribution. More precisely, assume that $a\in\mathscr{S}(\mathbf{R}^{2n})$. Then the Weyl quantization $a^w(x,D)$ for a is defined by the formula

$$a^w(x,D)f(x)=(2\pi)^{-n}\iint a((x+y)/2,\xi)f(y)e^{i\langle x-y,\xi\rangle}\,dyd\xi,$$

when $f\in\mathscr{S}(\mathbf{R}^n)$. The map $a^w(x,D)$ is continuous on $\mathscr{S}(\mathbf{R}^n)$, and extends to a continuous map from $\mathscr{S}'(\mathbf{R}^n)$ to $\mathscr{S}(\mathbf{R}^n)$. Moreover, the definition of $a^w(x,D)$ extends to any $a\in\mathscr{S}'(\mathbf{R}^{2n})$, and then $a^w(x,D)$ is continuous from $\mathscr{S}(\mathbf{R}^n)$ to $\mathscr{S}'(\mathbf{R}^n)$. (See [7].)

As a consequence of Fourier's inversion formula it follows that for any $g,h\in L^2(\mathbf{R}^n)$, we have

$$a^w(x,D)f(x)=(f,h)g(x)$$

for every $f\in L^2(\mathbf{R}^n)$ (e.i. $a^w(x,D)$ is a rank-one operator on L^2), if and only if $a=(2\pi)^{n/2}W_{g,h}$. Here $(\cdot,\cdot)$ denotes the usual scalar product on L^2. Consequently, Hudson's theorem is deeply related to certain properties for Weyl operators of rank one. In the last part of the next section we present some of these properties.

1. Nonnegative Wigner Distribution

In this section we give a proof of Hudson's theorem in several dimensions. We also discuss consequences for Weyl operators of rank one.

Before stating the main result, it is convenient to make the following definition.

Definition 1.1. The tempered distribution f on $\mathbf{R}^n$ is called *pseudo-Gaussian* if there is a subspace V of $\mathbf{R}^n$ and an element $y_0\in V^\perp$ such that $f(x_1,x_2)=f_0(x_1)\delta_{y_0}(x_2)$, where $x_1\in V$, $x_2\in V^\perp$ and $f_0(x_1)=Ce^{-Q(x_1)}$ for some complex number C and Q is a polynomial of degree 2 such that $\operatorname{Re} Q$ is bounded from below.

If in addition $V = \mathbf{R}^n$ here above, then f is called *semi-Gaussian.*

We note that f is Gaussian if and only if it is semi-Gaussian and that $f(x) \to 0$ as x turns to infinity. We also note that if f is pseudo-Gaussian, then there is a sequence f_j for $j \ge 1$ of Gaussians such that $f_j \to f$ in $\mathscr{S}'$ as j turns to infinity.

The main result is the following.

Theorem 1.2. *Assume that $f, g \in \mathscr{S}'(\mathbf{R}^n) \setminus \{0\}$. Then the following are true:*

1. *$W_{f,g}$ is a positive measure, if and only if f is pseudo-Gaussian and $g = c\,f$ for some constant $c > 0$;*
2. *if in addition $f, g \in L^1_{loc}$, then $W_{f,g}$ is a positive measure if and only if f is semi-Gaussian and $g = c\,f$ for some constant $c > 0$;*
3. *if in addition $f, g \in L^p(\mathbf{R}^n)$ for some $1 \le p < \infty$, then $W_{f,g}$ is a positive measure, if and only if f is Gaussian and $g = c\,f$ for some constant $c > 0$.*

Remark 1.3. In view of 3 in Theorem 1.2 we note that if $f \in L^p(\mathbf{R}^n)$ and $g \in L^{p'}(\mathbf{R}^n)$, where $1/p + 1/p' = 1$, then an application of Hölder's inequality shows that $W_{f,g}$ is a continuous function. If in addition $1 < p < \infty$, then $W_{f,g}$ vanishes at infinity. (See also [10].) Hence if $p = 2$, then statement 3 in Theorem 1.2 can be reformulated as

3. if in addition $f, g \in L^2(\mathbf{R}^n)$, then $W_{f,g}$ is a nonnegative function, if and only if f is Gaussian and $g = c\,f$ for some constant $c > 0$.

We need some preparations for the proof of Theorem 1.2. We start with the following lemma. Here and in what follows, we let the Fourier transform be defined as

$$\widehat{f}(\xi) = (\mathscr{F} f)(\xi) = (2\pi)^{-n/2} \int f(x) e^{-i\langle x, \xi\rangle}\, dx$$

when $f \in \mathscr{S}(\mathbf{R}^n)$. Then $\mathscr{F}$ is continuous on $\mathscr{S}$ which extends to a continuous mapping on $\mathscr{S}'$ and to a unitary mapping on L^2.

Lemma 1.4. *Let $f, g \in \mathscr{S}'(\mathbf{R}^n)$. Then*

$$W_{\widehat{f},\widehat{g}}(\xi, x) = W_{f,g}(-x, \xi). \tag{1.1}$$

If $a(x)$ and $b(\xi)$ are positive definite quadratic forms on $\mathbf{R}^n$ and

$$\phi_{a,b}(x) = e^{-a(x)} e^{-b(D)} \phi(x)$$

when $\phi \in L^2(\mathbf{R}^n)$, then

$$W_{f_{a,b}, g_{a,b}}(x, \xi) = e^{-2a(x)} e^{-a(D_\xi)/2} e^{-2b(\xi)} e^{-b(D_x)/2} W_{f,g}(x, \xi). \tag{1.2}$$

Proof. We may assume that $f, g \in \mathscr{S}$. Formula (1.1) is well-known and follows from straight-forward computations if we express f in terms of its Fourier transform in (0.1). We omit the details.

Let f_a and g_a be the functions obtained when f and g are multiplied by $e^{-a(x)}$. Since $a(x-y/2)+a(x+y/2)=2a(x)+a(y)/2$, it follows that

$$W_{f_a,g_a}(x,\xi)=(2\pi)^{-n/2}e^{-2a(x)}\int e^{i\langle y,\xi\rangle}e^{-a(y)/2}f(x-y/2)\overline{g(x+y/2)}\,dy,$$

i. e.

$$W_{f_a,g_a}(x,\xi)=e^{-2a(x)}e^{-a(D_\xi)/2}W_{f,g}(x,\xi). \tag{1.2$'$}$$

Since $e^{-b(D)}$ is multiplication by $e^{-b(\xi)}$ on the Fourier transform side it also follows that

$$W_{e^{-b(D)}f,e^{-b(D)}g}(x,\xi)=e^{-2b(\xi)}e^{-b(D_x)/2}W_{f,g}(x,\xi). \tag{1.2$''$}$$

The result now follows by combining (1.2)″ with (1.2)′. □

Next we recall that a function f is called *positive definite* when f is continuous, and that for each pairs of finite sets $\{x_i\}_{i\in I}\subseteq \mathbf{R}^n$ and $\{c_j\}_{j\in I}\subseteq \mathbf{C}$ we have

$$\sum_{j,k\in I} f(x_j-x_k)c_j\overline{c_k}\ge 0. \tag{1.3}$$

In particular, it follows from (1.3) that if f is positive definite, then

$$\overline{f(-x)}=f(x),\quad \text{and}\quad |f(x)|\le f(0). \tag{1.4}$$

Positive definite functions were completely characterized in [1] by Bochner who proved the following. (See also [11] or [17].)

Lemma 1.5. *Assume that f is continuous on $\mathbf{R}^n$. Then the following conditions are equivalent:*

1. *f is positive definite;*
2. *$\widehat{f}$ is a positive measure of finite mass.*

Proof of Theorem 1.2. It suffices to prove the necessity of 1. In the first part of the proof we assume that f and g belong to $L^2(\mathbf{R}^n)\cap \mathbf{A}$, where $\mathbf{A}$ is the space of all functions ϕ which extend to entire analytic functions on $\mathbf{C}^n$ and satisfy an inequality of the form

$$|\phi(z)|=|\phi(x+iy)|\le F(|y|)e^{-|x|},$$

for some continuous function F on $\mathbf{R}$. Thereafter we consider the general case and reduce ourself to the first part by considering $f_{a,b}/\|f_{a,b}\|_{L^2}$ and $g_{a,b}/\|g_{a,b}\|_{L^2}$.

Assume therefore that $f,g\in L^2\cap\mathbf{A}$ and that $W_{f,g}\ge 0$. It is no restriction to assume that $\|f\|_2=\|g\|_2=1$. The condition that $W_{f,g}\ge 0$ means that $y\mapsto f(x-y)\overline{g(x+y)}$ is a positive definite function for every x in view of Lemma 1.5. Hence (1.4) gives

$$f(x-y)\overline{g(x+y)}=\overline{f(x+y)}g(x-y) \tag{1.5}$$

and

$$|f(x-y)g(x+y)|\le f(x)\overline{g(x)}. \tag{1.6}$$

Since $\|f\|_2=\|g\|_2$ it is obvious that (1.5) implies that $g=cf$, where $c\in\mathbf{C}$ and $|c|=1$. Since the right-hand side of (1.6) is nonnegative it follows that $c=1$.

Hence $g = f$. Since f is entire, the function $y \mapsto f(x-y)f(x+y)$ is not identically equal to 0 when $x \in \mathbf{R}^n$. Hence it follows from (1.6) that $|f(x)| > 0$ for every x. If we take $x - y$ and $x + y$ as new independent variables in (1.6) we see that this inequality means that

$$|f(x)||f(y)| \leq |f((x+y)/2)|^2.$$

Hence

$$h((x+y)/2) \leq \frac{1}{2}h(x) + \frac{1}{2}h(y),$$

where $h(x) = -\log|f(x)|$. It follows that $h(x)$ is a smooth and convex function.

We shall use the decay and analyticity assumptions on f again. Let $f^*(z) = \overline{f(\bar{z})}$ be the entire analytic function which equals the conjugate of f on $\mathbf{R}^n$. Then $f^* \in \mathbf{A}$, and in the integral defining $W_{f,f}(x,\xi)$ we can make a change of contour of integration allowing us to replace y by $y + 2iy_0$ in the formula for the integrand, where y_0 may be an arbitrary real vector. This shows that

$$e^{2\langle y_0,\xi\rangle}W_{f,f}(x,\xi) = W_{f_{y_0},f_{y_0}}(x,\xi),$$

where $f_{y_0}(x) = f(x - iy_0)$. Since the left-hand side is nonnegative we draw the same conclusions for f_{y_0} as for f. Replacing y_0 by y in our notation and setting $h(z) = -\log|f(z)|$ we have proved that $h(z)$ is pluri-harmonic with the property that $x \mapsto h(x+iy)$ is convex for every y. We want to prove that this implies that h is a polynomial of degree ≤ 2.

Let $x \in \mathbf{R}^n$ and set $p(\sigma) = h(\sigma x)$ when $\sigma = s + it \in \mathbf{C}$. Then $p(\sigma)$ is harmonic and convex when σ is restricted to any line where t is constant. Hence $(d/ds)^2 p \geq 0$. But it follows from the maximum principle that every harmonic function ≥ 0 on $\mathbf{C}$ is constant. Hence $h(sx)$ is a polynomial in s of degree ≤ 2 for every x, which implies that $h(x)$ is a polynomial of degree ≤ 2. This shows that f is a Gaussian.

In the general case, when no additional assumptions are made on f and g, we let a and b be positive definite quadratic forms on $\mathbf{R}^n$ and define $f_{a,b}$ and $g_{a,b}$ as in Lemma 1.4. Since $e^{-b(D)}$ is convolution by a Gaussian it is true that $f_{a,b}, g_{a,b} \in \mathbf{A}$, and $W_{f_{a,b},g_{a,b}} \geq 0$ in view of (1.2). It follows from the first part of the proof therefore that $f_{a,b}$ is a Gaussian and $g = f$. Hence $e^{-b(D)}f$ is semi-Gaussian. Since the Fourier transform of a pseudo-Gaussian is a pseudo-Gaussian we know therefore that $e^{-b(\xi)}\widehat{f}(\xi)$ is a pseudo-Gaussian. Hence $\widehat{f}$ is a pseudo-Gaussian. The result now follows by applying the inverse Fourier transform. The proof is complete. □

Next we present some applications to rank one operators in Weyl calculus.

Proposition 1.6. *Assume that $a^w(x,D)$ is a rank one operator on L^2. Then $a \geq 0$ if and only if a is Gaussian.*

Note here that $a \geq 0$ in Proposition 1.6 makes sense, since a is a continuous function in view of the Remark 1.3.

Proof. It follows from the introduction that $a = W_{f,g}$ for some $f, g \in L^2$. Since $W_{f,g}(x,\xi)$ can be interpreted as a partial Fourier transform of $f(x-y/2)\overline{g(x+y/2)}$ with respect to the y-variable it follows that f and g are Gaussians if and only if a is Gaussian. Hence Theorem 1.2 and Remark 1.3 shows that $a \geq 0$ if and only if a is Gaussian. □

From Proposition 1.6 we may now derive the following.

Theorem 1.7. *Assume that $a^w(x,D)$ is an orthonormal projection of rank one on L^2. Then $\|a\|_{L^1} \geq 1$ with equality if and only if a is Gaussian.*

Proof. From the assumptions it follows that $a = (2\pi)^{n/2}W_{f,f}$ for some unit vector $f \in L^2$. We may assume that $a \in L^1$. By straight-forward computations it follows that

$$\begin{aligned} \iint a(x,\xi)\,dxd\xi &= \iiint f(x-y/2)\overline{f(x+y/2)}e^{i\langle y,\xi\rangle}\,dydxd\xi \\ &= \int |f(x)|^2\,dx = 1. \end{aligned}$$

Here the second equality follows from Fourier's inversion formula. Hence $\|a\|_{L^1} \geq 1$ with equality if and only if $a \geq 0$, or, equivalently, if and only if a is Gaussian. □

Acknowledgements I would like to express my sincere gratitude to my supervisor Professor Anders Melin for his invaluable support as well as fruitful and stimulating discussions during my time as Ph.D. student. I also thank the referee for valuable comments, leading to improvements of the paper.

References

[1] S. Bochner, *Vorlesungen über Fouriersche Integrale*, Akademie Verlag, 1932.

[2] N. G. De Bruijn, A theory of generalized functions, with applications to Wigner distribution and Weyl correspondence, *Nieuw Archief voor Wiskunde* **21** (1973), 205–280.

[3] G. B. Folland, *Harmonic Analysis in Phase Space*, Princeton University Press, Princeton, 1989.

[4] K. H. Gröchenig, *Foundations of Time-Frequency Analysis*, Birkhäuser, Boston, 2001.

[5] R. L. Hudson, When is the Wigner quasi-probability density non-negative, *Rep. Math. Phys.* **6** (1974), 249–252.

[6] L. Hörmander, The Weyl calculus of pseudo-differential operators, *Comm. Pure. Appl. Math.* **32** (1979), 359–443.

[7] ———, *The analysis of linear partial differential operators I, III*, Springer-Verlag, Berlin Heidelberg NewYork Tokyo, 1983, 1985.

[8] A. J. E. M. Janssen,A note on Hudson's theorem about functions with nonnegative Wigner distributions, *SIAM J. Math. Anal.* **15** (1984), 170–176.

[9] ———, Bilinear phase-plane distributions functions and positivity, *J. Math. Phys.* **26** (1985), 1986–1994.

[10] E. H. Lieb, Integral bounds for radar ambiguity functions and Wigner distributions, *J. Math. Phys.* **31** (1990), 594–599.

[11] M. Reed and B. Simon, *Scattering Theory*, Academic Press, London New York, 1979.

[12] B.-W. Schulze, *Boundary Value Problems and Singular Pseudo-Differential Operators*, Wiley, Chichester, 1998.

[13] M. A. Shubin, *Pseudodifferential Operators and Spectral Theory*, Springer-Verlag, Berlin, 1987.

[14] B. Simon, *Trace Ideals and Their Applications* I, London Math. Soc. Lecture Note Series, Cambridge University Press, Cambridge, London, New York, Melbourne, 1979.

[15] J. Toft, *Continuity and positivity problems in pseudo-differential calculus*, Thesis, Department of Mathematics, University of Lund, Lund, 1996.

[16] ———, Continuity properties for non-commutative convolution algebras with applications in pseudo-differential calculus, *Bull. Sci. Math.* **126** (2) (2002), 115–142.

[17] ———, Positivity properties for non-commutative convolution algebras with applications in pseudo-differential calculus, Bull. Sci. Math. **127** (2) (2003), 101–132.

[18] M. W. Wong, *Weyl Transforms*, Springer-Verlag, 1998.

Joachim Toft
Department of Mathematics and Systems Engineering
Växjö University
Sweden
e-mail: `joachim.toft@vxu.se`

Operator Theory:
Advances and Applications, Vol. 164, 161–172

Distributions and Pseudo-Differential Operators on Infinite-Dimensional Spaces with Applications in Quantum Physics

Andrei Khrennikov

To Marillia

Abstract. This paper is the review on theory of infinite-dimensional pseudo-differential operators(PDO) and their application to quantization of systems with the infinite number of degrees of freedom. There are considered various approaches to the theory of infinite-dimensional PDO (e.g., Berezin's approach based on polynomial operators). There is presented in details the calculus of PDO based on the theory of distributions on infinite-dimensional spaces (general locally convex spaces). This calculus is based on the "Feynman measure" on the phase space (introduced by Smolyanov in 80th). Symbols of the most important PDO in the representation of second quantization are calculated.

Mathematics Subject Classification (2000). Primary 47G30; Secondary 58D30.

Keywords. Distributions, pseudo-differential operators, Feynman measure, infinite-dimensional phase space, quantum physics.

1. Introduction

Since Weyl's book [1], pseudo-differential operators have been actively used in quantum mechanics. The procedure of operator quantization is based on the correspondence between classical observables, functions $a(q,p)$ on the phase space $Q \times P$, where $Q = P = R^n$, and quantum observables, pseudo-differential operators (PDO) $\hat{a}$ with symbols given by classical observables. In [1] there were considered PDO with Weyl's (or symmetric) symbols. Later there were considered

---

This paper was partly supported by EU-Network "QP and Applications" and Nat. Sc. Found., grant N PHY99-07949 at KITP, Santa-Barbara, and visiting professor fellowship at Russian State Humanitarian University.

various classes of symbols, qp, pq, Wick, anti-Wick and general τ-symbols, see, e.g., [2].

Quantum field theory is a theory with infinite number of degrees of freedom. It is natural to try to generalize Weyl's idea on quantization with the aid of PDO to systems with the infinite number of degrees of freedom. This is one of the reasons for the development of the theory of PDO on infinite-dimensional spaces. Of course, this problem is also very interesting from the purely mathematical viewpoint and infinite-dimensional PDO can be used in theories of distributions and partial differential equations on infinite-dimensional spaces (e.g., Hilbert or Banach spaces, or locally convex spaces of test functions or generalized functions), cf., [3]-[6]. However, personally I was motivated merely by quantum applications and this induces a rather special structure of this review.

We recall [2] that a PDO A on a space $L_2(R^n, dx)$ with τ symbol $A(q,p)$ is the integral operator

$$Af(q) = \frac{1}{(2\pi)^n} \int dpdq' \exp\{i(p, q - q')\} A((1-\tau)q + \tau q', p) f(q'). \tag{1.1}$$

There is no translation invariant measure ("Lebesgue" or "Haar" measure) in the infinite-dimensional case. The expression $dp\, dq$ can not be defined in the measure theoretic framework. Therefore PDO are usually introduced either as the limits of finite-dimensional operators [3], or on the basis of polynomial operators [7]–[9] or by means of the theory of distributions on infinite-dimensional spaces [10] (see also [11]–[14] and, at the physical level of rigor, [9]).

Finally, we remark that in theory of superfields and superstrings one needs PDO in spaces of functions (and distributions) depending on infinite number of commuting and anti-commuting variables: $x_i x_j = x_j x_i, \theta_i \theta_j = -\theta_j \theta_i$. Theory of such "super-PDO" was developed by author [14]. In theory of superstrings there are also used models over fields of p-adic numbers [15]. The corresponding theory of PDO was developed in [16], [17].

In the present paper, we also use the theory of distributions; in section 2, we construct functional spaces (consisting of the Fourier transforms of distributions) that possess a remarkable property:

Every formula of PDO theory in R^n (see, for example, [2]) is also true for infinite-dimensional PDO in these spaces when the measure

$$exp\{i\alpha(p, q)\}dpdq$$

is replaced by the "Feynman measure" in the phase space; in particular, the PDO with τ symbol is here defined by means of such a substitution in equation (1.1).

2. Feynman Measure in the Phase Space

If X is a locally convex space (LCS) and Y its dual, then for every $y \in Y$ the symbol f denotes the function on X defined by

$$f_y(x) = \exp\{i(y, x)\};$$

the symbol T denotes the linear hull of the set $\{f_y : y \in Y\}$, the symbol Ψ_X denotes LCS of functions on X containing T as a dense subspace and such that the space Ψ'_X dual to it separates the points of Ψ_X. The space Ψ'_X is equipped with a locally convex topology consistent with the duality between Ψ'_X and Ψ_X. By the Fourier transform of an element $\lambda \in \Psi'_X$ we define the function $\mathbf{F}(\lambda)$ on Y determined by the equation

$$\mathbf{F}(\lambda)(y) = (\lambda, f_y).$$

It is assumed that the space $\Phi_Y = \mathbf{F}(\Psi'_X)$, which consists of functions on Y that are the images under the mapping $\mathbf{F}$ of the elements in Ψ'_X, is equipped with a topology for which $\mathbf{F}$ is a homeomorphism.

Elements of the space dual to Φ_Y (which is denoted by the symbol $\mathbf{M}_Y$) are called distributions on Y. For a distribution $\mu \in \mathbf{M}_Y$ the Fourier transform is defined as the function $\mathbf{F}'(\mu)$ on X, where $\mathbf{F}' : \mathbf{M}_Y \to \Psi_X$ is the operator that is the adjoint to $\mathbf{F} : \Psi'_X \to \Phi_Y$; then (by definition) *Parseval's equation* holds:

If $\mu \in \mathbf{M}_Y$ and $\Psi \in \Phi_Y$, then $(\mu, \varphi) = (\mathbf{F}^{-1}(\varphi), \mathbf{F}'(\mu))$.

For $\lambda \in \Psi'_X$ and $\psi \in \Psi_X, \mu \in \mathbf{M}_Y$ and $\varphi \in \Phi_Y$, we set $\int d\lambda(x)\psi(x) = (\lambda, \psi), \int \varphi(y)\mu(dy) = (\mu, \varphi)$.

If E is an LCS, then by E_C we shall denote its complexification. A function $f : E_C \to C$ is said to be entire if it is continuous on every compact set of space E_C and for all $e_1, e_2 \in E_C$ the function $f_{e_1,e_2}(z) = f(e_1 + ze_2)$, $z \in C$, is entire. We denote by $\mathbf{A}(E_C)$ the space of entire functions on E_C equipped with the topology of uniform convergence on compact sets of the space E_C.

In all that follows below, it is assumed that the LCS X is complete and possesses the approximation property (this last means that the identity operator belongs to the closure in the compact convergence topology of the set of linear continuous operators with finite-dimensional images). In the scheme presented above, we set $\Psi_X = \mathbf{A}(X_C)$ (see [11], [12]).

Definition 2.1. (see [11], [12], cf. [18]–[24]). By the "Feynman measure" on the space Y with mean value $a \in Y$ and correlation functional $-2b$, where b is a complex-valued quadratic form on X that is continuous on compact sets, we mean the distribution $\gamma_{b,a} \in \mathbf{M}_Y$ having the Fourier transform

$$\exp\{b(x,x) + i(a,x)\}.$$

Everywhere below, P is a complete bornological space (e.g., Frechet space), $Q = P'$, the space Q being equipped with the Mackey topology; it is assumed that the spaces P and Q possesses the approximation property. For example, $P = \mathbf{S}(R^n)$ is the space of infinitely differentiable rapidly decreasing functions with the ordinary topology.

Definition 2.2. By the Smolyanov distribution ("Feynman measure" on the phase space) we shall mean the Feynman measure in the space $Q \times P$ with correlation functional b_α determined by the equation

$$b_\alpha(p \oplus q, p \oplus q) = -2\alpha i(q,p), \alpha \in C,$$

and mathematical expectation $a = q \oplus p$.

Such a distribution will be denoted by the symbol $v^\alpha_{q\oplus p}$. A "Feynman measure" in the phase space (the Smolyanov distribution) was introduced for the first time (at least in the mathematical literature) by means of Parseval's equation in [10].

Note (see [11], [12]) that the space Φ_Y of functions integrable with respect to the Feynman measure consists of the Fourier transforms of measures with compact supports on the space X_C (in particular, functions in Φ_Y are entire on Y_C of first order of growth). For the Smolyanov distribution, the class of integrable functions can be significantly extended by means of the following proposition:

Proposition 2.3. *Let* $\lambda \in \Psi'_P$, $\theta \in \Psi'_Q$, $a = \mathbf{F}(\lambda), b = \mathbf{F}(\theta)$, *and let* $I = \int a(q')b(p')v^\alpha_{q\oplus p}(dq'dp')$. *Then*

$$I = \int d\lambda(p') \exp\{i(q,p')\} a(\alpha p' + p); \tag{2.1}$$

$$I = \int d\theta(q') \exp\{i(p,q')\} b(\alpha q' + q). \tag{2.2}$$

Using (2.1), we extend the integral to functions of the form $f(q,p) = \sum_{k=1}^n a_k(q)b_k(p)$, where $a_k \in \Phi_Q$ and $b_k \in \mathbf{A}(P_C)$; if $a_k \in \mathbf{A}(Q_C), b_k \in \Phi_P$, then it is possible to use (2.2) (for these integrals, we shall use the same symbol as above).

Remark 2.4. The space $\mathbf{A}$ is chosen only for the sake of definiteness. One can integrate similarly functions $f(q,p) = a(q)b(p)$, where, for example, a is a smooth or continuous function, and b is the Fourier transform of a functional on a space of smooth or continuous functions (see [24]). Thus, by restrictions on one of the factors one can eliminate practically all the restrictions from the other.

3. Algebra of Infinite-Dimensional Pseudo-Differential Operators

The PDO A with τ symbol $A(q,p) \in \Phi_{Q\times P}$ on the space Φ_Q is determined by the equation ($\tau \in C$)

$$A_\psi(q) = \int A((1-\tau)q + \tau q', p)\varphi(q')v^1_{q\oplus 0}(dq'dp) \tag{3.1}$$

(if $\tau = 0, 1$, and $\frac{1}{2}$, then the τ symbol will be called, respectively, the qp, pq and Weyl symbol).

Proposition 3.1. *The PDO A is a continuous operator on the space* Φ_Q.

Everywhere below, for $\varphi = \mathbf{F}(\lambda)$ we set $\tilde{\varphi} = \lambda$.

Theorem 3.2. *(Formula of composition). If $A = A_1 \circ A_2$ and $A_1(q,p), A_2(q,p)$, and $A(q,p)$ are τ symbols of the PDO A_1, A_2, and A, respectively, then*

$$A(q,p) = \int\int A_1(q',p')A_2(q'',p'')v_{q\oplus p}^{-\tau}(dq'dp'')v_{q\oplus p}^{1-\tau}(dq''dp'). \qquad (3.2)$$

Proof. We denote the Feynman integral in (3.2) by $B(q,p)$. Calculating this integral, we find that

$$\begin{aligned} B(q,p) &= \int\int d\tilde{A}_1 \otimes \tilde{A}_2(p',q',p'',q'') \\ &\quad \times \exp\{-i\tau(q'',p') + i(1-\tau)(q',p'')\} + i(q'+q'',p) + i(q,p'+p'') \end{aligned}$$

(the direct product $\otimes$ of the elements $\lambda_1, \lambda_2 \in \Psi'$ is defined in the natural manner, see [11], [12], [24]). We introduce the linear continuous operator $L_\tau : \mathbf{A}(P_C \times Q_C) \to \mathbf{A}((P_C \times Q_C)^2)$:

$$L_{\tau,\varphi}(p_1,q_1,p_2,q_2) = \exp\{-i\tau(q_2,p_1) + i(1-\tau)(q_1,p_2)\}\varphi(p_1+p_2,q_1+q_2).$$

Then $B(q,p) = \mathbf{F}L'_\tau(\tilde{A}_1 \otimes \tilde{A}_2)(q,p)$, where L'_τ is the operator that is the adjoint of L_τ. Therefore, the function $B(q,p) \in \Phi_{Q\times P}$ and there is defined the PDO B with τ symbol $B(q,p)$. Note that

$$B\varphi(q) = \int dL'_\tau(\tilde{A}_1, \otimes\tilde{A}_2) \otimes \tilde{\varphi}(p_1,q_1,p_2)\exp\{i(q,p_1) + i\tau(q_1,p_1) + i\tau(q_1+q,p_2)\}$$

Similarly,

$$\begin{aligned} A\varphi(q) &= \int d\tilde{A}_1, \otimes\tilde{A}_2 \otimes \tilde{\varphi}(p_1,q_1,p_2,q_2,p_3) \\ &\quad \times \exp\{i(q,p_1) + i\tau(q_1,p_1) + i(q_1+q,p_2+p_3) + i\tau(q_2,p_2) + i(q_2,p_3)\} \\ &= \int dL'_\tau(\tilde{A}_1 \otimes \tilde{A}_2) \otimes \tilde{\varphi}(p_1,q_1,p_2)\exp\{i(q,p_1) + i\tau(q_1,p_1) + i(q_1+q,p_2)\} \\ &= B\varphi(q). \end{aligned}$$

□

We prove similarly

Theorem 3.3. *(Connection between symbols of an operator). If $A_\tau(q,p)$ and $A_t(q,p)$ are the τ and t symbols of the PDO A, then*

$$A_t(q,p) = \int A_\tau(q',p')v_{q\oplus p}^{\tau-t}(dq'dp'). \qquad (3.3)$$

It follows from (3.1) that for the PDO A with τ symbol $A(q,p)$ there holds the representation

$$A\varphi(q) = \int d\tilde{A} \otimes \tilde{\varphi}(p_1,q_1,p_2)\exp\{i(q,p_1) + i\tau(q_1,p_1) + i(q_1+q,p_2)\}. \qquad (3.4)$$

In particular, for PDO with pq and qp symbols we obtain

$$A\varphi(q) = \int d\tilde{A} \otimes \tilde{\varphi}(p_1,q_1,p_2)\exp\{i(q_1+q,p_1+p_2)\}; \qquad (3.5)$$

$$A\varphi(q) = \int d\tilde{\varphi}(p) \exp\{i(q,p)\} A(q,p). \tag{3.6}$$

Proposition 3.4. *If A is a PDO with qp symbol $A(q,p)$, then*

$$A(q,p) = \exp\{-i(p,q)\} A f_p(q). \tag{3.7}$$

For every $r \in P$ and every $s \in Q$ we define the operators $q(r)$ and $p(s)$ as the PDO with τ symbols (q,r) and (p,s), respectively. Then

$$q(r)\varphi(q) = (q,r)\varphi(q), \quad p(s)\varphi(q) = -i(\varphi'(q), s).$$

Theorem 3.5. *If A is the PDO with τ symbol $A(q,p)$, then for every $\varphi \in \Phi_Q$*

$$A\varphi = \int d\tilde{A}(r,s) \exp\{i\tau(s,r) + iq(r)\} \exp\{ip(s)\}\varphi;$$

$$A\varphi = \int d\tilde{A}(r,s) \exp\{i(\tau-1)(s,r) + ip(s)\} \exp\{iq(r)\}\varphi;$$

$$A\varphi = \int d\tilde{A}(r,s) \exp\{i(\tau-1/2)(s,r) + ip(s) + q(r))\}\varphi.$$

In the space of distribution $\mathbf{M}_Q$ the PDO with τ symbol $A(q,p) \in \Phi_{Q\times P}$ is introduced as the operator that is the adjoint to the PDO in the space of functions Φ_Q with $(1-\tau)$ symbol $A(q-p)$. It follows from the properties of adjoint operators that Theorems 3.2 and 3.3 hold for PDO on the space of distributions $\mathbf{M}_Q$.

4. Pseudo-Differential Operators on the Schrödinger Space

In all that follow below, it is assumed that the LCS P is continuously and densely embedded in the separable Hilbert space H and $H' \equiv H \subset Q$. By γ we denote the Gaussian Radon measure on Q with zero mean and covariation operator $\beta/2$, it being assumed that $\beta(P) \subset P$ and the operators $\beta, \beta^{-1} : P \to P$ are continuous. We denote the space $L_2(Q, d_\gamma)$ by $\mathbf{H}_Q$ and, following [25], shall call it the Schrödinger space (the field operators act on $\mathbf{H}_Q$ as multiplication operators). We embed the space Φ_Q in $\mathbf{H}_Q$ and we shall call the PDO A in Φ_Q with τ symbol $A(q,p) \in \Phi_{Q\times P}$, regarded as an operator on $\mathbf{H}_Q$, the PDO on the space $\mathbf{H}_Q$ with Schrödinger τ symbol ($\equiv S-\tau$ symbol) $A(q,p)$. On functions of the class Φ_Q, this operator is given by (3.1); by virtue of the definition, all propositions of section 3 are also valid for PDO with $S-\tau$ symbols.

Theorem 4.1. *(Symbol of adjoint operator). If A is the PDO with $S-\tau$ symbol $A(q_1,q_2) \in \Phi_{Q\times Q}(\subset \Phi_{Q\times P})$, then the Hilbert adjoint operator A^* is the PDO with $S-(1-\tau)$ symbol $A^*(q_1,q_2) \in \Phi_{Q\times Q}$, and*

$$\bar{A}^*(q_1,q_2) = \int d\tilde{A}(p_1,p_2) \exp\{(1-2\tau)(\beta^{-1}p_2,p_2) + i(q_1, p_1 - 2i\beta^{-1}p_2) + i(q_2,p_2)\}. \tag{4.1}$$

For the Weyl-Schrödinger symbol ($\tau = \frac{1}{2}$), formula (4.1) simplifies considerably: $\bar{A}^*(q_1, q_2) = A(q_1, q_2 - 2i(\beta^{-1})'q_1)$, where $(\beta^{-1})' : Q \to Q$ is the operator that is the adjoint to $\beta^{-1} : P \to P$.

Theorem 4.2. *(Normal symbol). If A is the PDO in the space $\mathbf{H}_Q$ with $S - qp$ symbol $A(q,p) \in \Phi_{Q\times P}$, then for the normal (Wick) symbol $A_n(\bar{v}, v), \bar{v}, v \in P_C$, the following equation holds:*

$$A_n(\bar{v}, v) = \int A(q, -i\sqrt{2}\beta^{-1}v)$$

$$\times \exp\{-2(\beta^{-1}\text{Re v}, \text{Re v}) + 2\sqrt{2}(\text{q}, \beta^{-1}\text{Re v})\}\gamma(\text{dq}). \tag{4.2}$$

Formula (4.1), defining the PDO, corresponds to decomposition of the operator in the space $\mathbf{H}_Q$ with respect to the field operators $q(r)$ and the operators $p(s) = -i\mathbf{D}_s$, see [25]. For every $s \in P$, the momentum operators $\pi(s)$ and the operators of annihilation $a(s)$ and creation $a^*(s)$ are determined by the equations

$$\pi(s) = p(s) + iq(\beta^{-1}s),$$

$$a(s) = [q(\beta^{-1/2}s) + i\pi(\beta^{-1/2}s)]/\sqrt{2}, \quad a^*(s) = [q(\beta^{-1/2}s) - i\pi(\beta_s^{1/2})]/\sqrt{2}.$$

We remark that $a(s) = ip(\beta_s^{1/2})/\sqrt{2}$. The $S - \tau$ symbols of these operators have the form

$$\pi(s)(q,p) = (s,p) + (q, \beta^{-1}s),$$

$$a(s)(q,p) = i(\beta^{-1/2}s, p)/\sqrt{2}, \quad a^*(s)(q,p) = [2(q, \beta^{-1/2}s) - i(\beta^{-1/2}s, p)]/\sqrt{2}.$$

A PDO theory in $\mathbf{H}_Q$ corresponding to decomposition of operators with respect to the *field operators* $q(r)$ and the *momentum operators* $\pi(s)$ can be presented in the following form. Note that the PDO A on the space $L_2(R^n, dx)$ (see(1)) is unitarily equivalent to the operator on $L_2(R^n, d\gamma)$ (which we also denote by A) determined by the equation

$$A\varphi(q) = \int d\tilde{A}(p_1, p_2) \exp\{i(q + \tau p_2, p_1) - (q, \beta^{-1}p_2) - (\beta^{-1}p_2, p_2)/2\}\varphi(q + p_2). \tag{4.3}$$

If $A(q_1, q_2) \in \Phi_{Q\times Q}$, then this formula is also well defined in the infinite-dimensional case (for $\varphi \in \Phi_Q$). We shall call the operator A determined by (4.3) the PDO on $\mathbf{H}_Q$ with τ symbol $A(q_1, q_2)$. For these PDO, formulas (3.2) and (3.3) and the following theorems hold.

Theorem 4.3. *(Adjoint operator). If A is the PDO on $\mathbf{H}_Q$ with τ symbol $A(q_1, q_2) \in \Phi_{Q\times Q}$, then the operator A^* is the PDO on $\mathbf{H}_Q$ with $1 - \tau$ symbol $\bar{A}(q_1, q_2)$.*

Theorem 4.4. *(Normal symbol). Let A be the PDO on $\mathbf{H}_Q$ with Weyl symbol $A(q_1, q_2) \in \Phi_{Q\times Q}$. Then for the normal symbol $A_n(\bar{v}, v)$ the following equation holds:*

$$\begin{aligned} A_n(\bar{v}, v) = \int\int A(q_1, (\beta^{-1})'q_2) \exp\{-2(\beta^{-1}v, \bar{v}) + 2\sqrt{2}((q_1, \beta^{-1}\text{Re v}) \\ +(q_2, \beta^{-1}\text{Im v}))\}\gamma\otimes\gamma(\text{dq}_1, \text{dq}_2). \end{aligned} \tag{4.4}$$

Using the composition formula (3.2), from Theorems 4.2 and 4.3 we can obtain formulas that express the normal symbol in terms of the $S-\tau$ symbol and τ symbol.

In [7], Weyl (symmetric) symbols were defined in terms of normal symbols by means of the formula that is the inverse of (4.4). This made it possible to extend the definition of a Weyl symbol to non-polynomial operators. In [8], Berezin posed the problem of extending the concept of qp and pq symbols in the infinite-dimensional case to nonpolynomial operators. The algebra of PDO on the space $\mathbf{H}_Q$ (which is isomorphic to the Fock space) with qp and pq symbols of the class $\Phi_{Q\times Q}$ can be regarded as one of the solutions to this problem.

We remark that the τ symbol of the regularization of a quantum-field Hamiltonian has the following form: quadratic form + constant. When the regularization is lifted, the constant tends to infinity.

Finally, we formulate a theorem on the connection between a symbol and the symbol of the Schrödinger operator.

Theorem 4.5. *Let $A_1(q_1,q_2)$ and $A_2(q_1,q_2)$ be, respectively, the Weyl symbol and Weyl-Schrödinger symbol of the operator A. Then*

$$A_2(q_1,q_2) = A_1(q_1,q_2 + i(\beta^{-1})'q_1).$$

5. On Extension of the Class of Operators Having Schrödinger Symbol

Let A be a linear operator on the Schrödinger space $\mathbf{H}_Q$, with the domain of definition of the operator A containing all functions $f_p(\cdot) = \exp\{i(\cdot,p)\}, p \in P_C$. We introduce the $S-qp$ symbol $A(q,p)$ of the operator A by means of (3.7) (by virtue of Proposition 3.4, this definition agrees with the one used above). For every $p \in P_C$, the function $A(\cdot,p) \in \mathbf{H}_Q$. Note that, as for PDO with symbols of the class $\Phi_{Q\times P,}$ the normal symbol of the operator A is related to the $S-qp$ symbol by formula (4.2).

The function $f: P_C \to \mathbf{H}_Q$ is said to be entire if it is continuous on compact sets and for any $h \in \mathbf{H}_Q$ the function $p \to (f(p),h)_{\mathbf{H}_Q}$ is entire.

Proposition 5.1. *If the operator A on the space $\mathbf{H}_Q$ is continuous, then the function $p \to A(\cdot,p)\exp\{i(\cdot,p)\}$ is entire and for any $\varphi \in \Phi_Q$ the integral equation* (9) *holds.*

The integral in formula (3.3) is understood in the weak sense: for every $h \in \mathbf{H}_Q$,

$$\int A\varphi(q)\bar{h}(q)\gamma(dq) = \int d\tilde{\varphi}(p)\int A(q,p)\exp\{i(q,p)\}\bar{h}(q)\gamma(dq).$$

The proof of the proposition consists of justifying the manipulation

$$A(\int d\tilde{\varphi}(p)f_p(q) = \int d\tilde{\varphi}(p)A(f_p)(q) = \int d\tilde{\varphi}(p)A(q,p)f_p(q).$$

We denote by $\mathbf{U}_{Q\times P}$ the space of measurable functions $A(q,p)$ such that for almost all q the function $(q,\cdot)$ is entire and for any positive finite measure μ on the space P_C

$$\int\int |A(q,p)|^2 \exp\{-2(q,\mathrm{Im\ p})\}\gamma\otimes\mu(\mathrm{dq\ dp}) < \infty.$$

If $A(q,p)\in\mathbf{U}_{Q\times P}$, then (3.6) defines a linear operator on the space $\mathbf{H}_Q$ with domain of definition Φ_Q. We shall call such operators PDO with S-qp symbols of the class $\mathbf{U}_{Q\times P}(\supset\Phi_{Q\times P})$.

Proposition 5.2. *Let $A = A_1\circ A_2$ and $A(q,p), A_1(q,p), A_2(q,p)$ be the $S-qp$ symbols of these operators. If A_1 is PDO with $S-qp$ symbol of the class $\mathbf{U}_{Q\times P}$ and for every $p\in P_C$ the function $A_2(\cdot,p)\in\Phi_Q$, then the symbols of the operators A, A_1, A_2 are related by equation* (3.2).

Proof.

$$\begin{aligned} A(f_p)(q) &= A_1(A_2(\cdot,p)f_p(\cdot))(q) = \int d\mathbf{F}_q^{-1}(A_2(\cdot,p)f_p(\cdot))(p')\times A(q,p')f_{p'}(q)\\ &= \int d\mathbf{F}_q^{-1}(A_2(\cdot,p))(p')A_1(q,p+p')f_{p+p'}(q), \end{aligned}$$

where we denote by symbol $\mathbf{F}_q^{-1}$ the inverse Fourier transformation with respect to the variable q. □

6. Second Quantization Operators

We denote by H_0 the completion of the space P in the norm $|p|^2 = (p,p) = (\beta p,p)$; let λ be a contractive operator on H_0. We calculate the S-qp symbol of the operator $\Gamma(\lambda)$ (for the definition of the operator $\Gamma(\lambda)$, see, for example, [26]. Using equations (1.16) and (1.33a) of [26], we obtain

$$\Gamma(\lambda)f_p(q) = \exp\{i(q,\lambda p) - < (1-\lambda*\lambda)p,p > /4\}.$$

Thus, for the $S-qp$ symbol of the operator $\Gamma(\lambda)$ the equation

$$\Gamma(\lambda)(q,p) = \exp\{i((q,(\lambda-1)p) - (\beta p,p)/4 + (\beta\lambda p,\lambda p)/4\}$$

holds.

The function $\Gamma(\lambda)(q,p)$ belongs to the space $\mathbf{U}_{Q\times P}$ not only for contractive operators λ but also for any continuous operator $\lambda : P\to P$ (and, therefore, defines a PDO on $\mathbf{H}_Q$; we shall also denote this PDO by $\Gamma(\lambda)$).

In all that follows below, we shall assume that the constant $c_0\in C$ is such that for (fixed) operator λ there is defined a function $u(t) = \exp\{c_0 t\lambda\}$ with values in the space of linear continuous operators on the LCS P_C, the function $u(t)$ being moreover strongly continuous.

Using the equation $\exp\{tc_0 d\Gamma(\lambda)\} = \Gamma(\exp\{c_0 t\lambda\})$ (where $d\Gamma(\lambda)$ is the second quantization of the operator λ, see [26], for the $S - qp$ symbol of the evolution operator $\exp\{c_0 t d\Gamma(\lambda)\}$ we obtain the formula

$$e^{c_0 t d\Gamma(\lambda)}(q,p) = \exp\{i(q,(e^{c_0 t\lambda} - 1)p) - < (1 - e^{2c_0 t\lambda})p, p > /4\}.$$

To find the $S - qp$ symbol of the operator $d\Gamma(\lambda)$, we note that

$$[d\Gamma(\lambda)]^n(q,p) = c_0^{-n}\frac{d^n}{dt^n}\exp\{c_0 t d\Gamma(\lambda)\}(q,p)|_{t=0}.$$

In particular, $d\Gamma(\lambda)(q,p) =< \lambda p, p > /2 + i(q, \lambda p)$.

The operators $d\Gamma(\lambda)$ and $\exp\{c_0 t d\Gamma(\lambda)\}$ are PDO on the Schrödinger space with $S - qp$ symbols of the class $\mathbf{U}_{Q\times P}$.

If $P = \mathbf{S}(R^3), \beta = \mu^{-1}$, where $\mu = (-\Delta + m_0^2)^{1/2}$, then the PDO $d\Gamma(\mu)$ is the Hamiltonian of the free field, $d\Gamma(1)$ is the particle number operator, and $\exp\{itd\Gamma(\mu)\}$ and $\exp\{-td\Gamma(\mu)\}$ are evolution operators [25], [26].

7. Pseudo-Differential Operators in Superanalysis and Non-Archimedean Analysis

Methods developed in the theory of infinite-dimensional PDO have unexpected applications in various domains of mathematics and theoretical physics. The crucial point is that in the infinite-dimensional case there was developed a calculus of PDO [10]–[13], [24] which is not based on any concrete measure. Instead of the Lebesgue measure on the phase space, there was considered the Smolyanov distribution ("Feynman measure" on the phase-space). This trick can be used to develop the theory of PDO on other algebraic structures in the cases when it is impossible to introduce an analogue of Lebesgue measure. One of such applications was done in superanalysis [14]. In so called functional approach to superanalysis [27], [28] and [29], [14], it is possible to define the Fourier transform in spaces of super analytic functions. In the framework of the correspoding theory of distributions there was defined the "Feynman measure" on the phase-superspace and the formalism of infinite-dimensional PDO [11], [12] was practically copied to the supercase [29]–[32], [14]. This formalism is especially interesting in the case of the infinite number of commuting and anticommuting variables (infinite-dimensional superspace) and it has applications to the theory of quantum superfileds and superstrings (including field theory for superstrings). Last years there was intensively developed p-adic and general non-Archimedean physics, see e.g. [15]–[17]. In particular, there were developed quantum models with wave functions taking values in non-Archimedean fields. Corresponding theory of PDO was proposed in [16], [17] and based on an analog of "Feynman measure" on the phase-space. Theory of non-Archimedean infinite-dimensional PDO was developed in [33]. In this paper this theory was applied to quantum field theory; there were defined quantum field Hamiltonians for interacting non-Archimedean quantum fields (see also [34] for recent development of p-adic dynamical systems).

Acknowledgment

Different aspects of the theory of infinite-dimensional PDO and their applications to quantum physics were discussed with O. G. Smolyanov, E. T. Shavgulidze, V. P. Maslov, S. Albeverio, L. Hörmander, L. Gärding, A. Melin; PDO in superanalysis – with V. S. Vladimirov, I. V. Volovich, B. De Witt, C. De Witt-Morette. I would like to thank all these people for fruitful discussions.

References

[1] H. Weyl, *The Theory of Groups and Quantum Mechanics,* Dover, New-York, 1931.

[2] M. A. Shubin, *Pseudo-Differential Operators and Spectral Theory,* English Transl., Springer Verlag, 1986; "Nauka", Moscow, 1978.

[3] P. M. Blekher and M. I. Vishik, On a class of pseudo-differential operators with infinitely many variables and applications, *Math. USSR Sb.* **15** (1971), 446–494.

[4] O. G. Smolyanov, Linear differential operators in spaces of measures and functions on Hilbert space, *Uspekhi Mat. Nauk.* **28** (1973), 251–252.

[5] Yu. M. Berezanskij, *Self Adjoint Operators in Spaces of Functions of an Infinite Number of Variables,* Amer. Math. Soc., Providence, R. I., 1986.

[6] A. Yu. Khrennikov, Equations with infinite-dimensional pseudo-differential operators, *Dokl. Akad. Nauk SSSR* **267** (1982), 1313–1318.

[7] F. A. Berezin, *The Method of Second Quantization,* Academic Press, 1966.

[8] F. A. Berezin, Representation of operators by means of functionals, *Trudy Moscow. Mat. Obschch.* **17** (1967), 117–196.

[9] G. S. Agarwal and E. Wolf, Calculus for functions of noncommuting operators and general phase-space methods in quantum mechanics I, II, III, *Phys. Rev. D* (3) **2** (1970), 2161–2186, 2187–2205, 2206–2225.

[10] O. G. Smolyanov, Infinite-dimensional pseudo-differential operators and Schrödinger quantization, *Dokl. Akad. Nauk SSSR* **263** (1982), 558–562.

[11] A. Yu. Khrennikov, The Feynman measure in phase space and symbols of infinite-dimensional pseudo-differential operators, *Math. Notes* **37** (1985), 734–742.

[12] A. Yu. Khrennikov, Second quantization and pseudo-differential operators, *Theoret. Math. Phys.* **66** (1986), 339–349.

[13] O. G. Smolyanov and A. Yu. Khrennikov, An algebra of infinite-dimensional pseudo-differential operators, *Soviet Math. Dokl.* **35** (1987), 1310–1314.

[14] A.Yu. Khrennikov, *Superanalysis,* Nauka, Fizmatlit, Moscow, 1997 (in Russian). English Translation: Kluwer, Dordreht, 1999.

[15] V. S. Vladimirov , I. V. Volovich and E. I. Zelenov, *p-adic Analysis and Mathematical Physics,* World Scientific, Singapore, 1993.

[16] A. Yu. Khrennikov, *p-Adic Valued Distributions and Their applications to the Mathematical Physics,* Kluwer, Dordreht, 1994.

[17] A. Yu. Khrennikov, *Non-Archimedean Analysis: Quantum Paradoxes, Dynamical Systems and Biological Models,* Kluwer, Dordreht, 1997.

[18] C. De Witt-Morette, Feynman's path integral. Definition without limiting procedure, *Comm. Math. Phys.* **28** (1972), 47–67.

[19] S. A. Albeverio and R. J. Höegh-Krohn, *Mathematical theory of Feynman path integrals,* Lecture Notes in Math., **523**, Springer-Verlag, Berlin-Heidelberg, 1976.

[20] A. A. Slavnov and L. D. Faddeev, *Introduction to the Quantum Theory of Gauge Fields,* Nauka, Moscow, 1978, English Transl., *Gauge Fields,* Benjamin/Cummings, Reading, Mass., 1980.

[21] A. V. Uglanov, Feynman measures with correlation operators of indefinite sign, *Soviet Math. Dokl.* **25** (1982), 37–40.

[22] O. G. Smolyanov and A. Yu. Khrennikov, The central limit theorem for generalized measures on infinite-dimensional spaces, *Soviet Math. Dokl.* **31** (1985), 279–283.

[23] A. Yu. Khrennikov, Integration with respect to generalized measures on linear topological spaces, *Trans. Moscow Math. Soc.* **49** (1987), 113–129.

[24] A. Yu. Khrennikov, Infinite-dimensional pseudo-differential operators, *Izv. Akad. Nauk USSR, ser. Math.* **51** (1987), 46–68.

[25] J. Glimm and A. Jaffe, Boson quantum field models, in *Mathematics of Contemporary Physics, Proc. Instructional Conf.*, Editor: R. F. Streater, Academic Press, 1972, 77–143.

[26] B. Simon, *The $P(\Phi)_2$ Euclidean (Quantum) Field Theory,* Princeton University Press, Princeton, N. J., 1974.

[27] V. S. Vladimirov and I. V. Volovich, Superanalysis I: Differential calculus, *Theoret. Math. Phys.* **59** (1984), 3–27.

[28] V. S. Vladimirov and I. V. Volovich, Superanalysis II: Integral calculus, *Theoret. Math. Phys.* **60** (1984), 169–198.

[29] A. Yu. Khrennikov, Superanalysis: theory of generalized functions and pseudo-differential operators, *Theoret. Math. Phys.* **72** (1987), 420–429.

[30] A. Yu. Khrennikov, Pseudo-differential equations in functional superanalysis, 1. Method of Fourier transform, *Diff. Equations* **24** (1988), 2144–2154.

[31] A. Yu. Khrennikov, Pseudo-differential equations in functional superanalysis, 2. Formula of Feynman-Kac, *Diff. Equations* **25** (1989) 314–324.

[32] A. Yu. Khrennikov, Equations on the superspace, *Izvestia Academii Nauk USSR, ser. Math.* **54** (1990), 556–606.

[33] S. Albeverio and A. Yu. Khrennikov, p-adic Hilbert space representation of quantum systems with an infinite number of degrees of freedom, *Int. J. of Modern Phys. B* **10** (1996), 1665–1673.

[34] A. Yu. Khrennikov and M. Nilsson, *p-Adic Deterministic and Random Dynamical Systems.* Kluwer, Dordreht, 2004.

Andrei Khrennikov
International Center
for Mathematical Modeling
in Physics and Cognitive Sciences
University of Växjö
Sweden
e-mail: `Andrei.Khrennikov@msi.vxu.se`

Operator Theory:
Advances and Applications, Vol. 164, 173–192

Ultradistributions and Time-Frequency Analysis

Nenad Teofanov

Abstract. The aim of the paper is to show the connection between the theory of ultradistributions and time-frequency analysis. This is done through time-frequency representations and modulation spaces. Furthermore, some classes of pseudo-differential operators are observed.

Mathematics Subject Classification (2000). Primary 46F05, 47G30; Secondary 35S05, 44A05.

Keywords. Ultradistributions, modulation spaces, pseudo-differential operators, time-frequency representations.

1. Introduction

The aim of the paper is to illustrate the natural connection between time-frequency analysis and the theory of (tempered) ultradistributions. This is done by the use of the so called time-frequency representations, Theorems 3.8 and 3.9. Another result is the definition of certain Gelfand-Shilov type spaces within the framework of modulation spaces, Proposition 4.3.

Pseudo-differential operators serve as yet another example which demonstrates how the combination of the techniques may lead to new results, Theorem 5.5. They are a traditional tool in micro-local analysis, [6], [35], [41], [53]. However, the use of the techniques of time-frequency analysis in the study of pseudo-differential operators lead to a new insight into their properties. Not only some classical results are generalized, e.g. the Calderon–Vaillancourt theorem, but also the scope of the theory has been enlarged. More concretely, operators whose symbols are non smooth, or the ones whose symbols may grow faster than polynomials are studied by the use of time-frequency analysis techniques, [21],[23], [24], [48], [49].

This research was supported by MNZŽS of Serbia.

The paper is organized as follows. In Section 2 we introduce the notation. A survey of test function spaces is given in Section 3. Subsection 3.3 contains main results of the first part of the paper. Modulation spaces and their connection to certain test function spaces is given in Section 4. In the last Section we study the action of pseudo-differential operators on modulation spaces.

2. Basic Notation

By $\mathbf{N}$, $\mathbf{Z}$, $\mathbf{R}$ and $\mathbf{C}$, we denote sets of positive integers, integers, real numbers and complex numbers respectively, and $\mathbf{N}_0 = \mathbf{N} \cup \{0\}$. By $\mathbf{R}^d$ ($\mathbf{N}_0^d$), $d \in \mathbf{N}$, we denote set of $d-$dimensional real numbers (nonnegative integers). Euclidean norm of $x \in \mathbf{R}^d$ is given by $|x| = \left(x_1^2 + \cdots + x_d^2\right)^{1/2}$, and $\langle x \rangle = (1+|x|^2)^{1/2}$. For multiindices $\alpha, \beta \in \mathbf{N}_0^d$, we have $|\alpha| = \alpha_1 + \cdots + \alpha_d$, $\alpha! = \alpha_1! \cdots \alpha_d!$, $x^\alpha = x_1^{\alpha_1} \cdots x_d^{\alpha_d}$, $x \in \mathbf{R}^d$, and, for $\beta \leq \alpha$, i.e. $\beta_j \leq \alpha_j$, $j \in \{1, 2, \ldots, d\}$, $\binom{\alpha}{\beta} = \binom{\alpha_1}{\beta_1} \cdot \cdots \cdot \binom{\alpha_d}{\beta_d}$. The letter C denotes a positive constant, not necessarily the same at every appearance. Dual pairing between a test function space $\mathcal{A}$ and its dual $\mathcal{A}'$ is denoted by $\langle \cdot, \cdot \rangle = {}_{\mathcal{A}'}\langle \cdot, \cdot \rangle_{\mathcal{A}}$. By $\check{f}$ we denote the reflection $\check{f}(x) = f(-x)$. The operator of partial differentiation D is given by

$$D^\alpha = D_x^\alpha = D_1^{\alpha_1} \cdots D_d^{\alpha_d} = \left(\frac{1}{2\pi i}\frac{\partial}{\partial x^1}\right)^{\alpha_1} \cdots \left(\frac{1}{2\pi i}\frac{\partial}{\partial x^d}\right)^{\alpha_d}$$

for all multiindices $\alpha \in \mathbf{N}_0^d$ and all $x = (x_1, \ldots, x_d) \in \mathbf{R}^d$. If f and g are smooth enough, then the Leibnitz formula holds

$$D^\alpha (fg)(x) = \sum_{\beta \leq \alpha} \binom{\alpha}{\beta} D^{(\alpha-\beta)} f(x) D^{(\beta)} g(x).$$

By Ω we denote an open subset of $\mathbf{R}^d$. The space of infinitely differentiable functions on Ω is denoted by $C^\infty(\Omega)$. Throughout the paper, the integrals are taken over $\mathbf{R}^d$, or $\mathbf{R}$, unless otherwise indicated. $L^2(\mathbf{R}^d)$ is the Hilbert space of square integrable functions with the inner product $(f,g) = \int f(x)\overline{g(x)}dx = \langle f, \overline{g} \rangle$. By $\|f\|$ we denote norm of $f \in L^2(\mathbf{R}^d)$. The Fourier transform of an integrable function f is defined by

$$\mathcal{F}f(\xi) = \hat{f}(\xi) = \int e^{-2\pi i x \xi} f(x)dx, \quad \xi \in \mathbf{R}^d.$$

Translation and modulation operators, T and M are defined by

$$T_x f(\cdot) = f(\cdot - x) \quad \text{and} \quad M_x f(\cdot) = e^{2\pi i x \cdot} f(\cdot), \quad x \in \mathbf{R}^d.$$

The following relations hold

$$M_y T_x = e^{2\pi i x \cdot y} T_x M_y, \quad (T_x f)\hat{} = M_{-x}\hat{f}, \quad (M_x f)\hat{} = T_x \hat{f}, \quad x, y \in \mathbf{R}^d, f, g \in L^2(\mathbf{R}^d).$$

3. Test Function Spaces

We suppose that a reader is familiar with the "classical" Schwartz test function spaces $\mathcal{D}(\Omega)$, $\mathcal{E}(\Omega)$, $\mathcal{S}(\mathbf{R}^d)$ and their duals, $\mathcal{D}'(\Omega)$, $\mathcal{E}'(\Omega)$ and $\mathcal{S}'(\mathbf{R}^d)$, [43].

We have $\mathcal{D}(\Omega) \hookrightarrow \mathcal{E}(\Omega) \hookrightarrow \mathcal{E}'(\Omega) \hookrightarrow \mathcal{D}'(\Omega)$, and

$$\mathcal{D}(\mathbf{R}^d) \hookrightarrow \mathcal{S}(\mathbf{R}^d) \hookrightarrow \mathcal{E}(\mathbf{R}^d) \hookrightarrow \mathcal{E}'(\mathbf{R}^d) \hookrightarrow \mathcal{S}'(\mathbf{R}^d) \hookrightarrow \mathcal{D}'(\mathbf{R}^d),$$

where "$A \hookrightarrow B$" means that A is dense subset of B and that the inclusion mapping is continuous.

$\mathcal{S}(\mathbf{R}^d)$ and $\mathcal{S}'(\mathbf{R}^d)$ play a particularly important role in various applications since the Fourier transform is a topological isomorphism between $\mathcal{S}(\mathbf{R}^d)$ and $\mathcal{S}(\mathbf{R}^d)$ and extends to a continuous linear transform from $\mathcal{S}'(\mathbf{R}^d)$ onto itself.

In order to deal with particular problems in applications various generalizations of the Schwartz type spaces were proposed. We give here only a very brief and incomplete list in order to make our aim more transparent. More precisely, we restrict our exposition to Gevrey classes and certain Gelfand-Shilov type spaces. Their duals are spaces of ultradistributions.

Starting with the work of Beurling [2] and Roumieu [42] several theories of ultradistributions are developed, e.g. Denjoy-Carleman-Komatsu [32], Beurling-Björck [3], Braun-Meise-Taylor [4], Cioranescu-Zsidó [11]. The notion of equivalence of ultradistribution theories allows to transfer results from one theory to another. This includes, for example, topological properties, repesentation and structure theorems, behavior under various integral transforms, hypoellipticity, see [10], [7], [19], [29], [32], [36], [41].

3.1. Gevrey Classes

Gevrey type spaces fill the gap between the space of analytic functions $\mathcal{A}(\Omega)$ and the space of infinitely differentiable functions $C^\infty(\Omega)$. This turns out to be particularly important in the study of operators whose behavior differs in C^∞ and analytic framework (local and microlocal analysis). Recall, the space of analytic functions is defined by

$$\mathcal{A}(\Omega) = \{f \in C^\infty(\Omega) \mid (\forall K \Subset \Omega)(\exists C > 0)(\exists h > 0) \quad \sup_{x \in K} |\partial^\alpha f(x)| \leq C h^{|\alpha|} |\alpha|!\}.$$

$K \Subset \Omega$ means that K is a compact subset of Ω. For $1 < s < \infty$ we define a Gevrey class by

$$G^s(\Omega) = \{f \in C^\infty(\Omega) \mid (\forall K \Subset \Omega)(\exists C > 0)(\exists h > 0) \quad \sup_{x \in K} |\partial^\alpha f(x)| \leq C h^{|\alpha|} |\alpha|!^s\}.$$

we denote by $G^s_0(\Omega)$ a subspace of $G^s(\Omega)$ which consists of compactly supported functions. We have $\mathcal{A}(\Omega) \hookrightarrow \cap_{s>1} G^s(\Omega)$ and $\cup_{s \geq 1} G^s(\Omega) \hookrightarrow C^\infty(\Omega)$.

For $1 < s < \infty$ the Gevrey class $\mathcal{E}^{\{s\}}(\Omega)$ (resp. $\mathcal{E}^{(s)}(\Omega)$) consists of all $f \in C^\infty(\Omega)$ such that for any compact set K in Ω there are constants h and C (resp. for any $h > 0$ there is a constant C) such that

$$\sup_{x \in K} |\partial^\alpha f(x)| \leq C h^{|\alpha|} |\alpha|!^s.$$

$\mathcal{D}^{\{s\}}(\Omega)$ (resp. $\mathcal{D}^{\{s\}}(\Omega)$) is the space of all $f \in \mathcal{E}^{\{s\}}(\Omega)$ (resp. $f \in \mathcal{E}^{(s)}(\Omega)$) with compact support under a natural locally convex topology. The corresponding dual spaces $(\mathcal{E}^{(s)}(\Omega))'$ and $(\mathcal{D}^{(s)}(\Omega))'$ are spaces of Beurling ultradistributions, while $(\mathcal{E}^{\{s\}}(\Omega))'$ and $(\mathcal{D}^{\{s\}}(\Omega))'$ are spaces of Roumieu type ultradistributions, see[33]. Common notation for both projective and inductive limit is $*$ in the upper index (instead of (s) or $\{s\}$). The ultradistributions spaces are a generalization of the corresponding Schwartz distribution spaces since

$$\mathcal{E}^*(\Omega) \hookrightarrow \mathcal{E}(\Omega) \quad \text{and} \quad \mathcal{D}^*(\Omega) \hookrightarrow \mathcal{D}(\Omega)$$

imply $\mathcal{D}'(\Omega) \hookrightarrow (\mathcal{D}^*(\Omega))'$ and $\mathcal{E}'(\Omega) \hookrightarrow (\mathcal{E}^*(\Omega))'$. Also, $(\mathcal{E}^*(\Omega))' \hookrightarrow (\mathcal{D}^*(\Omega))'$.

Remark 3.1.
- Global definition, with $\Omega = \mathbf{R}^d$, is obtained by taking the supremum over $\mathbf{R}^d$.
- One may use $|\alpha|^{s|\alpha|}$ or $\Gamma(s|\alpha|+1)$ instead of $|\alpha|!^s$, [41, Proposition 1.4.2].
- Weighted ultradistribution spaces $(\mathcal{D}^*_{L^s}(\mathbf{R}^d))'$, $s \geq 1$, are studied in [7].
- Likewise $\mathcal{D}(\mathbf{R}^d)$ and $\mathcal{E}(\mathbf{R}^d)$, spaces $\mathcal{D}^*(\mathbf{R}^d)$ and $\mathcal{E}^*(\mathbf{R}^d)$ are not invariant under the Fourier transform. Their behavior under the Fourier transform is given by Paley-Wiener type theorems, see e.g. [41, Paragraph 1.6].

The last remark is a motivation for the study of other generalizations. The idea is to obtain Beurling and Roumieu type spaces invariant under the Fourier transform, i.e. to define a natural counterpart of the space of tempered distributions in the framework of ultradistributions.

3.2. Gelfand-Shilov Type Spaces

For the sake of simplicity we skip the definition of Gelfand-Shilov type spaces S_α, S^β, and $W_M^{M^\times}$, and refer the reader to [18], [25], [27].

Definition 3.2. Let there be given $\alpha, \beta \geq 0$, and $A, B > 0$. Gelfand-Shilov type space $S^{\alpha,B}_{\beta,A} = S^{\alpha,B}_{\beta,A}(\mathbf{R}^d)$ is defined by

$$S^{\alpha,B}_{\beta,A} = \{f \in C^\infty(\mathbf{R}^d) \mid (\exists C > 0) \sup_{x \in \mathbf{R}^d} |x^p \partial^q f(x)| \leq C A^p p^{p\alpha} B^q q^{q\beta}, \ \forall p, q \in \mathbf{N}_0^d\}.$$

We introduce the following notation for projective and inductive limits:

$$\Sigma^\alpha_\beta = S^{\alpha,0}_{\beta,0} = \operatorname{proj} \lim_{A>0,B>0} S^{\alpha,B}_{\beta,A}; \quad S^\alpha_\beta := \operatorname{ind} \lim_{A>0,B>0} S^{\alpha,B}_{\beta,A}.$$

S^α_β is nontrivial if and only if $\alpha+\beta > 1$, or $\alpha+\beta = 1$ and $\alpha\beta > 0$, [18]. Spaces Σ^α_α, $\alpha > 1/2$, were introduced and studied in [36]. Note that $\Sigma^{1/2}_{1/2}$, if defined as above, is trivial. An alternative definition of $\Sigma^{1/2}_{1/2}$ based on the Hermite expansions is given in [36]. We have

$$\Sigma^\alpha_\beta \hookrightarrow S^\alpha_\beta \hookrightarrow \mathcal{S}.$$

For the proof of the following theorem we refer to [18]. See also [10], [29] and [36].

Theorem 3.3. *Let there be given $\alpha, \beta \geq 0$. The Fourier transform is a topological isomorphism between S^{β}_{α} and S^{α}_{β} ($\mathcal{F}(S^{\beta}_{\alpha}) = S^{\alpha}_{\beta}$) and extends to a continuous linear transform from $(S^{\beta}_{\alpha})'$ onto $(S^{\alpha}_{\beta})'$. In particular, if $\alpha = \beta$ and $\alpha \geq 1/2$ then $\mathcal{F}(S^{\alpha}_{\alpha}) = S^{\alpha}_{\alpha}$.*

Similar assertions hold for Σ^{α}_{β}.

Therefore we obtain a family of Fourier transform invariant spaces which are contained in $\mathcal{S}$. Corresponding dual spaces are, due to this, called tempered ultradistributions (of Beurling or Roumieu type).

Further on we will use the notation $\mathcal{S}^{(\alpha)} = \Sigma^{\alpha}_{\alpha}$, $\mathcal{S}^{\{\alpha\}} = S^{\alpha}_{\alpha}$, and $\mathcal{S}^*$, $*$ means (α) or $\{\alpha\}$.

The relation between Gevrey type spaces and Gelfand-Shilov type spaces is given by the inclusions

$$\mathcal{D}^*(\mathbf{R}^d) \hookrightarrow \mathcal{S}^*(\mathbf{R}^d) \hookrightarrow \mathcal{E}^*(\mathbf{R}^d).$$

In particular, $G^{\alpha}_0(\mathbf{R}^d) \hookrightarrow S^{\alpha}_{\alpha}(\mathbf{R}^d) \hookrightarrow G^{\alpha}(\mathbf{R}^d)$, $\alpha > 1$.

Apart from being invariant under the Fourier transform, spaces $\mathcal{S}^*$ have another important property. Namely, $S^{\alpha}_{\alpha}(\mathbf{R}^d)$ (respectively $\Sigma^{\alpha}_{\alpha}(\mathbf{R}^d)$) are subspaces of the space of entire functions in $\mathbf{C}^d$ restricted to $\mathbf{R}^d$ provided $1/2 \leq \alpha < 1$ (respectively $1/2 < \alpha \leq 1$) while $S^1_1(\mathbf{R}^d)$ is a subspace of the set of real analytic functions in $\mathbf{R}^d$ (hence the spaces do not contain compactly supported functions except $f \equiv 0$). However, the spaces are "sufficiently rich" in the sense of [18]. On contrary, Gevrey classes can not jump in the quasi-analytic case.

Remark 3.4.
- Quasi-analytic Gelfand-Shilov type spaces are essential for the ultradistributional approach to hyperfunctions, [9]. For example, S^1_1 is known as the Sato test function space for a space of Fourier hyperfunctions.
- For an abstract harmonic analysis approach to Gelfand-Shilov type spaces we refer to [28].
- For a relation between the spaces of type $W^{M^{\times}}_{M}$ (studied in [27]) and Gelfand-Shilov type spaces, see [29].

3.3. Time-Frequency Representations

In this subsection we give a characterization of $\mathcal{S}^*$ by means of time-frequency representations. Although Theorems 3.8 and 3.9 are not completely new results we present them not only to make the exposition self contained, but also in order to emphasize the role of time-frequency analysis techniques in the study of ultradistributions.

The following theorem gives a beautiful symmetric characterization of $\mathcal{S}^*$. It has been reinvented several times, [10], [25], [29], [39].

Theorem 3.5. *Let there be given $s \geq 1/2$. The following conditions are equivalent:*

a) *$f \in \mathcal{S}^{\{s\}}$ (resp. $f \in \mathcal{S}^{(s)}$, $s > 1/2$).*

b) *$\sup_{x \in \mathbf{R}^d} |x^{\alpha} f(x)| \leq C h^{|\alpha|} \alpha!^s$ and $\sup_{\xi \in \mathbf{R}^d} |\xi^{\beta} \hat{f}(\xi)| \leq C k^{|\beta|} \beta!^s$ for some (resp. every) $h, k > 0$.*

c) $\sup_{x\in\mathbf{R}^d} |x^\alpha f(x)| \leq Ch^{|\alpha|}\alpha!^s$ *and* $\sup_{x\in\mathbf{R}^d} |\partial^\beta f(x)| \leq Ck^{|\beta|}\beta!^s$ *for some (resp. every)* $h,k>0$.

d) $\sup_{x\in\mathbf{R}^d} |f(x)|e^{h|x|^{1/s}} < \infty$ *and* $\sup_{\xi\in\mathbf{R}^d} |\hat{f}(\xi)|e^{k|\xi|^{1/s}} < \infty$, *for some (resp. every)* $h,k>0$.

Proof. For the proof of the inductive limit case see [10]. Projective limit case is observed in [29]. □

The following definition collects some of the most important time-frequency representations.

Definition 3.6. Let $g \in L^2(\mathbf{R}^d)\setminus\{0\}$. The short-time Fourier transform of a signal $f \in L^2(\mathbf{R}^d)$, known also as the Gabor transform, is given by

$$V_g f(x,\xi) = \int e^{-2\pi i t\xi}\overline{g(t-x)}f(t)dt, \quad x,\xi \in \mathbf{R}^d. \tag{3.1}$$

The cross ambiguity function (the Fourier–Wigner transform) of f and g is

$$A(f,g)(x,\xi) = \int e^{2\pi i\xi t} f(t+\frac{x}{2})\overline{g(t-\frac{x}{2})}dt, \quad x,\xi \in \mathbf{R}^d \tag{3.2}$$

and the cross Wigner distribution of f and g is

$$W(f,g)(x,\xi) = \int e^{-2\pi i\xi t} f(x+\frac{t}{2})\overline{g(x-\frac{t}{2})}dt, \quad x,\xi \in \mathbf{R}^d. \tag{3.3}$$

The quadratic expressions $Af := A(f,f)$ and $Wf := W(f,f)$ are called the (radar) ambiguity function and the Wigner distribution of f.

For various applications of the time-frequency representations in signal analysis and in harmonic analysis see [17], [21], [22], [24], [26], [27], [29], [51] and the references given there.

For $f,g \in L^2(\mathbf{R}^d)$, following relations hold:

$$A(f,g)(x,\xi) = e^{\pi i x\xi} V_g f(x,\xi),$$
$$W(f,g)(x,\xi) = 2^d e^{4\pi i x\xi} V_{\check{g}} f(2x,2\xi),$$
$$W(f,g)(x,\xi) = (\mathcal{F}A(f,g))(x,\xi), \quad x,\xi \in \mathbf{R}^d.$$

Thus, for our purpose, it will be enough to choose one of the representations. We consider the cross Wigner distribution. The choice is made due to the role which it plays in the theory of pseudo-differential and localization operators, see the next section.

Proposition 3.7. *We list here some of the properties of the cross Wigner distribution.*

a) $W(f,g)$ *maps* $\mathcal{S}(\mathbf{R}^d)\times\mathcal{S}(\mathbf{R}^d)$ *into* $\mathcal{S}(\mathbf{R}^d\times\mathbf{R}^d)$ *and extends to a map from* $\mathcal{S}'(\mathbf{R}^d)\times\mathcal{S}'(\mathbf{R}^d)$ *into* $\mathcal{S}'(\mathbf{R}^d\times\mathbf{R}^d)$.

b) *For* $f_1, f_2, g_1, g_2 \in L^2(\mathbf{R}^d)$, *the Moyal identity holds:*

$$\langle W(f_1, g_1), W(f_2, g_2)\rangle = \langle f_1, f_2\rangle\overline{\langle g_1, g_2\rangle}.$$

c) $\|W(f, g)\|_\infty \leq 2^d\|f\|\|g\|$.

d) $W(\hat{f}, \hat{g})(x, \xi) = W(f, g)(-\xi, x)$.

e) $W(f, g)(x, \xi) = \overline{W(g, f)}(x, \xi)$.

For the proofs we refer to [17], [21], [51].

The space $\Sigma_{1/2}^{1/2} = \mathcal{S}^{(1/2)}$ is not included in the following theorem.

Theorem 3.8. *Let* $f, g \in \mathcal{S}^*(\mathbf{R}^d)$. *Then* $W(f, g)(x, \xi) \in \mathcal{S}^*(\mathbf{R}^d \times \mathbf{R}^d)$. *The same is true for the Fourier–Wigner transform and the short-time Fourier transform.*

Proof. The proof for the case $f = g$, i.e. for the Wigner distribution $W(f, f)$ is given in [8] for $\mathcal{S}^{\{s\}}$ and in [29] for $\mathcal{S}^*$. We follow the proof of Proposition 3.7 a) given in [51] and give the proof in two steps.

Step 1. Obviously, $f, g \in \mathcal{S}^*(\mathbf{R}^d)$ implies that $f(x)g(t) \in \mathcal{S}^*(\mathbf{R}^d \times \mathbf{R}^d)$, that is

$$\sup_{x,t\in\mathbf{R}^d} |x^\alpha t^\beta \partial_x^\gamma \partial_t^\delta f(x)g(t)| \leq Ch^{|\alpha|+|\beta|+|\gamma|+|\delta|}|\alpha|^{s|\alpha|} \cdot \beta^{s|\beta|} \cdot |\gamma|^{s|\gamma|} \cdot \delta^{s|\delta|}, \quad (3.4)$$

for some (resp. all) $h > 0$. Here we use $|\cdot|^{s|\cdot|}$ instead of $|\cdot|!^s$ for the convenience, see Remark 3.1. Recall, $\mathcal{S}^*$ denotes $\mathcal{S}^{(s)}$ or $\mathcal{S}^{\{s\}}$, $s \geq 1/2$. Let us show that, under the same assumptions $\varphi(x, t) := f(x + \frac{t}{2})g(x - \frac{t}{2}) \in \mathcal{S}^*(\mathbf{R}^d \times \mathbf{R}^d)$. By Theorem 3.5 c) we need to show

$$\sup_{x,t\in\mathbf{R}^d} |x^\alpha t^\beta \varphi(x, t)| \leq Ch^{|\alpha|+|\beta|}|\alpha|^{s|\alpha|} \cdot \beta^{s|\beta|},$$

and

$$\sup_{x,t\in\mathbf{R}^d} |\partial_x^\alpha \partial_t^\beta \varphi(x, t)| \leq Ck^{|\alpha|+|\beta|}|\alpha|^{s|\alpha|} \cdot \beta^{s|\beta|}$$

for some (resp. every) $h, k > 0$. The first inequality easily follows from assumptions on f and g and a change of variables:

$$\begin{aligned}
&\sup_{x,t\in\mathbf{R}^d} |x^\alpha t^\beta f(x + \frac{t}{2})g(x - \frac{t}{2})| = \sup_{y,t\in\mathbf{R}^d} |(y - t/2)^\alpha t^\beta f(y)g(y - t)| \\
&\leq 2^{-|\alpha|} \sup_{y,t\in\mathbf{R}^d} |(y - (t - y))^\alpha(-1)^{|\beta|}((y - t) - y)^\beta f(y)g(y - t)| \\
&= 2^{-|\alpha|} \sup\nolimits_{y,z\in\mathbf{R}^d} |(y - z)^\alpha(z - y)^\beta f(y)g(z)| \\
&\leq C_{\alpha,\beta} \sup\nolimits_{y,z\in\mathbf{R}^d} |(z - y)^{\alpha+\beta} f(y)g(z)|.
\end{aligned}$$

Now, (3.4) implies $\sup_{x,t\in\mathbf{R}^d} |x^\alpha t^\beta \varphi(x, t)| \leq Ch^{|\alpha|+|\beta|}|\alpha|^{s|\alpha|} \cdot \beta^{s|\beta|}$ for some (resp. every) $h > 0$.

In order to prove the second inequality recall that in the definition of $\mathcal{S}^*$ we may take $|\alpha|^{s|\alpha|}$ instead of $\alpha!^s$. The Leibniz formula gives

$$|\partial_x^\alpha \partial_t^\beta \varphi(x,t)| = |\sum_{\delta\le\alpha}\sum_{\gamma\le\beta}\binom{\alpha}{\delta}\binom{\beta}{\gamma}\frac{1}{2^{|\alpha|+|\beta|}}\partial_x^\delta\partial_t^\gamma f(x+t/2)\partial_x^{\alpha-\delta}\partial_t^{\beta-\gamma}g(x-t/2)|$$

$$\le C_{\alpha,\beta}\sup_{x,t\in\mathbf{R}^d}|\partial_x^\delta\partial_t^\gamma f(x+t/2)\partial_x^{\alpha-\delta}\partial_t^{\beta-\gamma}g(x-t/2)|.$$

Now, $m^{sm}n^{sn}\le(m+n)^{s(m+n)}$ and (3.4) imply

$$\sup_{x,t\in\mathbf{R}^d}|\partial_x^\alpha\partial_t^\beta\varphi(x,t)|\le Ck^{|\alpha|+|\beta|}|\alpha|^{s|\alpha|}\cdot\beta^{s|\beta|}$$

for some (resp. all) $k>0$.

Therefore, $\varphi(x,t)\in\mathcal{S}^*(\mathbf{R}^d\times\mathbf{R}^d)$.

Step 2. Now, we show that $\Phi(x,\xi)=\int e^{-2\pi it\xi}\varphi(x,t)dt\in\mathcal{S}^*(\mathbf{R}^d\times\mathbf{R}^d)$ if $\varphi\in\mathcal{S}^*(\mathbf{R}^d\times\mathbf{R}^d)$.

Again, we use Theorem 3.5 c) and show that

$$\sup_{x,\xi\in\mathbf{R}^d}|x^\alpha\xi^\beta\Phi(x,\xi)|\le Ch^{|\alpha|+|\beta|}\alpha!^s\beta!^s \text{ and } \sup_{x,\xi\in\mathbf{R}^d}|\partial_x^\alpha\partial_\xi^\beta\Phi(x,\xi)|\le Ck^{|\beta|}\alpha!^s\beta!^s$$

for some (resp. every) $h,k>0$. Integration by parts gives

$$|x^\alpha\xi^\beta\int e^{-2\pi it\xi}\varphi(x,t)dt| = |\int(-1)^\beta D_t^\beta e^{-2\pi it\xi}x^\alpha\varphi(x,t)dt|$$

$$= |\int e^{-2\pi it\xi}x^\alpha D_t^\beta\varphi(x,t)dt|\le C\int(1+t^2)^{-d}(1+t^2)^d x^\alpha\partial_t^\beta\varphi(x,t)dt|.$$

Since $\varphi\in\mathcal{S}^*(\mathbf{R}^d\times\mathbf{R}^d)$ we have $|(1+t^2)^d x^\alpha\partial_t^\beta\varphi(x,t)|\le C\tilde{h}^{2d+\alpha+\beta}(2d)^s\alpha!^s\beta!^s$, for some (any) $\tilde{h}>0$. Then, an easy case studies examination shows that for some (resp. for all) $h>0$, $\sup_{x,\xi\in\mathbf{R}^d}|x^\alpha\xi^\beta\Phi(x,\xi)|\le Ch^{|\alpha|+|\beta|}\alpha!^s\beta!^s$.

For the second equality, we have

$$|\partial_x^\alpha\partial_\xi^\beta\int e^{-2\pi it\xi}\varphi(x,t)dt| = |\int(-2\pi it)^\beta e^{-2\pi it\xi}\partial_x^\alpha\varphi(x,t)dt|$$

$$= |\int(2\pi i)^{|\beta|}e^{-2\pi it\xi}t^\beta\partial_x^\alpha\varphi(x,t)dt|\le|2\pi i|^{|\beta|}|\int e^{-2\pi it\xi}(1+t^2)^{-d}t^\beta\partial_x^\alpha\psi(x,t)dt|,$$

where $\psi=(1+t^2)^d\varphi$. Now $\varphi\in\mathcal{S}^*$ implies that $\psi\in\mathcal{S}^*$ hence

$$|t^\beta\partial_x^\alpha\psi(x,t)|\le Ch^{|\alpha|+|\beta|}\alpha!^s\beta!^s\Rightarrow\sup_{x,\xi\in\mathbf{R}^d}|\partial_x^\alpha\partial_\xi^\beta\Phi(x,\xi)|\le Ck^{|\beta|}\alpha!^s\beta!^s.$$

Theorem 3.8 is therefore proved in a general case for transforms of the type $\int e^{-2\pi it\xi}\varphi(x,t)dt$ with $\varphi\in\mathcal{S}^*$. In particular, the assertion holds for the cross Wigner distribution as claimed. Proofs for the Fourier Wigner transform and the short-time Fourier transform are analogous.

Note that Theorem 3.8 for the inductive limit case and the short-time Fourier transform is proved in [25], see also [21, Section 11.2]. The main difference is that we use Theorem 3.5 c) which allows to estimate actions of multiplication and differentiation separately. □

Theorem 3.9. *Let $g \in \mathcal{S}^*(\mathbf{R}^d)$ and let $f \in \mathcal{S}'^*(\mathbf{R}^d)$. If the cross Wigner distribution $W(f,g)(x,\xi)$ (resp. the cross ambiguity function or the short-time Fourier transform) belongs to $\mathcal{S}^*(\mathbf{R}^d \times \mathbf{R}^d)$ then $f \in \mathcal{S}^*(\mathbf{R}^d)$.*

Proof. We combine the ideas of the proof of similar assertions given in [25] and [8]. Firstly, we show the inversion formula for the cross Wigner distribution in the spirit of [21, Section 3.2]. First, note that, for given $g_1, g_2 \in L^2(\mathbf{R}^d)$ and $\langle g_1, g_2\rangle \neq 0$,

$$l(h) = \frac{1}{\langle g_1, g_2\rangle} 2^d \int\int e^{4\pi i x\xi} W(f,g_1)(x,\xi)\langle h, M_{2\xi}T_{2x}\check{g}_2\rangle dx d\xi, \quad h \in L^2(\mathbf{R}^d),$$

is a bounded functional on $L^2(\mathbf{R}^d)$ and therefore it defines a unique function $\tilde{f} \in L^2(\mathbf{R}^d)$,

$$\tilde{f}(t) = 2^d \int\int e^{4\pi i x\xi} W(f,g)(x,\xi) M_{2\xi}T_{2x}\check{g}(t) dx d\xi \tag{3.5}$$

such that $l(h) = \langle \tilde{f}, h\rangle$, for all $h \in L^2(\mathbf{R}^d)$.

Let us show that for all $f \in L^2(\mathbf{R}^d)$ the following inversion formula holds

$$f = \frac{1}{\langle g_1, g_2\rangle} 2^d \int\int e^{4\pi i x\xi} W(f,g_1)(x,\xi) M_{2\xi}T_{2x}\check{g}_2 dx d\xi.$$

Actually, (3.5), the Moyal formula and

$$W(f,g)(x,\xi) = \langle f, 2^d e^{4\pi i x\xi} M_{2\xi}T_{2x}\check{g}\rangle$$

imply

$$\langle \tilde{f}, h\rangle = \frac{1}{\langle g_1, g_2\rangle} \int\int W(f,g_1)(x,\xi) W(h,g_2) dx d\xi = \langle f, h\rangle.$$

Thus $\tilde{f} = f$ in $L^2(\mathbf{R}^d)$. Moreover, it can be shown that $f \in \mathcal{S}(\mathbf{R}^d)$, [21, Proposition 11.2.4].

We are now in a position to show that $f \in \mathcal{S}^*(\mathbf{R}^d)$. We will again use Theorem 3.5. Now it is convenient to use part d) of the Theorem. Therefore, we need to show that $\sup_{x\in\mathbf{R}} |f(x)| e^{h|x|^{1/s}} < \infty$ and $\sup_{\omega\in\mathbf{R}} |\hat{f}(\omega)| e^{k|\omega|^{1/s}} < \infty$, for some (resp. for every) $h, k > 0$.

Note that $e^{h|t|^{1/s}} M_{2\xi}T_{2x}\check{g}(t) = M_{2\xi}T_{2x} e^{h|t+2x|^{1/s}}\check{g}(t)$. The inversion formula implies

$$\begin{aligned}
&\sup_{t\in\mathbf{R}^d} e^{h|t|^{1/s}} \frac{1}{\langle g_1, g_2\rangle} |2^d \int\int e^{4\pi i x\xi} W(f,g_1)(x,\xi) M_{2\xi}T_{2x}\check{g}_2 dx d\xi| \\
&\leq \sup_{t\in\mathbf{R}^d} |2^d \int\int e^{4\pi i x\xi} W(f,g_1)(x,\xi) e^{h|2x|^{1/s}} M_{2\xi}T_{2x} e^{h|t|^{1/s}} \check{g}(t) dx d\xi \\
&\leq C \int\int e^{-\tilde{h}|x|^{1/s}} e^{-k|2x|^{1/s}} e^{h|2x|^{1/s}} dx d\xi < \infty,
\end{aligned}$$

since $W(f,g_1) \in \mathcal{S}^*(\mathbf{R}^d \times \mathbf{R}^d)$ and $g \in \mathcal{S}^*$. In order to prove

$$\sup_{\omega \in \mathbf{R}} |\hat{f}(\omega)| e^{k|\omega|^{1/s}} < \infty$$

we use $(M_{2\xi} T_{2x} \breve{g}(t))\hat{}(\omega) = T_{2\xi} M_{-2x} \hat{\breve{g}}(\omega)$ and the same arguments as above. □

Remark 3.10. Theorems 3.8 and 3.9 are proved in [25] for the inductive limit case and the short-time Fourier transform. The proof for the Wigner distribution $W(f,f)$ is given in [8].

3.4. Tempered Ultradistributions

This subsection is given for the sake of completeness, having in mind that Gevrey and (non-quasianalytic) Gelfand-Shilov type spaces might be considered as special cases of the following construction [32].

Let $(M_p) = (M_p)_{p \in \mathbf{N}_0}$ be a sequence of positive numbers which satisfies some of the following conditions:
$(M.1)$ (logarithmic convexity) $M_p^2 \leq M_{p-1} M_{p+1}, \quad p \in \mathbf{N}$;
$(M.2)'$ (stability under the action of differential operators) There exist positive constants A, H such that $M_{p+1} \leq AH^p M_p$;
$(M.2)$ (stability under the action of ultradifferential operators) There exist positive constants A, H such that

$$M_{p+1} \leq AH^p \min{}_{0 \leq q \leq p} M_{p-q} M_q, \quad p, q \in \mathbf{N}_0;$$

$(M.3)'$ (non-quasi-analicity) $\sum_{p \in \mathbf{N}} \frac{M_{p-1}}{M_p} < \infty$;
$(M.3)$ (strong non-quasi-analicity) There exists $A > 0$, such that

$$\sum_{p=q+1}^{\infty} \frac{M_{p-1}}{M_p} < Aq \frac{M_{q+1}}{M_q}, \quad q \in \mathbf{N}.$$

We assume $M_0 = 1$. Obviously $(M.2) \Rightarrow (M.2)'$, $(M.3) \Rightarrow (M.3)'$.

The so-called *associated function* for a sequence (M_p) is defined by

$$M(\rho) = \sup_{p \in \mathbf{N}_0} \ln \frac{\rho^p M_0}{M_p}, \quad 0 < \rho < \infty.$$

Properties $(M.1) - (M.3)$ can be expressed via the associated function. It plays an important role in the proofs of properties of ultradistributions, see [32, page 29].

We give the definition in one dimension for the sake of simplicity only.

Definition 3.11. Let there be given $h \geq 0$, and a sequence (M_p) such that $(M.1) - (M.3)'$ holds. The space $\mathcal{S}_h^{(M_p)} = \mathcal{S}_h^{(M_p)}(\mathbf{R})$ is the space of smooth functions f on $\mathbf{R}$ such that

$$\sup_{x \in \mathbf{R}} \sup_{\alpha, \beta \in \mathbf{N}_0} \frac{h^{\alpha+\beta} |x^\alpha f^{(\beta)}(x)|}{M_\alpha M_\beta} < \infty.$$

It is a Banach space with the norm

$$\sigma_{m,\infty}(f) = \sup_{\alpha,\beta\in\mathbf{N}_0} \frac{h^{\alpha+\beta}}{M_\alpha M_\beta}\|x^\alpha f^{(\beta)}(x)\|_\infty, \quad f \in \mathcal{S}_h^{(M_p)}.$$

Let $\mathcal{S}^{(M_p)}$ and $\mathcal{S}^{\{M_p\}}$ be a projective (when $h \to \infty$) and an inductive (when $h \to 0$) limit of $\mathcal{S}_h^{(M_p)}$, and $\mathcal{S}_h^{\{M_p\}}$, respectively. Dual spaces of $\mathcal{S}^{(M_p)}$ and $\mathcal{S}^{\{M_p\}}$, denoted by $\mathcal{S}'^{(M_p)}$ and $\mathcal{S}'^{\{M_p\}}$, are then spaces of tempered ultradistributions of Beurling and Roumieu type respectively. They are studied in [27], [10], [29], [32], [33], [36], where one can find expansion of the elements of $\mathcal{S}^{(M_p)}$, $\mathcal{S}^{\{M_p\}}$ and their duals via Hermite functions, as well as their characterization through the Fourier transform, the Wigner distribution and the Bargman transform. Structural theorems, continuity of ultradifferential operators, the Hilbert transform and the convolution are studied there as well.

The following inclusions hold

$$\mathcal{D}^{(M_p)}(\mathbf{R}) \hookrightarrow \mathcal{S}^{(M_p)}(\mathbf{R}) \hookrightarrow \mathcal{S}(\mathbf{R}) \hookrightarrow L^2(\mathbf{R}) \hookrightarrow \mathcal{S}'(\mathbf{R}) \hookrightarrow \mathcal{S}'^{(M_p)}(\mathbf{R}) \hookrightarrow \mathcal{D}'^{(M_p)}(\mathbf{R}).$$

Remark 3.12.
- If $s > 1$ examples of (M_p) sequences which satisfy some of the above conditions are $M_p = p!^s$, $M_p = p^{sp}$, $M_p = \Gamma(sp+1)$, where Γ denotes the Gamma function. Therefore spaces $\mathcal{S}^*$ of the preceding paragraph are examples of spaces obtained by the above construction.
- Spaces $\mathcal{D}^*$ and $\mathcal{E}^*$ and corresponding weighted versions are introduced and studied within the above approach in [7], [32], [33], [37], [38].

4. Modulation Spaces

Modulation spaces consist of functions or distributions whose short-time Fourier transform satisfies some prescribed decay at infinity as well as some integrability conditions. They are recognized as the most important spaces in time-frequency analysis, [21], [14], [1]. The decay of the short-time Fourier transform is controlled by a weight function, i.e. a non negative locally integrable function on $\mathbf{R}^{2d}$. It is often sufficient to suppose that weights have (at most) polynomial growth. However, in order to deal with ultradistribution, almost exponential growth should be allowed.

Definition 4.1. Let $\gamma \in [0,1)$. A strictly positive and continuous function w_γ on $\mathbf{R}^d \times \mathbf{R}^d$ is called an exp–type weight if there exist $h \geq 0$ and $C > 0$ such that

$$w_\gamma(x+y,\xi+\eta) \leq Ce^{h(|x|^\gamma+|\xi|^\gamma)}w_\gamma(y,\eta), \quad x,y,\xi,\eta \in \mathbf{R}^d$$

and $w_\gamma(x,\epsilon_1\cdot\xi_1,\dots,\epsilon_d\cdot\xi_d) = w_\gamma(x,\xi)$, $\epsilon = (\epsilon_1,\dots,\epsilon_d) \in \{-1,1\}^d$.

A typical example of an exp–type weight is

$$w_\gamma(x,\xi) = e^{h_1|x|^\gamma+h_2|\xi|^\gamma}, \quad x,\xi \in \mathbf{R}^d,\ h_1,h_2 \geq 0. \tag{4.1}$$

More generally, $(1+|x|+|\xi|)^a e^{b|x|^{\gamma_1}+c|\xi|^{\gamma_2}}$ is an exp-type weight for any $a,b,c \geq 0$ and $\gamma_1,\gamma_2 \in [0,1)$.

Definition 4.2. Let there be given $\gamma \in [0,1)$, an exp-type weight w_γ, $t \in \mathbf{R}$, and $1 \le p, q < \infty$. Let $0 \not\equiv g \in \mathcal{S}^{(1/\gamma)}$ (if $\gamma = 0$, then $0 \not\equiv g \in \mathcal{S}$). Then

$$M_{p,q}^{w_\gamma,t} = \left\{ f \in \mathcal{S}'^{(1/\gamma)} : \|f\|_{M_{p,q}^{w_\gamma,t}} < \infty \right\}, \tag{4.2}$$

$$\|f\|_{M_{p,q}^{w_\gamma,t}} = \left[\int \left(\int |V_g f(x,\xi)|^p \, w_\gamma^p(x,\xi) e^{t(|x|^\gamma + |\xi|^\gamma)} dx \right)^{q/p} d\xi \right]^{1/q}$$

is called the modulation space.

$M_{p,q}^{w_\gamma,t}$ is a Banach space [21, Theorem 11.3.5.]. If $t = 0$, we write $M_{p,q}^{w_\gamma} = M_{p,q}^{w_\gamma,0}$, for short. The above definition is independent of the choice of g, $0 \not\equiv g \in \mathcal{S}^{(1/\gamma)}$, in the sense that different functions define the same modulation space and equivalent norms [14].

Modulation spaces $M_{p,q}^w$, where $1 \le p, q < \infty$ and $w(x+y, \xi+\eta) \le C(1+|x|+|\xi|)^s w(y,\eta)$, for some $C > 0$ and all $x, y, \xi, \eta \in \mathbf{R}^d$, are studied in [1], [16], [21] and [45]. Obviously, every weight defined in such a way is an exp–type weight. In particular, for $w_s(x,\xi) = (1+|x|+|\xi|)^s$, $x, \xi \in \mathbf{R}^d$, $s \ge 0$ properties of pseudo-differential operators whose symbols belong to the corresponding modulation spaces are given in [24], see also [16], [21], [31], [48], [49] for some generalizations.

Inclusion relations between modulation spaces and other spaces of functions or distributions (L^p, Besov and Sobolev spaces) are given in [14], [21], [30], [48]. Note that for $w_s(x,y) = (1+|x|+|y|)^s$, $x, y \in \mathbf{R}^d$,

$$\mathcal{S} = \operatorname{proj} \lim_{s \to \infty} M_{p,q}^{w_s}, \quad 1 \le p, q < \infty,$$

[15], [49]. Definition 4.2 and Theorems 3.8 and 3.9 imply that an analogous statement should hold for Gelfand-Shilov type spaces if polynomial weights are replaced by weights of almost exponential growth. Indeed,

Proposition 4.3. *Let $1 \le p < \infty$ and fix $\gamma \in (0,1)$. Put $w_{\gamma,s}(x,\xi) = e^{s(|x|^\gamma + |\xi|^\gamma)}$, $s \in \mathbf{R}$. We have*

$$\mathcal{S}^{(1/\gamma)} = \operatorname{proj} \lim_{s \to \infty} M_{2,2}^{w_{\gamma,s}}, \quad \mathcal{S}^{\{1/\gamma\}} = \operatorname{ind} \lim_{s \to \infty} M_{p,p}^{w_{\gamma,s}}.$$

For the proof of projective limit case we refer to [39]. The inductive limit case is proved in [25].

5. Pseudo-Differential Operators

Pseudo-differential operators in Gevrey classes are studied in [6], [19], [35], [41], [53]. For operators which map $G_0^s(\Omega)$ to $G^s(\Omega)$, $s > 1$, almost exponential (subexponential) growth of corresponding symbols is allowed. This gives rise to operators of "infinite order" in contrast to "finite order" symbols which are slowly increasing at infinity. Depending on the problem at hand various classes of symbols are introduced. Just to give an example, a class of symbols $S_{\varrho,\delta}^{\infty,\theta}(\mathbf{R}^{2d})$ consists of all functions $\sigma \in C^\infty(\mathbf{R}^{2d})$ for which there exist constants $C > 0$ and $B \geq 0$ such that for every $\varepsilon > 0$ there is a constant $c_\varepsilon > 0$ such that

$$\sup_{x\in\mathbf{R}^d} \left| D_x^\alpha D_\xi^\beta \sigma(x,\xi) \right| \leq c_\varepsilon C^{|\alpha+\beta|} \alpha!^{\theta(\varrho-\delta)} \beta! (1+|\xi|)^{\delta|\alpha|-\varrho|\beta|} e^{(\varepsilon|\xi|)^{1/\theta}} \tag{5.1}$$

for every $\alpha, \beta \in \mathbf{N}_0^d$ and every $|\xi| \geq B|\beta|^\theta$, [53].

In order to deal with Gelfand-Shilov type spaces, we use their definition via modulation spaces. In the context of modulation spaces with polynomial weights smooth symbols with at most polynomial growth are studied in [13], [45], [46]. The following class, which includes functions of almost exponential growth, is studied in [40].

Definition 5.1. Let $\gamma \in (0,1)$. Let there be given $L_1, L_2 \geq 0$, and $\lambda, \tau \in \mathbf{R}$. We define the symbol class $S_{L_1,L_2}^{\lambda,\tau,\gamma}(\mathbf{R}^{2d}) = S_{L_1,L_2}^{\lambda,\tau,\gamma}$ as the set of $\sigma \in C^\infty(\mathbf{R}^{2d})$ satisfying

$$\left|\frac{L_1^{|\alpha|}}{\alpha!^{1/\gamma}} \frac{L_2^{|\beta|}}{\beta!^{1/\gamma}} D_x^\alpha D_\xi^\beta \sigma(x,\xi)\right| \leq C e^{\lambda|x|^\gamma + \tau|\xi|^\gamma}, \quad \alpha,\beta \in \mathbf{N}_0^d, x,\xi \in \mathbf{R}^d$$

for some positive constant $C = C_\sigma$ depending on L_1, L_2, λ, τ and γ. The infimum of such constants C_σ will be denoted by $\|\sigma\|_{L_1,L_2}^{\lambda,\tau,\gamma}$.

$S_{L_1,L_2}^{\lambda,\tau,\gamma}$ contains polynomial symbols as well as some ultrapolynomial symbols. For example, if $|a_n| \leq C\frac{k^n}{n!^{1/\gamma}}$, for some positive constants C and k, and all $n \in \mathbf{N}_0$ then

$$\sigma(x,\xi) := \sum_{n=0}^{\infty} a_n \left(1+|x|^2+|\xi|^2\right)^{n/2} \in S_{L_1,L_2}^{\lambda,\tau,\gamma}$$

for all $L_1, L_2 > 0$ and

$$\lambda, \tau \geq (k^\gamma 2^{\gamma/2}(1+d(L_1^2+L_2^2))^{\gamma/2})/\gamma.$$

For a relation between $S_{L_1,L_2}^{\lambda,\tau,\gamma}$ and classes of symbol introduced in [6], [45] and [53], as well as for the continuity properties of the corresponding operators we refer to [40]. As an illustration we give the following result from [40]. The notion of Weyl symbol is given in Definition 5.3.

Proposition 5.2. *Let $\gamma \in (0,1)$. Let there be given $L_1, L_2 \geq 0$, and $\lambda, \tau \in \mathbf{R}$, and $L_1 \geq \frac{-\tau}{2^{\gamma+1/\gamma}}$ if $\tau < 0$.*

Observe a pseudo-differential equation $\sigma(x,D)u = f$, where the Weyl symbol of $\sigma(x,D)$ belongs to the class $S_{A,B}^{\lambda,\tau,\gamma}$, with

$$A > 2\tilde{\gamma}L_1, \quad B > 2\tilde{\gamma}L_2, \quad \textit{where} \quad \tilde{\gamma} = \left(\frac{4}{1-\gamma^2}\right)^{1/\gamma}.$$

If $u \in \mathcal{S}^{(1/\gamma)}$, then $f \in \mathcal{S}^{(1/\gamma)}$ as well. Moreover, the mapping $\sigma(x,D) : \mathcal{S}^{(1/\gamma)} \mapsto \mathcal{S}^{(1/\gamma)}$ is continuous.

Another consequence is the boundedness of the operator $\sum_{|\alpha|\leq n} a_\alpha D^\alpha$, on $\mathcal{S}^{(1/\gamma)}$. It follows from the fact that its Weyl symbol belongs to the class $S^{\lambda,\tau,\gamma}_{A,B}$ for any choice of $A, B > 0$.

Definition 5.3. Let $\mathcal{A}$ be a class of symbols. We say that $\sigma \in \mathcal{A}$ is the Weyl symbol of the operator $\sigma(x, D)$ if and only if

$$\sigma(x,D)f(x) = \int\int \sigma\left(\frac{x+y}{2}, \xi\right) e^{2\pi i(x-y)\xi} f(y)dyd\xi, \quad f \in \mathcal{S}^{(1/\gamma)}(\mathbf{R}^d). \qquad (5.2)$$

We also say that $\sigma(x, D)$ is the Weyl transforms of the symbol $\sigma(x, \xi)$ and refer to (5.2) as the Weyl correspondence between an operator and a symbol.

The Weyl correspondence can be defined by the means of the cross Wigner distribution [51]. Namely, for $f, g \in L^2(\mathbf{R}^d)$ and $\sigma \in L^2(\mathbf{R}^{2d})$

$$(\sigma(x,D)f, g) = (\sigma, W(g,f)) = \langle \sigma, \overline{W(g,f)} \rangle = \int\int \sigma(x,\xi) W(f,g)(x,\xi) dx d\xi.$$

Remark 5.4.
- The cross Wigner distribution and the short-time Fourier transform play important role in the study of localization operators, [1], [12], [52]. For given "windows" $\varphi_1, \varphi_2 \in L^2(\mathbf{R}^d)$ and a suitable symbol a, the time-frequency localization operator $A_a^{\varphi_1,\varphi_2}$ is defined by

  $$A_a^{\varphi_1,\varphi_2} f = \int\int a(x,\xi) V_{\varphi_1} f(x,\xi) M_\xi T_x \varphi_2 dx d\xi.$$

  For various choices of windows and symbols we refer to [12]. We are interested here in a representation of localization operator through the cross Wigner distribution. It can be shown that $A_a^{\varphi_1,\varphi_2}$ is operator whose Weyl symbol σ is given by $\sigma = a * W(\varphi_1, \varphi_2)$.
- The so called anti-Wick operators arise as a special case of localization operators, [1, Chapters 8,9]. Recall, an anti-Wick symbol σ determines Weyl symbol $\sigma * (2^{2d} e^{-|\cdot|^2})$. Let $\sigma(x,\xi) \in \mathcal{S}'^{(\gamma)}$. Then there exists $s > 0$ such that $\sigma * \left(2^{2d} e^{-|\cdot|^2}\right) \in S^{\lambda,\tau,\gamma}_{L_1,L_2}$ for any $\lambda, \tau \geq s$, and $0 \leq L_1, L_2 \leq 2^{-1/2}$, [40].
- Another time-frequency representation, Rihaczek distribution is closely related to the Kohn-Nirenberg correspondence between symbols and operators, see [21], [23].

Let us now consider a class of symbols $\sigma \in C^\infty(\mathbf{R}^d \times \mathbf{R}^d)$ satisfying:

(S1) $\sigma(z) \geq 1$, $z = (x,\xi) \in \mathbf{R}^d \times \mathbf{R}^d$.

(S2) $(\exists C > 0)$ $(\exists \eta > 0)$ such that

$$\sigma(z+w) \leq C e^{\eta |z|^\gamma} \sigma(w), \quad z, w \in \mathbf{R}^{2d}.$$

(S3) $(\forall h \geq 0)$ $(\exists C > 0)$ $(\exists \tilde{s} \geq 0)$ such that

$$\sup_{\alpha \in \mathbf{N}_0^{2d}, |\alpha| \geq 1} \left| \frac{h^{|\alpha|}}{\alpha!^{1/\gamma}} D^\alpha \sigma(z) \right| \leq C \frac{\sigma(z)}{(1+|z|)^{\tilde{s}}}, \quad z \in \mathbf{R}^{2d}.$$

(S4) $\sigma(x,\xi) \leq \sigma(x,\xi')$ for all $\xi, \xi' \in \mathbf{R}^d$ such that $|\xi| \leq |\xi'|$.

Note, by (S2), (S3) and $|z| \leq |x|^\gamma + |\xi|^\gamma$, we have $\sigma(x,\xi) \in S^{\eta,\eta}_{L_1,L_2}$ for every $L_1, L_2 \geq 0$. Also, condition (S2) implies that σ is an exp-type weight and $\sigma(z) \geq \frac{\sigma(0)}{C} e^{-\eta|z|^\gamma}$ for all $z \in \mathbf{R}^{2d}$.

Example. The following functions satisfy conditions (S1)–(S4).

a) $\sigma(z) = \sum_{k=0}^n a_k \langle z\rangle^{2k}$, $z = (x,\xi) \in \mathbf{R}^d \times \mathbf{R}^d$, where $a_0 \geq 1$, $a_k > 0$.
b) $\sigma(x,\xi) = (1 + |x|^2 + |\xi|^2)^{s/2}$, $s \geq 0$, $x, \xi \in \mathbf{R}^d$. In particular,

$$\sigma(\xi) = (1 + |\xi|^2)^{s/2}, \quad s \geq 0,\ \xi \in \mathbf{R}^d.$$

c) $\sigma(x,\xi) = |\xi|^2 + V(x)$, $x, \xi \in \mathbf{R}^d$, where
(V1) $V \in C^\infty(\mathbf{R}^d)$, $V \geq 1$, $V(x) \to \infty$ when $|x| \to \infty$;

(V2) $(\exists C > 0)$ $(\exists \eta > 0)$ such that $V(x+y) \leq Ce^{\eta|x|^\gamma} V(y)$, $x, y \in \mathbf{R}^d$;

(V3) $(\forall h \geq 0)$ $(\exists C > 0)$ such that

$$\sup_{\alpha \in \mathbf{N}_0^d, |\alpha| \geq 1} \left| \frac{h^{|\alpha|}}{\alpha!^{1/\gamma}} D^\alpha V(x) \right| \leq CV(x), \quad x \in \mathbf{R}^d.$$

d) $\sigma(x,\xi) = e^{(1+|x|^2+|\xi|^2)^{\gamma/2}}$, $x, \xi \in \mathbf{R}^d$.

Example c) implies that we may observe the Schrödinger operator with potential $V(x)$ which may have sub-exponential growth.

Theorem 5.5. *Assume that* $\sigma(x,\xi)$ *satisfies* (S1)–(S4). *Then*

a) *Let* $1 \leq p, q < \infty$ *and* $s \geq 0$. *For every* $f \in M^{\sigma,s}_{p,q}$ *there exist positive constants* C_1, C_2 *and* C_3 *such that*

$$C_1 \|f\|_{M^{\sigma,s}_{p,q}} \leq \|\sigma(x,D)f\|_{M^{1,s}_{p,q}} + C_2\|f\|_{M^{1,0}_{p,q}} \leq C_3 \|f\|_{M^{\sigma,s}_{p,q}}. \tag{5.3}$$

b) *If, additionally,* $\sigma(z) \geq Ce^{\mu|z|^\gamma}$ *for* $|z| \geq K$, *for some positive constants* C, μ *and* K, *and if* $\sigma(x,D)f \in M^{1,s}_{p,q}$ *for* $f \in M^{1,0}_{p,q}$, *then* f *belongs to* $M^{1,s+\mu}_{p,q}$.
c) *Let*

$$\sigma(x,\xi) \sim e^{\mu(|x|^\gamma + |\xi|^\gamma)} \quad \textit{for some} \quad \mu > 0 \quad \textit{when} \quad |x| + |\xi| \to \infty. \tag{5.4}$$

Then $\sigma(x,D)|_{\mathcal{S}^{(1/\gamma)}}$ *is essentially self-adjoint in* L^2 *and the domain of its unique self adjoint extension* L *is* $M^{\sigma,0}_{2,2}$. *Furthermore, the self adjoint operator* L *is compact operator and its eigenfunctions belong to* $\mathcal{S}^{(1/\gamma)}$.

Proof. Part a) is proved in [40], see also [46]. Part b) is a corollary of a). For simplicity we prove b) for $p = q = 2$. Let $B(0,K) = \{z \in \mathbf{R}^{2d} \mid |z| \leq K\}$.

$$\begin{aligned}
\|f\|^2_{M^{1,s+\mu}_{2,2}} &= \int\int |V_g f(x,y)|^2 e^{2(s+\mu)(|x|^\gamma + |y|^\gamma)} dxdy \\
&\leq \iint_{B(0,K)} |V_g f(x,y)|^2 e^{2(s+\mu)(|x|^\gamma + |y|^\gamma)} dxdy \\
&\quad + \iint_{\mathbf{R}^d \backslash B(0,K)} |V_g f(x,y)|^2 e^{2(s+\mu)(|x|^\gamma + |y|^\gamma)} dxdy \\
&\leq C + |\iint_{\mathbf{R}^d \backslash B(0,K)} |V_g f(x,y)|^2 \sigma^2(x,y) e^{2s(|x|^\gamma + |y|^\gamma)} dxdy| \leq C + \|f\|_{M^{\sigma,s}_{2,2}} \\
&\leq C + \|\sigma(x,D)f\|_{M^{1,s}_{2,2}} + C_2 \|f\|_{M^{1,0}_{2,2}} < \infty,
\end{aligned}$$

where we used a) together with the fact that $f \in M^{1,0}_{2,2}$ implies $V_g f \in M^{1,0}_{2,2}$.

c) We prove that $\sigma(x,D)|_{\mathcal{S}^{(1/\gamma)}}$ is essentially self-adjoint by using [44, Theorem 26.1]. To that end we need to show the $\mathcal{S}^{(1/\gamma)}$ hypoellipticity. This follows from b) since $\mathcal{S}^{(1/\gamma)} = \bigcap_{s \geq 0} M^{1,s}_{2,2}$. It remains to show that L is compact.

Now we need the following fact from time-frequency analysis. Recall, a set of functions $\{g_{k,n},\ k,n \in \mathbf{Z}\}$, is called a *Gabor system* if

$$g_{k,n}(x) = e^{2\pi bnx} g(x - ak) = M_{bn} T_{ak} g, \quad k,n \in \mathbf{Z},$$

for a fixed function g and time-frequency shift parameters $a, b > 0$. A Gabor system is a *Gabor frame* in $L^2(\mathbf{R})$ if there exists $0 < A \leq B < \infty$ such that

$$A\|f\|^2 \leq \sum_{k,n \in \mathbf{Z}} |\langle g_{k,n}, f\rangle|^2 \leq B\|f\|^2.$$

If $A = B$ the frame is tight. For a tight Gabor frame $\{g_{k,n},\ k,n \in \mathbf{Z}\}$, we have

$$\|f\|^2 = \sum_{k \in \mathbf{N}^d, n \in \mathbf{Z}^d} |\langle g_{k,n}, f\rangle|^2. \tag{5.5}$$

To show that L is compact it is sufficient to prove that the set

$$S = \{f \in L^2 \ : \ \sum_{k \in \mathbf{N}^d, n \in \mathbf{Z}^d} |\langle g_{k,n}, f\rangle|^2 \sigma_{k,n} \leq 1\}$$

where $\sigma_{k,n} = \sigma(n + \frac{\kappa}{2}, l)$ for $k = 2l + \kappa \in \mathbf{N}^d$, $\kappa \in \{0,1\}^d$, is compact in $L^2(\mathbf{R}^d)$.

Let $(f_n)_{n \in \mathbf{N}}$ be a sequence in S. We need to show that it contains a convergent subsequence. Since $(f_n)_{n \in \mathbf{N}}$ is bounded in $L^2(\mathbf{R}^d)$ it contains a weakly convergent subsequence $(f_{l_n})_{l_n \in \mathbf{N}}$, i.e. there exists $f \in S$ such that

$$\lim_{n \to \infty} (f_{l_n} - f, \phi) = 0, \quad \forall \phi \in L^2(\mathbf{R}^d). \tag{5.6}$$

Let us show that this convergence is strong, i.e. $\lim_{n\to\infty}\|f_{l_n}-f\|=0$, or, equivalently

$$\lim_{n\to\infty}\sum_{k\in\mathbf{N}^d,m\in\mathbf{Z}^d}|\langle f_{l_n}-f,g_{k,m}\rangle|^2=0. \tag{5.7}$$

Since $\lim_{|k|+|m|\to\infty}\sigma_{k,m}=\infty$, for arbitrary $\varepsilon/4>0$ there is a finite number of indices (k,m) such that $\dfrac{1}{\sigma_{k,m}}\le\dfrac{\varepsilon}{4}$. Denote the set of indices by I_1. Then we have

$$\sum_{k,m)\notin I_1}|\langle f_{l_n}-f,g_{k,m}\rangle|^2\frac{\sigma_{k,m}}{\sigma_{k,m}}$$
$$<2\frac{\varepsilon}{4}\Big(\sum_{(k,m)\notin I_1}|\langle f_{l_n},g_{k,m}\rangle|^2\sigma_{k,m}+\sum_{(k,m)\notin I_1}|\langle f,g_{k,m}\rangle|^2\sigma_{k,m}\Big)<\varepsilon.$$

Furthermore, since (5.6) implies that

$$\lim_{n\to\infty}\sum_{(k,m)\in I_1}|\langle f_{l_n}-f,g_{k,m}\rangle|^2=0,$$

by taking e.g. $\varepsilon=2^{-l_n}$ we obtain (5.7), hence S is compact in $L^2(\mathbf{R}^d)$.

The fact that eigenfunctions of L belong to $\mathcal{S}^{(1/\gamma)}$ is derived by part b) and $\cap_{s\ge0}M^{1,s}_{2,2}=\mathcal{S}^{(1/\gamma)}$. □

An immediate consequence of Theorem 5.5 is the continuity of the mapping $\sigma : M^{\sigma,s}_{p,q}\mapsto M^{1,s}_{p,q}$. Moreover, $\sigma(M^{\sigma,s}_{p,q})$, the image of $M^{\sigma,s}_{p,q}$ under σ, is a Banach subspace of $M^{1,s}_{p,q}$.

For spectral asymptotics we refer to [13],[40], [46].

We mention that regularity of solutions in Gelfand Shilov spaces $\mathcal{S}^{(1/\gamma)}$ to different classes of (pseudo)differential operators have been studied in [5], [6], [20].

Acknowledgment I am grateful to Professor Stevan Pilipović for many valuable discussions during the preparation of the paper. My thanks also goes to the Erwin Schrödinger Institute (ESI), Wien. On the base of their Junior Research Fellowship I visited ESI and Professor Hans Georg Feichtinger and his Numerical Harmonic Analysis Group in Wien, November 2004, when a part of the paper is prepared.

References

[1] R. Ashino, P. Boggiatto and M. W. Wong, *Advances in Pseudo-Differential Operators*, Operator Theory, Advances and Applications **155**, Birkhäuser, 2004.

[2] A. Beurling, *Quasi-analyticity and general distributions*, Lectures 4 and 5, AMS Summer Inst. Stanford, 1961.

[3] G. Björk, Linear partial differential operators and generalized distributions, *Ark. Mat.* **6** (1966), 351–407.

[4] R. W. Braun, R. Meise and B. A. Taylor, *Ultradifferentiable functions and Fourier analysis, Results in Mathematics* **17** (1990), 206–237.

[5] M. Cappiello, Pseudodifferential operators and spaces of type S, in "Progress in Analysis", Proceedings 3rd International ISAAC Congress, Editors: H. G. W. Begeher, R. P. Gilbert and M. W. Wong, **1** (2003), 681–688.

[6] M. Cappiello and L. Rodino SG-pseudodifferential operators and Gelfand-Shilov spaces, *Rocky Mount. J. Math.*, to appear.

[7] R. D. Carmichael, A. Kamiński and S. Pilipović, *Notes on Boundary Values in Ultradistribution Spaces*, Lecture Notes Series, **49**, Seoul National University, 1999.

[8] J. Cho, A characterization of Gelfand–Shilov space based on Wigner distribution, *Comm. Korean Math. Soc.* **14** no. **4** (1999), 761–767.

[9] J. Chung, S.-Y. Chung and D. Kim, A characterization for Fourier hyperfunctions, *Publ. RIMS. Kyoto Univ.* **30** (1994), 203–208.

[10] J. Chung, S.-Y. Chung and D. Kim, Characterization of the Gelfand–Shilov spaces via Fourier transforms, *Proc. Amer. Math. Soc.* **124** (7) (1996), 2101–2108.

[11] I. Cioranescu and L. Zsidó, ω-ultradistributions and their application to operator theory, *Spectral Theory, Banach Center Publ.* **8** (1982), 77–220.

[12] E. Cordero and K. Gröchenig, Time-frequency analysis of localization operators, *J. Funct. Anal.* **205** (2003), 107–131.

[13] W. Czaja and Z. Rzeszotnik, Pseudodifferential operators and Gabor frames: spectral asymptotics, *Math. Nachr.* **233-234** (2002), 77–88.

[14] H. G. Feichtinger, Modulation Spaces on Locally Compact Abelian Groups, in *Wavelets and Their Applications*, Editors: M. Krishna, R. Radha and S. Thangavelu, Allied Publishers, 2003, 99–140.

[15] H. G. Feichtinger and K. Gröchenig, Gabor frames and time-frequency analysis of distributions, *J. Funct. Anal.* **146** (1997), 464–495.

[16] H. G. Feichtinger and T. Strohmer, *Advances in Gabor Analysis*, Birkhäuser, 2003.

[17] G. B. Folland, *Harmonic Analysis in Phase Space* Annals of Mathematical Studies **122**, Princeton University Press, 1989.

[18] I. M. Gelfand and G. E. Shilov, Generalized Functions II, III Academic Press, 1967.

[19] T. Gramchev, Classical pseudodifferential operators and Egorov's theorem in Gevrey classes G^s, $s > 1$, *Uspekhi Mat. Nauk* **40** (6) (1985), 137–138.

[20] T. Gramchev, Pertubartive methods in scales of Banach spaces: Applications for Gevrey reguularity of solutions to semilinear partial diffeential equations, *Rend. Sem. Mat. Univ. Pol. Torino* **61** (2) (2003), 101–134.

[21] K. Gröchenig, *Foundations of Time-Frequency Analysis*, Birkhäuser, 2001.

[22] K. Gröchenig, Uncertainty principles for time-frequency representations, in "Advances in Gabor Analysis", Editors: H. G. Feichtinger and T. Strohmer, Birkhäuser, 2003, 11–30.

[23] K. Gröchenig, A pedestrian approach to pseudodifferential operators, in "Harmonic Analysis and Applications, Volume in the honor of J. J. Benedetto" Editor: C. Heil, Birkhäuser, to appear.

[24] K. Gröchenig and C. Heil, Modulation spaces and pseudodifferential operators, Integral Equations Operator Theory **34** (1999), 439–457.

[25] K. Gröchenig and G. Zimmermann, Spaces of test functions via the STFT, *J. Funct. Spaces Appl.* **2** (1) (2004), 25–53.

[26] A. J. E. M. Janssen, *Application of the Wigner Distribution to Harmonic Analysis of Generalized Stochastic Processes*, Mathematisch Centrum, Amsterdam, 1979.

[27] A. J. E. M. Janssen and S. J. L. Van Eijndhoven, Spaces of type W, growth of Hermite coefficients, Wigner distribution and Bargmann transform, *J. Math. Anal. Appl.* **152** (1990), 368–390.

[28] N. Kaiblinger, *Metaplectic Representation, Eigenfunctions of Phase Space Shifts, and Gelfand-Shilov Spaces for LCA Groups* Ph.D. Thesis, Institut für Mathematik, Wien, 1999.

[29] A. Kamiński, D. Perišić and S. Pilipović, On various integral transformations of tempered ultradistributions, *Demonstratio Math.* **33(3)** (2000), 641-655.

[30] K. A. Okoudjou, Embeddings of some classical Banach spaces into modulation spaces, *Proc. Amer. Math. Soc.* **132** (2004), 1639-1647.

[31] D. Labate, Pseudodifferential operators on modulation spaces, *J. Math. Anal. Appl.* **262** (1) (2001), 242–255.

[32] H. Komatsu, Ultradistributions, I, structure theorems and a characterization, *J. Fac. Sci. Univ. Tokyo Sect. IA* **20** (1973), 25–105.

[33] H. Komatsu, Microlocal analysis in Gevrey classes and in complex domains, *Lecture Notes in Math., Springer* **1726** (1989), 426–493.

[34] H. Komatsu, *Introduction to the Theory of Generalized Functions*, Science University of Tokyo, 1999.

[35] T. Matsuzawa, Hypoellipticity in ultradistribution spaces, *J. Fac. Sci. Univ. Tokyo Sect. IA* **34** (1987), 779–790.

[36] S. Pilipović, Tempered ultradistributions, *Boll. Un. Mat. Ital.* **7** (2-B) (1988), 235–251.

[37] S. Pilipović, Characterization of bounded sets in spaces of ultradistributions, *Proc. Amer. Math. Soc.* **120** (1994), 1191–1206.

[38] S. Pilipović, *Microlocal analysis of ultradistributions*, *Proc. Amer. Math. Soc.* **126** (1) (1998), 105–113.

[39] S. Pilipović and N. Teofanov, Wilson bases and ultra-modulation spaces, *Math. Nachr.* **242** (2002), 179–196.

[40] S. Pilipović and N. Teofanov, Pseudodifferential operators on ultra-modulation spaces, *J. Funct. Analy.* **208** (2004), 194–228.

[41] L. Rodino, *Linear Partial Differential Operators in Gevrey Spaces,* World Scientific, 1993.

[42] C. Roumieu, Sur quelques extensions de la notion de distribution, *Ann. Sci. École Norm, Sup.* **77** (3) (1960), 41–121.

[43] L. Schwartz *Théorie des distributions,* Hermann, Paris, 1950–1951.

[44] M. A. Shubin, *Pseudodifferential Operators and Spectral Theory,* 2nd Edition, Springer, 2001.

[45] K. Tachizawa, The boundedness of pseudodifferential operators on modulation spaces, *Math. Nachr.* **168** (1994), 263–277.

[46] K. Tachizawa, *The Pseudodifferential operators and Wilson bases, J. Math. Pures Appl.* **75** (1996), 509–529.

[47] N. Teofanov, *A note on ultrapolynomials and the Wigner distribution*, *Novi Sad J. Math.* **30** (1) (2000), 165–175.

[48] J. Toft, Continuity properties for modulation spaces, with applications to pseudo-differential calculus, I, *J. Funct. Anal.* **207** (2004), 399–429.

[49] J. Toft, Continuity properties for modulation spaces, with applications to pseudo-differential calculus, II, *Ann. Glob. Anal. Geom.* **26** (2004), 73–106.

[50] M. W. Wong, *An Introduction to Pseudo-Differential Operators*, 2nd Edition, World Scientific, 1999.

[51] M. W. Wong, *Weyl Transforms*, Springer-Verlag, 1998.

[52] M. W. Wong, *Wavelet Transforms and Localization Operators*, Birkhäuser, 2002.

[53] L. Zanghirati, Pseudodifferential operators of infinite order and Gevrey classes, *Ann. Univ. Ferrara - Sez. VII - Sc. Mat.* **31** (1985), 197–219.

Nenad Teofanov
Department of Mathematics and Informatics
Trg D. Obradovića 4
21000 Novi Sad
Serbia and Montenegro

e-mail: `tnenad@im.ns.ac.yu`

Operator Theory:
Advances and Applications, Vol. 164, 193–209

Frames and Generalized Shift-Invariant Systems

Ole Christensen

Abstract. With motivation from the theory of Hilbert–Schmidt operators we review recent topics concerning frames in $L^2(\mathbb{R})$ and their duals. Frames are generalizations of orthonormal bases in Hilbert spaces. As for an orthonormal basis, a frame allows each element in the underlying Hilbert space to be written as an unconditionally convergent infinite linear combination of the frame elements; however, in contrast to the situation for a basis, the coefficients might not be unique. We present the basic facts from frame theory and the motivation for the fact that most recent research concentrates on tight frames or dual frame pairs rather than general frames and their canonical dual. The corresponding results for Gabor frames and wavelet frames are discussed in detail.

Mathematics Subject Classification (2000). Primary 42C15; Secondary 42C40.

Keywords. Frames, Hilbert–Schmidt integral operators, Gabor systems, wavelets, generalized shift-invariant spaces.

1. Introduction

An orthonormal basis in $L^2(\mathbb{R})$ is a useful tool in order to obtain expansions of functions in $L^2(\mathbb{R})$. However, the conditions defining an orthonormal basis are very restrictive, so often it is impossible to construct an orthonormal basis having extra prescribed properties. It was observed early that some of the limitations can be removed by considering Riesz bases rather than orthonormal bases. The purpose of this paper is to go one step further and study the advantages of constructing *frames* rather than bases. We give the formal definition in the next section. For now, we just mention that a frame is some kind of overcomplete basis: via a frame for $L^2(\mathbb{R})$, one can express each function $f \in L^2(\mathbb{R})$ as an unconditionally convergent infinite linear combination of the frame elements, exactly as we are used to for bases. However, the coefficients in the expansion are not necessarily unique. Theoretically this opens for the opportunity to *choose* the most convenient coefficients, and we will actually demonstrate that this is particulary useful for wavelet frames.

The overcompleteness of frames has already proved useful in the context of noise compression, and its use is currently investigated in several areas of signal processing. In this paper we concentrate on the mathematical properties. Section 2 presents the general theory for frames in Hilbert spaces, and discusses the reasons for the fact that most recent research concentrates on tight frames or dual pairs of frames, rather than general frames and their canonical dual. Section 3 concentrates on frames having the Gabor structure. Finally, Section 4 discusses wavelet frames, in particular, the recent extension principles for construction of tight frames.

As motivation for our presenting the results here, we mention an application of frames to the study of integral operators. Our approach follows the paper in [14] by Chris Heil closely.

Let $K \in L^2(\mathbb{R}^2)$ be a kernel function and define the associated integral operator $A_K : L^2(\mathbb{R}) \to L^2(\mathbb{R})$ by

$$(A_K f)(x) = \int_{-\infty}^{\infty} K(x, y) f(y) dy.$$

Letting $\{e_k\}_{k=1}^{\infty}$ be an orthonormal basis for $L^2(\mathbb{R}^2)$, we can write

$$K = \sum_{k=1}^{\infty} \langle K, e_k \rangle e_k$$

and thus

$$A_K = \sum_{k=1}^{\infty} \langle K, e_k \rangle A_{e_k}. \tag{1.1}$$

Suppose that $\{e_k\}_{k=1}^{\infty}$ is a tensor product,

$$e_k(x, y) = f_{m(k)}(x) \overline{f_{n(k)}(y)},$$

where $\{f_k\}_{k=1}^{\infty}$ is an orthonormal basis for $L^2(\mathbb{R})$; then

$$A_{e_k} f = \int_{-\infty}^{\infty} f_{m(k)}(x) \overline{f_{n(k)}(y)} f(y) dy = \langle f, f_{n(k)} \rangle f_{m(k)},$$

i.e., A_{e_k} is a rank-one operator. Thus, (1.1) decomposes A_K as a superposition of rank-one operators. Finite-dimensional approximations of A_K are obtained via the operators

$$A_K^j = \sum_{k=1}^{j} \langle K, e_k \rangle A_{e_k}, \ j \in \mathbb{N}.$$

We notice that A_K is a Hilbert Schmidt operator and therefore compact. Thus, it has a countable sequence of non-negative singular values,

$$s_1(A_K) \geq s_2(A_K) \geq \cdots$$

The eigenvalues are given by

$$s_j(A_K) = \sqrt{\lambda_j(A_K^* A_K)},$$

where $\lambda_j(A_K^* A_K)$ are the eigenvalues of the positive self-adjoint operator $A_K^* A_K$. Alternatively,

$$s_j(A_K) = \inf\{||A_K - T|| \ : \ \mathrm{rank}(T) \le j - 1.\}$$

Since $\mathrm{rank}(A_K^j) \le j$, this implies that

$$s_j(A_K) \le ||A_K - A_K^{j-1}||.$$

Using the above technique, Heil [14] obtains estimates for $s_j(A_K)$, as well as a sufficient condition for A_K being a trace class operator. In his paper, $\{e_k\}_{k=1}^{\infty}$ is formed as a tensor product of families of the form

$$g_{m,n}(x) = e^{2\pi i m b x} g(x - na), \quad m, n \in \mathbb{Z}, \tag{1.2}$$

where $g \in L^2(\mathbb{R})$ and $a, b > 0$. However, for technical reasons he has to relax the condition that the functions in (1.2) form an orthonormal basis. In fact, in the proof, it turns out to be necessary that g as well as $\hat{g}$ have fast decay. According to the famous *Balian-Low Theorem* this can not be combined with the requirement that the functions in (1.2) forms an orthonormal basis:

Theorem 1.1. *Assume that* $\{e^{2\pi i m b x} g(x - na)\}_{m,n\in\mathbb{Z}}$ *is an orthonormal basis for* $L^2(\mathbb{R})$. *Then*

$$\left(\int_{\mathbb{R}} |xg(x)|^2 dx\right) \left(\int_{\mathbb{R}} |\gamma \hat{g}(\gamma)|^2 d\gamma\right) = \infty.$$

In [14], Heil solves this problem by replacing the condition that the functions in (1.2) forms an orthonormal basis with the condition that they form a so-called *tight frame*. As explained above, this condition leads to an expansion similar to the one obtained via an orthonormal basis.

2. General Frame Theory

Let $\mathbf{H}$ be a separable Hilbert space with the inner product $\langle \cdot, \cdot \rangle$ linear in the first entry.

Definition 2.1. A countable family of elements $\{f_k\}_{k\in I}$ in $\mathbf{H}$ is a

(i) *Bessel sequence* if there exists a constant $B > 0$ such that

$$\sum_{k\in I} |\langle f, f_k \rangle|^2 \le B||f||^2, \ \forall f \in \mathbf{H};$$

(ii) *frame* for $\mathbf{H}$ if there exist constants $A, B > 0$ such that

$$A||f||^2 \le \sum_{k\in I} |\langle f, f_k \rangle|^2 \le B||f||^2, \ \forall f \in \mathbf{H}; \tag{2.1}$$

The numbers A, B in (2.1) are called *frame bounds.*

(iii) *Riesz basis* for $\mathbf{H}$ if $\overline{\text{span}}\{f_k\}_{k\in I} = \mathbf{H}$ and there exist constants $A, B > 0$ such that

$$A\sum |c_k|^2 \leq \left|\left|\sum c_k f_k\right|\right|^2 \leq B\sum |c_k|^2. \tag{2.2}$$

for all finite sequences $\{c_k\}$.

Every orthonormal basis is a Riesz basis, and every Riesz basis is a frame (the bounds A, B in (2.2) are frame bounds). That is, Riesz bases and frames are natural tools to gain more flexibility than possible with an orthonormal basis. For an overview of the general theory for frames and Riesz bases we refer to [4]; a deeper treatment is given in the book [5]. Here, we just mention that the difference between a Riesz basis and a frame is that the elements in a frame might be dependent. More precisely, a frame $\{f_k\}_{k\in I}$ is a Riesz basis if and only if

$$\sum_{k\in I} c_k f_k = 0, \ \{c_k\} \in \ell^2(I) \Rightarrow c_k = 0, \ \forall k \in I.$$

Given a frame $\{f_k\}_{k\in I}$, the associated *frame operator* is a bounded invertible operator on $\mathbf{H}$, defined by

$$Sf = \sum_{k\in I} \langle f, f_k\rangle f_k.$$

The series defining the frame operator converges unconditionally for all $f \in \mathbf{H}$. Via the frame operator we obtain the *frame decomposition*, representing each $f \in \mathbf{H}$ as an infinite linear combination of the frame elements:

$$f = SS^{-1}f = \sum_{k\in I} \langle f, S^{-1}f_k\rangle f_k. \tag{2.3}$$

The family $\{S^{-1}f_k\}_{k\in I}$ is itself a frame, called the *canonical dual frame*. In case $\{f_k\}_{k\in I}$ is a frame but not a Riesz basis, there exist other frames $\{g_k\}_{k\in I}$ which satisfy that

$$f = \sum_{k\in I} \langle f, g_k\rangle f_k, \ \forall f \in \mathbf{H}; \tag{2.4}$$

each family $\{g_k\}_{k\in I}$ with this property is called a *dual frame*. All dual frames associated to a given frame have been characterized by Li [19]:

Theorem 2.2. *Let $\{f_k\}_{k\in I}$ be a frame for $\mathbf{H}$. The dual frames of $\{f_k\}_{k\in I}$ are precisely the families*

$$\{g_k\}_{k\in I} = \left\{ S^{-1}f_k + h_k - \sum_{j\in I} \langle S^{-1}f_k, f_j\rangle h_j \right\}_{k\in I}, \tag{2.5}$$

where $\{h_k\}_{k\in I}$ is a Bessel sequence.

Note that if $\{f_k\}_{k\in I}$ is a Riesz basis, then $g_k = S^{-1}f_k$ for all choices of $\{h_k\}$. Note also that if $\{g_k\}_{k\in I}$ is *any* dual of a frame $\{f_k\}_{k\in I}$, then this particular dual arises from (2.5) by letting $h_k = g_k$.

Formula (2.3) is the main reason for considering frames, but it also immediately reveals one of the fundamental problems with frames. In fact, in order for (2.3) to be practically useful, one has to invert the frame operator, which is difficult when $\mathbf{H}$ is infinite-dimensional. One way to avoid this difficulty is to consider *tight* frames, i.e., frames $\{f_k\}_{k\in I}$ for which

$$\sum_{k\in I} |\langle f, f_k\rangle|^2 = A||f||^2, \ \forall f \in \mathbf{H} \tag{2.6}$$

for some $A > 0$. For a tight frame, $\langle Sf, f\rangle = A||f||^2$, which implies that $S = AI$, and therefore

$$f = \frac{1}{A}\sum_{k\in I}\langle f, f_k\rangle f_k, \ \forall f \in \mathbf{H}. \tag{2.7}$$

We note that there is a procedure for constructing a tight frame associated to an arbitrary frame $\{f_k\}_{k\in I}$: in fact, $\{S^{-1/2}f_k\}_{k\in I}$ is a tight frame with frame bound $A = 1$. However, in order to construct this frame, we need to find the operator $S^{-1/2}$, which is a nontrivial matter.

3. Gabor Frames

Given parameters $a, b \in \mathbb{R}$ and a function $g \in L^2(\mathbb{R})$, a frame for $L^2(\mathbb{R})$ of the form $\{e^{2\pi imbx}g(x-na)\}_{m,n\in\mathbb{Z}}$ is called a *Gabor frame*. The word "Weyl-Heisenberg frame" is also used. Notice that this is exactly the type of functions Heil used in the motivating example from Section 1. Using the operators "translation" resp. "modulation" acting on functions $g \in L^2(\mathbb{R})$ by

$$(T_a g)(x) = g(x-a), \ a > 0, \ \text{resp.} \ (E_b g)(x) = e^{2\pi ibx}g(x), \ b > 0,$$

a Gabor frame can be written

$$\{e^{2\pi imbx}g(x-na)\}_{m,n\in\mathbb{Z}} = \{E_{mb}T_{na}g\}_{m,n\in\mathbb{Z}}.$$

The origin of Gabor frames goes back to the paper [16], where Gabor propose to expand signals f as a series $f(x) = \sum c_{m,n}e^{2\pi imbx}g(x-na)$, where g is the Gaussian. The idea was to use the expansion for communications: instead of transmitting the function f, one could send the coefficients $c_{m,n}$. The original idea has apparently not been developed very far, but Gabor frames have proved very useful in many other contexts. See the research articles about Gabor systems (theory and applications) contained in the books [13],[14]. Another important source of information is the book [17] by Gröchenig, where Gabor frames are used in the context of time-frequency analysis.

It is important to notice that the frame operator S for a Gabor frame $\{E_{mb}T_{na}g\}_{m,n\in\mathbb{Z}}$ commutes with the inherent modulation and translation operators:

Lemma 3.1. *Let $g \in L^2(\mathbb{R})$ and $a, b > 0$ be given, and assume that $\{E_{mb}T_{na}g\}_{m,n\in\mathbb{Z}}$ is a Bessel sequence with frame operator S. Then*

$$SE_{mb}T_{na} = E_{mb}T_{na}S, \ \forall m, n \in \mathbb{Z}.$$

As a consequence of Lemma 3.1, also S^{-1} commutes with the operators $E_{mb}T_{na}$. Since $S^{-1/2}$ is a limit of polynomials in S^{-1} in the strong operator topology, this operator also commutes with $E_{mb}T_{na}$. Thus, Lemma 3.1 has the following important consequence:

Theorem 3.2. *Let $g\in L^2(\mathbb{R})$ and $a, b>0$ be given, and assume that $\{E_{mb}T_{na}g\}_{m,n\in\mathbb{Z}}$ is a Gabor frame. Then the canonical dual also has the Gabor structure and is given by $\{E_{mb}T_{na}S^{-1}g\}_{m,n\in\mathbb{Z}}$. The canonical tight frame associated with $\{E_{mb}T_{na}g\}_{m,n\in\mathbb{Z}}$ is $\{E_{mb}T_{na}S^{-1/2}g\}_{m,n\in\mathbb{Z}}$.*

Corollary 3.3. *There exists a tight Gabor frame $\{E_{mb}T_{na}g\}_{m,n\in\mathbb{Z}}$, for which g as well as $\hat{g}$ decay exponentially.*

Proof. Let $h(x) = e^{-x^2}$. Then $\{E_{mb}T_{na}h\}$ is a Gabor frame for any $a, b \in]0, 1[$, and by Theorem 3.2,

$$\{S^{-1/2}E_{mb}T_{na}h\} = \{E_{mb}T_{na}S^{-1/2}h\},$$

which is a tight frame. It has been proved by Janssen and Bölcskei [3] that the generator $g := S^{-1/2}h$ as well as its Fourier transform decay exponentially. □

The frames in Corollary 3.3 have very nice properties, but they are not given explicitly. Except the constructions given already in [11] and variations thereof [1], only few Gabor frames are known for which the generator as well as a dual are known explicitly, e.g., as finite linear combinations of elementary functions.

Theorem 3.2 is very important for computations of the inverse frame associated to a Gabor frame: instead of calculating the *double infinite* family

$$\{S^{-1}E_{mb}T_{na}g\}_{m,n\in\mathbb{Z}},$$

it is enough to find $S^{-1}g$ and then apply the modulation and translation operators. It also gives a reason that even if $\{E_{mb}T_{na}g\}_{m,n\in\mathbb{Z}}$ contains a Riesz basis as a subfamily, it might not be an advantage to remove elements from $\{E_{mb}T_{na}g\}_{m,n\in\mathbb{Z}}$: the computational benefits from the lattice structure of $\{(na, mb)\}_{m,n\in\mathbb{Z}}$ will be lost, the operators $E_{mb}T_{na}$ will in general no longer commute with the frame operator, and the dual frame will be more complicated to find.

The function $S^{-1}g$ is often called the *dual window function,* and Theorem 3.2 is just one good reason to stick to the frame it generates when working with a Gabor frame. Despite the many nice properties of the canonical dual, we now discuss other duals associated to a given frame $\{E_{mb}T_{na}g\}_{m,n\in\mathbb{Z}}$. There are several reasons to do this. First, one might be interested in duals minimizing other norms than the ℓ^2-norm of the coefficients in the frame expansions (the coefficients in the canonical frame expansion minimizes the ℓ^2-norm among all possible expansion coefficients). Second, there are cases where a function g generating a tight Gabor frame is badly localized; in this case one might prefer to search for a dual which is better localized than the canonical dual.

The general characterization of all dual frames in Theorem 2.2 of course also applies to Gabor frames, but if $\{E_{mb}T_{na}g\}_{m,n\in\mathbb{Z}}$ is an overcomplete frame, not all of these duals have the Gabor structure. The duals with Gabor structure are characterized in the famous *Wexler–Raz Theorem* [22]:

Theorem 3.4. *Let $g, h \in L^2(\mathbb{R})$ and $a, b > 0$ be given. Then, if the two Gabor systems $\{E_{mb}T_{na}g\}_{m,n\in\mathbb{Z}}$ and $\{E_{mb}T_{na}h\}_{m,n\in\mathbb{Z}}$ are Bessel sequences, they are dual frames if and only if*

$$\langle h, E_{m/a}T_{n/b}g\rangle = 0 \text{ for all } (m,n) \neq (0,0) \text{ and } \langle h, g\rangle = ab. \tag{3.1}$$

Another characterization of the Gabor duals can be obtained via Theorem 2.2:

Proposition 3.5. *Let $g \in L^2(\mathbb{R})$ and $a, b > 0$ be given, and assume that $\{E_{mb}T_{na}g\}_{m,n\in\mathbb{Z}}$ is a frame for $L^2(\mathbb{R})$. Then the duals having Gabor structure are precisely the families $\{E_{mb}T_{na}h\}_{m,n\in\mathbb{Z}}$ where*

$$h = S^{-1}g + f - \sum_{m,n\in\mathbb{Z}} \langle S^{-1}g, E_{mb}T_{na}g\rangle E_{mb}T_{na}f \tag{3.2}$$

for some function $f \in L^2(\mathbb{R})$ for which $\{E_{mb}T_{na}f\}_{m,n\in\mathbb{Z}}$ is a Bessel sequence.

Proof. It is proved in [19] that all functions h of the given form generate Gabor duals. On the other hand, if $\{E_{mb}T_{na}k\}_{m,n\in\mathbb{Z}}$ is a Gabor dual, let $f = k$, and the formula in (3.2) gives the function $h = k$, i.e., it returns the frame $\{E_{mb}T_{na}k\}_{m,n\in\mathbb{Z}}$. □

4. Wavelet Frames

Given a function $\psi \in L^2(\mathbb{R})$ and parameters $a > 1, b > 0$, the associated *wavelet system* is the collection of functions $\{a^{j/2}\psi(a^jx - kb)\}_{j,k\in\mathbb{Z}}$. A frame of this type is called a *wavelet frame*. The definition shows that all the functions in the wavelet frame are generated by certain scalings and translations of the single function ψ, a feature which is very important in computations. A slight generalization is to consider frames generated by scaling and translating of a *finite* collection of functions $\psi_1, \dots, \psi_n$; a frame $\{a^{j/2}\psi_\ell(a^jx - kb)\}_{j,k\in\mathbb{Z},\ell=1,\dots,n}$ is called a *multiwavelet frame*.

Wavelet frames are typically defined in terms of properties of the Fourier transform of the generating functions.

In this paper we will concentrate on wavelet frames with scaling parameter $a = 2$ and translation parameter $b = 1$. Several results will be formulated in terms of the translation and scaling operators defined on $L^2(\mathbb{R})$ by

$$(T_k f)(x) = f(x-k), \ k \in \mathbb{Z}, \quad (Df)(x) = 2^{1/2} f(2x).$$

A frame $\{2^{j/2}\psi(2^j x - k)\}_{j,k\in\mathbb{Z}}$ is said to be *dyadic*; in terms of the operators D^j, T_k it can be written as $\{D^j T_k \psi\}_{j,k\in\mathbb{Z}}$. We will frequently use the short notation $\{\psi_{j,k}\}_{j,k\in\mathbb{Z}}$. Correspondingly, a multiwavelet will be written $\{D^j T_k \psi_\ell\}_{j,k\in\mathbb{Z},\ell=1,\dots,n}$ or $\{\psi_{\ell;j,k}\}_{j,k\in\mathbb{Z},\ell=1,\dots,n}$.

Letting S denote the frame operator associated with a frame $\{\psi_{j,k}\}_{j,k\in\mathbb{Z}}$, we have already noted that the frame decomposition

$$f = \sum_{j,k\in\mathbb{Z}} \langle f, S^{-1}\psi_{j,k}\rangle \psi_{j,k} \tag{4.1}$$

is difficult to apply due to the presence of the operator S^{-1}. Another annoying fact is that the canonical dual frame $\{S^{-1}\psi_{j,k}\}_{j,k\in\mathbb{Z}}$ usually does not have the wavelet structure; see the example below, which appeared in [9]:

Example. Let $\{\psi_{j,k}\}_{j,k\in\mathbb{Z}}$ be a wavelet orthonormal basis for $L^2(\mathbb{R})$. Given $\epsilon \in]0,1[$, we define a function θ by

$$\theta = \psi + \epsilon D\psi.$$

One can prove that $\{\theta_{j,k}\}_{j,k\in\mathbb{Z}}$ is a Riesz basis, but the dual frame does *not* have the wavelet structure. There are other properties which are not inherited by the dual. For example, if we assume that the function ψ has compact support, then also θ has compact support, and all the functions $\{\theta_{j,k}\}_{j,k\in\mathbb{Z}}$ have compact support. However, $S^{-1}\theta_{j,0}$ is not compactly supported. □

The above example provides us with a good reason to consider tight wavelet frames, because the dual frame automatically has the right structure in this case. However, for general frames this is the point where it is natural to exploit the potential overcompleteness of frames. In other words, if $\{\psi_{j,k}\}_{j,k\in\mathbb{Z}}$ is a frame but not a Riesz basis, we know that for given $f \in L^2(\mathbb{R})$ there exist coefficients $\{c_{j,k}\}_{j,k\in\mathbb{Z}} \neq \{\langle f, S^{-1}\psi_{j,k}\rangle\}_{j,k\in\mathbb{Z}}$ such that $f = \sum_{j,k\in\mathbb{Z}} c_{j,k}\psi_{j,k}$. Thus, it is natural to ask if we can find a function $\tilde{\psi}$ such that

$$f = \sum_{j,k\in\mathbb{Z}} \langle f, \tilde{\psi}_{j,k}\rangle \psi_{j,k}, \ \forall f \in L^2(\mathbb{R}).$$

Definition 4.1. Consider two sequences of functions $\psi_1, \dots, \psi_n \in L^2(\mathbb{R})$, resp. $\tilde{\psi}_1, \dots, \tilde{\psi}_n \in L^2(\mathbb{R})$. Then $\{D^j T_k \psi_\ell\}_{j,k\in\mathbb{Z},\ell=1,\dots,n}$ and $\{D^j T_k \tilde{\psi}_\ell\}_{j,k\in\mathbb{Z},\ell=1,\dots,n}$ are called a pair of dual wavelet frames if both are Bessel sequences and

$$f = \sum_{\ell=1}^{N} \sum_{j,k\in\mathbb{Z}} \langle f, D^j T_k \psi_\ell\rangle D^j T_k \tilde{\psi}_\ell, \ \forall f \in L^2(\mathbb{R}). \tag{4.2}$$

That Bessel sequences $\{D^j T_k \psi_\ell\}_{j,k\in\mathbb{Z},\ell=1,\dots,n}$ and $\{D^j T_k \tilde{\psi}_\ell\}_{j,k\in\mathbb{Z},\ell=1,\dots,n}$ are frames if they satisfy (4.2) follows from Cauchy-Schwartz' inequality applied to (4.2): the lower frame bound for one of the sets of functions is implied by the upper bound for the other family. A pair of dual wavelet frames is also called *sibling frames* or *bi-frames.*

Bownik and Weber gave in [2] an example of a wavelet system $\{\psi_{j,k}\}_{j,k\in\mathbb{Z}}$ for which the canonical dual does not have the wavelet structure; however, in this particular case, there exist infinitely many functions $\tilde{\psi}$ for which $\{\tilde{\psi}_{j,k}\}_{j,k\in\mathbb{Z}}$ is a dual frame.

It is not clear which conditions we need to put on a function ψ in order to ensure that ψ generates a wavelet frame having a dual of the same structure. We can of course apply Theorem 2.2 in order to find all duals associated to a given wavelet frame $\{\psi_{j,k}\}_{j,k\in\mathbb{Z}}$: if a wavelet dual exists, it will appear this way. Note that when doing so, it is actually enough to let the Bessel sequence $\{h_k\}_{k\in I}$ run through all families of the form $\{D^jT_k\tilde{\psi}\}_{j,k\in mz}$ (this follows from the comment after Theorem 2.2).

A characterization of all pairs of dual wavelet frames was obtained by Frazier et al. [15]; for convenience we remove the specification of the indices $j,k\in\mathbb{Z},\ell=1,\dots,n$:

Theorem 4.2. *Let $\psi_1,\dots,\psi_n,\tilde{\psi}_1,\dots,\tilde{\psi}_n\in L^2(\mathbb{R})$ and assume that $\{D^jT_k\psi_\ell\}$ and $\{D^jT_k\tilde{\psi}_\ell\}$ are Bessel sequences. Then $\{D^jT_k\psi_\ell\}$ and $\{D^jT_k\tilde{\psi}_\ell\}$ are a pair of dual wavelet frames if and only if*

$$\sum_{\ell=1}^{n}\sum_{j\in\mathbb{Z}}\hat{\psi}_\ell(2^j\gamma)\overline{\hat{\tilde{\psi}}_\ell(2^j\gamma)}=1$$

and

$$\sum_{\ell=1}^{n}\sum_{j=0}^{\infty}\hat{\psi}_\ell(2^j\gamma)\overline{\hat{\tilde{\psi}}_\ell(2^j(\gamma+q))}=0 \text{ for all odd integers } q.$$

We now present results by Ron and Shen, which enable us to construct tight multiwavelet frames. The generators $\psi_1,\dots,\psi_n$ will be constructed on the basis of a function which satisfies a refinement equation, and since we will work with all those functions simultaneously it is convenient to change our previous notation slightly and denote the refinable function by ψ_0.

General Setup: Let $\psi_0\in L^2(\mathbb{R})$. Assume that $\lim_{\gamma\to 0}\hat{\psi}_0(\gamma)=1$ and that there exists a function $H_0\in L^\infty(\mathbb{T})$ such that

$$\hat{\psi}_0(2\gamma)=H_0(\gamma)\hat{\psi}_0(\gamma). \tag{4.3}$$

Let $H_1,\dots,H_n\in L^\infty(\mathbb{T})$, and define $\psi_1,\dots,\psi_n\in L^2(\mathbb{R})$ by

$$\hat{\psi}_\ell(2\gamma)=H_\ell(\gamma)\hat{\psi}_0(\gamma),\ \ell=1,\dots,n. \tag{4.4}$$

Finally, let H denote the $(n+1)\times 2$ matrix-valued function defined by

$$H(\gamma)=\begin{pmatrix} H_0(\gamma) & T_{1/2}H_0(\gamma)\\ H_1(\gamma) & T_{1/2}H_1(\gamma)\\ \cdot & \cdot \\ \cdot & \cdot \\ H_n(\gamma) & T_{1/2}H_n(\gamma)\end{pmatrix}. \tag{4.5}$$

We will frequently suppress the dependence on γ and simply speak about the matrix H. The purpose is to find $H_1,\dots,H_n$ such that

$$\{D^jT_k\psi_1\}_{j,k\in\mathbb{Z}}\cup\{D^jT_k\psi_2\}_{j,k\in\mathbb{Z}}\cup\cdots\cup\{D^jT_k\psi_n\}_{j,k\in\mathbb{Z}} \tag{4.6}$$

constitute a tight multiwavelet frame. The *unitary extension principle* by Ron and Shen shows that a condition on the matrix H will imply that the multiwavelet system in (4.6) is a tight frame for $L^2(\mathbb{R})$:

Theorem 4.3. *Let $\{\psi_\ell, H_\ell\}_{\ell=0,\dots,n}$ be as in the general setup, and assume that the 2×2 matrix $H(\gamma)^*H(\gamma)$ is the identity for a.e. γ. Then the multi-wavelet system $\{D^jT_k\psi_\ell\}_{j,k\in\mathbb{Z},\ell=1,\dots,n}$ constitutes a tight frame for $L^2(\mathbb{R})$ with frame bound equal to one.*

Ron and Shen showed in [21] how one can construct compactly supported tight spline frames based on Theorem 4.3:

Example. Fix any $m=1,2,\dots,$ and consider the function ψ_0 defined by

$$\hat{\psi}_0(\gamma)=\frac{\sin^{2m}(\pi\gamma)}{(\pi\gamma)^{2m}}.$$

ψ_0 is know as the *B-spline of order* $2m$, and

$$\psi_0=\chi_{[-\frac{1}{2},\frac{1}{2}]}*\chi_{[-\frac{1}{2},\frac{1}{2}]}*\cdots*\chi_{[-\frac{1}{2},\frac{1}{2}]}\quad (2m \text{ factors}).$$

Let $\begin{pmatrix}2m\\ \ell\end{pmatrix}$ denote the binomial coefficients $\frac{(2m)!}{(2m-\ell)!\ell!}$ and define the 1-periodic bounded functions $H_1,H_2,\dots,H_{2m}$ by

$$H_\ell(\gamma)=\sqrt{\begin{pmatrix}2m\\ \ell\end{pmatrix}}\sin^\ell(\pi\gamma)\cos^{2m-\ell}(\pi\gamma).$$

Then one can show that $M:=H^*H$ is the identity on $\mathbb{C}^2$ for all γ; by Theorem 4.3 this implies that the $2m$ functions $\psi_1,\dots,\psi_{2m}$ defined by

$$\begin{aligned}\hat{\psi}_\ell(\gamma) &= H_\ell(\gamma/2)\psi_0(\gamma/2)\\ &= \sqrt{\begin{pmatrix}2m\\ \ell\end{pmatrix}}\frac{\sin^{2m+\ell}(\pi\gamma/2)\cos^{2m-\ell}(\pi\gamma/2)}{(\pi\gamma/2)^{2m}}\end{aligned}$$

generate a multiwavelet frame for $L^2(\mathbb{R})$.

Frequently one takes a slightly different choice of H_ℓ, namely,

$$H_\ell(\gamma) = i^\ell \sqrt{\binom{2m}{\ell}} \sin^\ell(\pi\gamma)\cos^{2m-\ell}(\pi\gamma).$$

Inserting this expression in $\hat{\psi}_\ell(\gamma) = H_\ell(\gamma/2)\hat{\psi}_\ell(\gamma/2)$ and using the commutator relations for the operators $\mathbf{F}$ (the Fourier transform), D, T_k shows that ψ_ℓ is a finite linear combination with real coefficients of the functions

$$DT_k\psi_0, \; k = -m, \dots, m.$$

It follows that ψ_ℓ is a real-valued spline with support in $[-m, m]$, degree $2m-1$, smoothness class C^{2m-2}, and knots at $\mathbb{Z}/2$. Note in particular that we obtain smoother generators by starting with higher order splines, but that the price to pay is that the number of generators increases as well. □

An important reformulation of Theorem 4.3 was obtained simultaneously by Daubechies, Han, Ron and Shen in [12] and Chui, He and Stöckler in [8]. It is called the *oblique extension principle:*

Theorem 4.4. *Let $\{\psi_\ell, H_\ell\}_{\ell=0}^n$ be as in the general setup. Assume that there exists a strictly positive function $\theta \in L^\infty(\mathbb{T})$ for which $\lim_{\gamma\to 0}\theta(\gamma) = 1$ and such that for a.e. γ,*

$$H_0(\gamma)\overline{H_0(\gamma+\nu)}\theta(2\gamma) + \sum_{\ell=1}^{n} H_\ell(\gamma)\overline{H_\ell(\gamma+\nu)} = \begin{cases} \theta(\gamma) & \text{if } \nu = 0 \\ 0 & \text{if } \nu = \frac{1}{2} \end{cases}. \tag{4.7}$$

Then the functions $\{D^jT_k\psi_\ell\}_{j,k\in\mathbb{Z},\ell=1,\dots,n}$ constitute a tight frame for $L^2(\mathbb{R})$ with frame bound equal to one.

By taking $\theta = 1$ in Theorem 4.4 we obtain Theorem 4.3. From the extra freedom in Theorem 4.4 concerning the choice of θ, one could expect it to be a more general result that Theorem 4.3, but the proof given in [12] shows that the class of frames which can be constructed is the same for the two Theorems. However, in practice Theorem 4.4 gives more flexibility because it naturally leads to some constructions one would not expect from Theorem 4.3. Let us explain this in more details. Suppose that ψ_0 is a compactly supported function satisfying (4.3) for some 1-periodic function $H_0 \in L^\infty(\mathbb{T})$, and that the functions H_ℓ, θ are chosen as trigonometric polynomials satisfying the conditions in Theorem 4.4. Writing $H_\ell(\gamma) = \sum_k c_{k\ell}e^{2\pi ik\gamma}$ (a finite sum), we see that

$$\hat{\psi}_\ell(2\gamma) = H_\ell(\gamma)\hat{\psi}_0(\gamma) = \mathbf{F}\sum_k c_{k\ell}T_{-k}\psi_0(\gamma).$$

This shows that the frame $\{D^jT_k\psi_\ell\}$ is generated by functions having compact support. The proof of Theorem 4.4 in [12] shows that the same frame can be constructed via Theorem 4.3, based on the function $\tilde{\psi}_0$ which might not be compactly supported; the fact that the resulting frame $\{D^jT_k\psi_\ell\}$ is generated by compactly supported functions is somewhat miraculous and could certainly not be predicted in advance. In short, this shows that there are constructions which appear naturally via Theorem 4.4, but one would not even think about constructing them via Theorem 4.3.

5. Generalized Shift-Invariant Systems

So far we have treated Gabor systems and wavelet systems separately. However, both are in fact special cases of the so-called *generalized shift-invariant systems,* i.e., families of the form $\{T_{C_jk}\phi_j\}_{j\in J,k\in\mathbb{Z}^d}$, where $\{C_j\}_{j\in J}$ is a countable collection of real invertible $d\times d$ matrices, and $\{\phi_j\}_{j\in J}\subset L^2(\mathbb{R}^d)$. Generalized shift-invariant systems were introduced in [18] and [20].

In this final section we present results from [6] concerning expansions in terms of such systems. To be more precise, our purpose is to obtain expansions of the type

$$f=\sum_{j\in J}\sum_{k\in\mathbb{Z}^d}\langle f,T_{C_jk}\tilde{\phi_j}\rangle T_{C_jk}\phi_j,\ \forall f\in\overline{\text{span}}\{T_{C_jk}\phi_j\}.$$

The novelty in [6] compared to the results in the literature is that it is not assumed that the generalized shift-invariant system is total; that is the space $\overline{\text{span}}\{T_{C_jk}\phi_j\}$ might be a proper subspace of $L^2(\mathbb{R})$.

Let us first consider the special case of a single generator. For an invertible matrix C, let $C^\sharp:=(C^{-1})^T$.

Proposition 5.1. *Let C be a real and invertible $d\times d$ matrix and $\phi,\tilde{\phi}\in L^2(\mathbb{R}^d)$. Assume that $\{T_{Ck}\phi\}_{k\in\mathbb{Z}^d}$ and $\{T_{Ck}\tilde{\phi}\}_{k\in\mathbb{Z}^d}$ are Bessel sequences. Then the following are equivalent:*

(i) $f=\sum_{k\in\mathbb{Z}^d}\langle f,T_{Ck}\tilde{\phi}\rangle T_{Ck}\phi,\ \forall f\in\overline{\mathit{span}}\{T_{Ck}\phi\}_{k\in\mathbb{Z}^d}$.

(ii) *On* $\{\gamma:\ \sum_{n\in\mathbb{Z}^d}|\hat{\phi}(\gamma+C^\sharp n)|^2\neq 0\}$, *we have*

$$\sum_{n\in\mathbb{Z}^d}\hat{\phi}(\gamma+C^\sharp n)\overline{\hat{\tilde{\phi}}(\gamma+C^\sharp n)}=|\det\ C|,\ a.e.$$

If the conditions are satisfied, then $\{T_{Ck}\phi\}_{k\in\mathbb{Z}^d}$ is a frame for $\overline{\mathit{span}}\{T_{Ck}\phi\}_{k\in\mathbb{Z}^d}$; furthermore, the orthogonal projection of $\{T_{Ck}\tilde{\phi}\}_{k\in\mathbb{Z}^d}$ onto $\overline{\mathit{span}}\{T_{Ck}\phi\}_{k\in\mathbb{Z}^d}$ is a frame for $\overline{\mathit{span}}\{T_{Ck}\phi\}_{k\in\mathbb{Z}^d}$.

The multigenerator case is similar, but technically and notationally more involved:

Theorem 5.2. *Assume that $\{T_{C_jk}\phi_j\}_{k\in\mathbb{Z}^d,j\in J}$ and $\{T_{C_jk}\tilde{\phi_j}\}_{k\in\mathbb{Z}^d,j\in J}$ are Bessel sequences. Then the following are equivalent:*

(i) $f=\sum_{j\in J}\sum_{k\in\mathbb{Z}^d}\langle f,T_{C_jk}\tilde{\phi_j}\rangle T_{C_jk}\phi_j,\ \forall f\in\mathcal{W}$.

(ii) *For all* $\ell \in J, m \in \mathbb{Z}^d$,

$$\hat{\phi}_\ell(\gamma) = \sum_{j\in J} \frac{1}{|\det\, C_j|}\hat{\phi}_j(\gamma)\left(\sum_{n\in\mathbb{Z}^d} e^{-2\pi i C_\ell m\cdot C_j^\sharp n}\hat{\phi}_\ell(\gamma + C_j^\sharp n)\overline{\hat{\tilde{\phi}}_j(\gamma + C_j^\sharp n)}\right).$$

If the conditions are satisfied, then $\{T_{C_j k}\phi_j\}_{k\in\mathbb{Z}^d, j\in J}$ *and* $\{P_{\mathcal{W}}T_{C_j k}\tilde{\phi}_j\}_{k\in\mathbb{Z}^d, j\in J}$ *are dual frames for* $\overline{span}\{T_{C_j k}\phi_j\}_{k\in\mathbb{Z}^d, j\in J}$.

In the summation over $n \in \mathbb{Z}^d$ in condition (ii), the indices j, n always appear in the combination $C_j^\sharp n$; let us consider all possible outcomes, i.e., let

$$\Lambda = \{C_j^\sharp n \ : \ j \in J, n \in \mathbb{Z}^d\}. \tag{5.1}$$

Given $\alpha \in \Lambda$, there might exist several pairs $(j,n) \in J \times \mathbb{Z}^d$ for which $\alpha = C_j^\sharp n$; let

$$J_\alpha = \{j \in J \ : \ \exists n \in \mathbb{Z}^d \text{ such that } \alpha = C_j^\sharp n\}.$$

If we are allowed to reorder the sum in condition (ii) of Theorem 5.2, we can rewrite the condition in terms of the index sets Λ and J_α:

Corollary 5.3. *Assume the setup from Theorem* 5.2 *and define* Λ *and* J_α *as above. Furthermore, assume that the series*

$$\sum_{\alpha\in\Lambda}\sum_{j\in J_\alpha} \frac{1}{|\det\, C_j|}\hat{\phi}_j(\gamma)e^{-2\pi i C_\ell m\cdot\alpha}\hat{\phi}_\ell(\gamma+\alpha)\overline{\hat{\tilde{\phi}}_j(\gamma+\alpha)}, \tag{5.2}$$

converges unconditionally for all $\ell \in J, m \in \mathbb{Z}$. *Then the following are equivalent:*

(i) $f = \sum_{j,k}\langle f, T_{C_j k}\tilde{\phi}_j\rangle T_{C_j k}\phi_j,\ \forall f \in \mathcal{W}$.

(ii) *For all* $\ell \in J$ *and all* $m \in \mathbb{Z}^d$,

$$\hat{\phi}_\ell(\gamma) = \sum_{\alpha\in\Lambda}\left(\sum_{j\in J_\alpha}\frac{1}{|\det\, C_j|}\hat{\phi}_j(\gamma)\overline{\hat{\tilde{\phi}}_j(\gamma+\alpha)}\right) e^{-2\pi i C_\ell m\cdot\alpha}\hat{\phi}_\ell(\gamma+\alpha).$$

In particular, the conditions (i) *and* (ii) *are satisfied if*

$$\sum_{j\in J_\alpha}\frac{1}{|\det\, C_j|}\hat{\phi}_j(\gamma)\overline{\hat{\tilde{\phi}}_j(\gamma+\alpha)} = \delta_{\alpha,0}$$

a.e. on $\{\gamma :\ \sum_{j\in J}|\hat{\phi}_j(\gamma)|^2 \neq 0\}$.

In the rest of this section we show how this general result simplifies in the case of Gabor systems and wavelet systems. For notational convenience we restrict our attention to the one-dimensional setting, i.e., systems in $L^2(\mathbb{R})$.

In general, a Gabor system does not have the shift-invariant structure. However, $\{E_{jb}T_{ka}\tilde{\phi}\}$ and $\{E_{jb}T_{ka}\phi\}$ are duals if and only if the systems $\{T_{ka}E_{jb}\tilde{\phi}\}$ and $\{T_{ka}E_{jb}\phi\}$ are duals, and these systems have the shift-invariant structure with

$$a_j = a, \phi_j = E_{jb}\phi, \tilde{\phi}_j = E_{jb}\tilde{\phi}.$$

It can be shown that Corollary 5.3 yields the following:

Corollary 5.4. *Assume that* $\{E_{jb}T_{ka}\tilde{\phi}\}_{j,k\in\mathbb{Z}}$ *and* $\{E_{jb}T_{ka}\phi\}_{j,k\in\mathbb{Z}}$ *are Bessel sequences, and that the series*

$$\sum_{j\in\mathbb{Z}}\sum_{n\in\mathbb{Z}}\hat{\phi}(\gamma+jb)\hat{\phi}(\gamma+n/a)\overline{\hat{\tilde{\phi}}(\gamma+n/a+jb)}$$

converges unconditionally. Then the following are equivalent:

(i) $f=\sum_{j\in\mathbb{Z}}\sum_{k\in\mathbb{Z}}\langle f,E_{jb}T_{ka}\tilde{\phi}\rangle E_{jb}T_{ka}\phi,\ \forall f\in\overline{span}\{E_{jb}T_{ka}\phi\}_{j,k\in\mathbb{Z}}$.

(ii)

$$\hat{\phi}(\gamma)=\sum_{n\in\mathbb{Z}}\left(\frac{1}{a}\sum_{j\in\mathbb{Z}}\hat{\phi}(\gamma+jb)\overline{\hat{\tilde{\phi}}(\gamma+jb+n/a)}\right)\hat{\phi}(\gamma+n/a).$$

In particular, the condition (ii) *is satisfied if*

$$\sum_{j\in J}\hat{\phi}(\gamma+jb)\overline{\hat{\tilde{\phi}}(\gamma+jb+n/a)}=a\delta_{n,0}$$

a.e. on $\{\gamma:\ \sum_{j\in J}|\hat{\phi}(\gamma+jb)|^2\neq 0\}$.

Note that, compared with Corollary 5.3, the dependence on the two parameters $\ell\in J$ and $m\in\mathbb{Z}$ is eliminated.

A wavelet system $\{D^jT_k\phi\}_{j,k\in\mathbb{Z}}$ has the structure of a generalized shift-invariant system with

$$a_j=2^{-j},\phi_j=D^j\phi.$$

Let

$$\Lambda=\{2^jn\ :\ j\in\mathbb{Z},n\in\mathbb{Z}\},$$

and for $\alpha\in\Lambda$,

$$J_\alpha=\{j\in\mathbb{Z}\ :\ \exists n\in\mathbb{Z}^d \text{ such that } \alpha=2^jn\}.$$

Using that $\mathbf{F}D^j=D^{-j}\mathbf{F}$, we can now formulate the condition (5.3) as

$$\hat{\phi}(2^{-\ell}\gamma)=\sum_{\alpha\in\Lambda}\left(\sum_{j\in J_\alpha}\hat{\phi}(2^{-j}\gamma)\overline{\hat{\tilde{\phi}}(2^{-j}(\gamma+\alpha))}\right)e^{-2\pi i2^{-\ell}m\alpha}\hat{\phi}(2^{-\ell}(\gamma+\alpha)).\quad(5.3)$$

This identity has to hold for all $\ell,m\in\mathbb{Z}$. As for Gabor systems, a simplification occurs, i.e., it is enough to check the condition for $\ell=0$ and $m\in\mathbb{Z}$:

Corollary 5.5. *The condition* (5.3) *is satisfied for all* $m,\ell\in\mathbb{Z}$ *if and only if for all* $m\in\mathbb{Z}$

$$\hat{\phi}(\gamma)=\sum_{\alpha\in\Lambda}\left(\sum_{j\in J_\alpha}\hat{\phi}(2^{-j}\gamma)\overline{\hat{\tilde{\phi}}(2^{-j}(\gamma+\alpha))}\right)e^{-2\pi im\alpha}\hat{\phi}(\gamma+\alpha).$$

In particular, this condition is satisfied if

$$\sum_{j\in J_\alpha} \hat{\phi}(2^{-j}\gamma)\overline{\hat{\tilde{\phi}}(2^{-j}(\gamma+\alpha))} = \delta_{\alpha,0}$$

a.e. on $\{\gamma : \sum_{j\in\mathbb{Z}} |\hat{\phi}(2^{-j}\gamma)|^2 \neq 0\}$.

6. Generalized Duals for Riesz Bases of Translates

Most of the results in the present paper are derived with the aim of constructing series expansions of the type (2.3) which are convenient to use in practice. This implies that we want $\{f_k\}_{k=1}^\infty$ and $\{g_k\}_{k=1}^\infty$ to have a common structure (Gabor structure, wavelet structure,..) and that the generators have compact support and are given explicitly. Recently, such expansions have been obtained for Riesz bases consisting of translates of a single function, i.e., in the setting of shift-invariant systems:

Theorem 6.1. *Assume that* $\phi \in L^2(\mathbb{R})$ *has support on an interval* $[0, N]$ *and that* $\{T_k\phi\}_{k\in\mathbb{Z}}$ *is a Riesz basis for its closed linear span. Then the following are equivalent:*

(i) *There exists a function* $\tilde{\phi}$ *with supp* $\tilde{\phi} \subseteq [0,1]$, *and such that*

$$f = \sum_{k\in\mathbb{Z}} \langle f, T_k\tilde{\phi}\rangle \phi, \ \forall f \in \overline{span}\{T_k\tilde{\phi}\}_{k\in\mathbb{Z}}. \tag{6.1}$$

(ii) *If* $\sum_{k=0}^{N-1} c_k\phi(x+k) = 0$ *for all* $x \in [0,1]$, *then* $c_0 = 0$.

In case the conditions are satisfied, we can choose $\tilde{\phi}$ *of the form*

$$\tilde{\phi}(x) = \left(\sum_{k=0}^{N-1} d_k\phi(x+k)\right) \chi_{[0,1]}(x) \tag{6.2}$$

for some scalar coefficients d_k. $\{T_k\tilde{\phi}\}_{k\in\mathbb{Z}}$ *is an orthogonal sequence.*

For the proof we refer to [7]. That paper contains several variations on the theme: for example, the condition (ii) implies that $\tilde{\phi}$ in (6.1) can be chosen as smooth as desired, even at the points $x = 0$ and $x = 1$. Regardless of the structure of the generator function ϕ, one can furthermore choose $\tilde{\phi}$ to be a polynomial on $[0,1]$.

We note that the function $\tilde{\phi}$ satisfying (6.1) is not required to belong to $\overline{\text{span}}\{T_k\phi\}_{k\in\mathbb{Z}}$, so $\{T_k\tilde{\phi}\}_{k\in\mathbb{Z}}$ is not a dual frame in the traditional sense; it is called an *oblique dual.*

The B-splines are natural candidates for the function ϕ. Recall that they are defined inductively by

$$B_1(x) = \chi_{[0,1]}, \ B_{N+1}(x) = (B_N * B_1)(x) = \int_0^1 B_N(x-t)dt.$$

With our definition, we have $\text{supp}B_N = [0, N]$. It is well-known that $\{T_k B_N\}_{k\in\mathbb{Z}}$ is a Riesz sequence for each $N \in \mathbb{N}$. Our results can be applied to all B-splines:

Proposition 6.2. *Let $N \in \mathbb{N}$. Then the functions $B_N(\cdot + k)$, $k = 0, \cdots, N-1$, are linearly independent on the interval $[0, 1]$. Thus, the sequence $\{T_k B_N\}_{k\in\mathbb{Z}}$ has an oblique dual generator of the form*

$$\tilde{\phi}(x) = \left(\sum_{k=0}^{N-1} d_k B_N(x+k) \right) \chi_{[0,1]}(x).$$

Acknowledgment: The author thanks the reviewer for several remarks improving the presentation.

References

[1] J. Benedetto and D. Walnut, Gabor frames for L^2 and related spaces, in *Wavelets: mathematics and applications,* Editors: J. Benedetto and M. Frazier, CRC Press, Boca Raton, 1993, 97–162.

[2] M. Bownik and E. Weber, Affine frames, GMRA's and the canonical dual, *Studia Math.* **159** no.3 (2003), 453–479.

[3] H. Bölcskei and A.J.E.M. Janssen, Gabor frames, unimodularity, and window decay, *J. Fourier Anal. Appl.* **6** no. 3 (2000), 255–276.

[4] O. Christensen, Frames, bases, and discrete Gabor/wavelet expansions, *Bull. Amer. Math. Soc.* **38** no. 3 (2001), 273–291.

[5] O. Christensen, *An Introduction to Frames and Riesz Bases,* Birkhäuser, Boston, 2003.

[6] O. Christensen and Y. C. Eldar, Generalized shift-invariant systems and frames for subspaces, *J. Fourier Anal. Appl.* **11** no. 3 (2005), 299–313.

[7] O. Christensen, H. O. Kim, R. Y. Kim and J. K. Lim, Riesz sequences of translates and oblique duals with support on $[0, 1]$, Preprint 2005.

[8] C. Chui, W. He and J. Stöckler, Compactly supported tight and sibling frames with maximum vanishing moments, *Appl. Comp. Harm. Anal.* **13** (2002), 224–262.

[9] I. Daubechies, The wavelet transform, time-frequency localization and signal analysis, *IEEE Trans. Inform. Theory* **36** (1990), 961–1005.

[10] I. Daubechies and B. Han, The canonical dual of a wavelet frame, *Appl. Comp. Harm. Anal.* **12** no.3 (2002), 269–285.

[11] I. Daubechies, A. Grossmann and Y. Meyer, Painless nonorthogonal expansions, *J. Math. Phys.* **27** (1986), 1271–1283.

[12] I. Daubechies, B. Han, A. Ron and Z. Shen, Framelets: MRA-based constructions of wavelet frames, *Appl. Comp. Harm. Anal.* **14** no. 1 (2003), 1–46.

[13] H.G. Feichtinger and T Strohmer, *Gabor Analysis and Algorithms: Theory and Applications,* Birkhäuser, Boston, 1998.

[14] H.G. Feichtinger and T. Strohmer, *Advances in Gabor Analysis,* Birkhäuser, Boston, 2002.

[15] M. Frazier, G. Garrigos, K. Wang and G. Weiss, A characterization of functions that generate wavelet and related expansion, *J. Fourier Anal. Appl.* **3** (1997), 883–906.

[16] D. Gabor, Theory of communications, *J. IEE (London)* **93** no. 3 (1946), 429–457.

[17] K. Gröchenig, *Foundations of Time-Frequency Analysis,* Birkhäuser, 2000.

[18] E. Hernandez, D. Labate and G. Weiss, A unified characterization of reproducing systems generated by a finite family II, *J. Geom. Anal.* **12** no. 4 (2002), 615–662.

[19] S. Li, On general frame decompositions, *Numer. Funct. Anal. and Optimiz.* **16** no. 9 & 10 (1995), 1181–1191.

[20] A. Ron and Z. Shen, Generalized shift-invariant systems, *Const. Approx.* **22** no.1 (2005), 1–45.

[21] A. Ron and Z. Shen, Compactly supported tight affine spline frames in $L_2(R^d)$, *Math. Comp.* **67** (1998), 191–207.

[22] J. Wexler and S. Raz, Discrete Gabor expansions, *Signal Processing* **21** (1990), 207–220.

Ole Christensen
Department of Mathematics
Technical University of Denmark
Building 303
Denmark
e-mail: `Ole.Christensen@mat.dtu.dk`

Operator Theory:
Advances and Applications, Vol. 164, 211–226

The Wigner Distribution of Gaussian Weakly Harmonizable Stochastic Processes

Patrik Wahlberg

Abstract. The paper treats the Wigner distribution of scalar-valued stochastic processes defined on $\mathbb{R}^d$. We show that if the process is Gaussian and weakly harmonizable then a stochastic Wigner distribution is well defined. The special case of stationary processes is studied, in which case the Wigner distribution is weakly stationary in the time variable and the variance is equal to the deterministic Wigner distribution of the covariance function.

Mathematics Subject Classification (2000). Primary 60G15, 60G12; Secondary 46G10, 60G57.

Keywords. Wigner distribution, Gaussian weakly harmonizable stochastic processes.

1. Introduction

The Wigner distribution is an important tool in several fields of study, in particular time-frequency analysis, the theory of pseudo-differential operators and quantum mechanics [8, 9, 11]. Within time-frequency analysis the Wigner distribution is one of several methods to represent an object (function, distribution) simultaneously in the time and frequency variables.

The Wigner distribution of deterministic functions and distributions is a well explored area. Within engineering, physical signals are however often modeled as random objects, ie stochastic processes or generalized stochastic processes. Therefore it is of interest for applications of time-frequency analysis within engineering to study the Wigner distribution of stochastic processes. The Wigner distribution is then a stochastic process or a generalized stochastic process.

A brief overview of the literature on the Wigner distribution of stochastic processes relevant to this paper follows. In [14] a test function space for generalized stochastic processes is introduced and the Wigner distribution discussed. In [18] a sufficient condition is given for the stochastic Wigner distribution of a stochastic process to exist as a stochastic Riemann integral. This condition is shown to be fulfilled if the process is Gaussian and the covariance function belongs to Feichtinger's algebra $S_0(\mathbb{R}^{2d})$ in [24]. In [7], [12] and [17] a theory of generalized stochastic processes using S_0 as a test function space is developed, and the expectation of the Wigner distribution of such a generalized stochastic process is studied.

Our contribution consists of showing that for a process defined on $\mathbb{R}^d$ a stochastic Wigner distribution $W : \mathbb{R}^d \times \mathcal{B}(\mathbb{R}^d) \ni (t, A) \mapsto H$, where H denotes the Hilbert space of zero mean second order stochastic variables, can be defined under the assumptions that the process is scalar-valued Gaussian and weakly harmonizable. In the complex-valued case we also impose *symmetry*, in a sense defined in Section 2, which gives a simplification compared to the real-valued case. The tools we use for the proof are some results for Gaussian processes, the theory of weakly harmonizable processes [15, 22] which involves bimeasure theory and integration with respect to vector-valued measures [6] for the spectral representation of such processes, and a result of R. C. Blei [2] which implies that a product of two bimeasures on $\mathbb{R}^d \times \mathbb{R}^d$ can be extended to a bimeasure on $\mathbb{R}^{2d} \times \mathbb{R}^{2d}$. We also specialize to the case of stationary processes and show that W is weakly stationary in the t variable then. The variance turns out to equal a constant times the deterministic Wigner distribution of the covariance function.

2. Gaussian Stochastic Processes on $\mathbb{R}^d$

We consider complex-valued stochastic processes [5] on $\mathbb{R}^d$, i.e. maps

$$X : \mathbb{R}^d \times \Omega \mapsto \mathbb{C} \tag{2.1}$$

where $(\Omega, \mathcal{B}, \mathbb{P})$ is a probability triple. Here Ω is the so called *sample space*, $\mathcal{B}$ is a σ-algebra of subsets of Ω (the "events"), and $\mathbb{P}$ is a measure $\mathbb{P} : \mathcal{B} \mapsto [0, 1]$ such that $\mathbb{P}(\Omega) = 1$ which assigns a probability to each event. The integral with respect to the measure $\mathbb{P}$ is denoted E for "expectation". For fixed $t \in \mathbb{R}^d$ $X(t, \cdot)$ is a random variable, and for fixed $\omega \in \Omega$ $X(\cdot, \omega)$ is a function on $\mathbb{R}^d$, called realization or trajectory, which depends on the random ω. When $d > 1$ a stochastic process is sometimes called a stochastic field. We restrict to zero mean real- or complex-valued Gaussian processes X [5, 13, 19]. Then by definition $E(X(t, \cdot)) \equiv 0$, and for any finite set $T = \{t_j\}_{j=1}^n \subset \mathbb{R}^d$, the $2n$-dimensional extended vector consisting of the real and the imaginary part of the"sampling vector" of the process $X_T = (X(t_1, \cdot), \ldots, X(t_n, \cdot))$, $X_T^e = (\Re X_T, \Im X_T)$, $X_T^e : \Omega \mapsto \mathbb{R}^{2n}$, has a $2n$-dimensional real Gaussian probability density. In the case when the covariance matrix

$$C = E\left((X_T^e)^t X_T^e\right) \in \mathbb{R}^{2d \times 2d} \tag{2.2}$$

is non-singular this means that for any $A \in \mathcal{B}(\mathbb{R}^{2n})$ (the Borel σ-algebra of $\mathbb{R}^{2n}$), the probability for $X_T^e \in A$ is given by

$$(2\pi)^{-n} (\det C)^{-1/2} \int_A e^{-\frac{1}{2} x^t C^{-1} x} dx. \tag{2.3}$$

More generally, Gaussianity can be defined also when the covariance matrix is singular using characteristic functions, by requiring that for any finite set $T = \{t_j\}_{j=1}^n \subset \mathbb{R}^d$ and any real vector $y = (y_1, \ldots, y_{2n}) \in \mathbb{R}^{2n}$ we have

$$E \exp\left(iy^t X_T^e\right) = \exp\left(-\frac{1}{2} y^t C y\right). \tag{2.4}$$

If the process X is zero mean Gaussian then $X(t,\cdot) \in H = \{x \in L^2(\mathbb{P}), E(x) = 0\}$, where H is a Hilbert space, for all $t \in \mathbb{R}^d$. The inner product in H is

$$(x,y)_H = E(x\overline{y}). \tag{2.5}$$

Modifying notation slightly we can consider the process to be a map

$$X : \mathbb{R}^d \mapsto H. \tag{2.6}$$

The covariance function is the function $\mathbb{R}^{2d} \mapsto \mathbb{C}$ defined by

$$r(t,u) = r_X(t,u) = E(X(t)\overline{X(u)}), \ t \in \mathbb{R}^d, \ u \in \mathbb{R}^d, \tag{2.7}$$

and it characterizes completely a Gaussian zero mean process.

In the case of complex-valued processes we assume that the process is *symmetric*, i.e. $\lambda X(t)$ and $X(t)$ have identical probability distribution laws for each $\lambda \in \mathbb{C}$ such that $|\lambda| = 1$ [13, 19]. This is equivalent to $E(X(t)) \equiv 0$ and $E(X(t)X(u)) \equiv 0$ [19].

Let $x_1, \ldots, x_n$ be complex-valued zero mean jointly Gaussian stochastic variables. Then according to Wick's theorem [13]

$$E(x_1 \cdots x_n) = \sum \prod_k E(x_{i_k} x_{j_k}) \tag{2.8}$$

where the sum is over all partitions of $\{1, \ldots, n\}$ into disjoint pairs $\{i_k, j_k\}$. Thus for $n = 4$ we have

$$E(x_1x_2x_3x_4) = E(x_1x_2)E(x_3x_4) + E(x_1x_3)E(x_2x_4) + E(x_1x_4)E(x_2x_3). \tag{2.9}$$

We define the zero mean process $X^* : \mathbb{R}^{2d} \mapsto H$ by

$$X^*(t) = X^*(t_1,t_2) = X(t_1)\overline{X(t_2)} - r(t_1,t_2), \ \mathbb{R}^{2d} \ni t = (t_1,t_2), \ t_1 \in \mathbb{R}^d, \ t_2 \in \mathbb{R}^d. \tag{2.10}$$

The process X^* takes its values in H due to (2.9), which in the case X is complex-valued symmetric implies

$$E(X(t_1)\overline{X(t_2)}\ \overline{X(u_1)}X(u_2)) = r(t_1,t_2)\overline{r(u_1,u_2)} + r(t_1,u_1)\overline{r(t_2,u_2)} \tag{2.11}$$

since the last term of (2.9) vanishes due to $E(X(t)X(u)) \equiv 0$. For a complex-valued symmetric Gaussian process X it follows from (2.11) and (2.7) that X^* has covariance

$$r_{X^*}(t,u) = r_{X^*}(t_1,t_2;u_1,u_2) = E(X^*(t_1,t_2)\overline{X^*(u_1,u_2)}) = r(t_1,u_1)\overline{r(t_2,u_2)}. \tag{2.12}$$

In the case of a real-valued Gaussian process X we obtain from (2.9)

$$r_{X^*}(t_1,t_2;u_1,u_2) = r(t_1,u_1)r(t_2,u_2) + r(t_1,u_2)r(t_2,u_1). \tag{2.13}$$

3. Scalar-Valued Weakly Harmonizable Stochastic Processes

A process X on $\mathbb{R}^d$ is *weakly harmonizable* [15, 22], denoted $X \in WH(\mathbb{R}^d, H)$, if its covariance function can be represented as

$$r(t,u) = \iint_{\mathbb{R}^{2d}} e^{i(t\xi - u\eta)} m(d\xi, d\eta), \tag{3.1}$$

where the *spectral measure* m is a non-negative definite *bimeasure*, denoted $m \in BM(\mathbb{R}^d \times \mathbb{R}^d)$. A bimeasure is a map $m : \mathcal{B}(\mathbb{R}^d) \times \mathcal{B}(\mathbb{R}^d) \mapsto \mathbb{C}$ which is a measure of bounded variation in each argument separately, i.e. $m(A, \cdot)$ and $m(\cdot, B)$ are measures of bounded variation for fixed $A, B \in \mathcal{B}(\mathbb{R}^d)$ [15, 20, 21, 22, 23, 25]. A bimeasure m has automatically finite *Fréchet variation* [15, 25], i.e.

$$\|m\|_F := \sup \left| \sum_{j,k} \alpha_j \overline{\beta_k} m(A_j, B_k) \right| < \infty, \tag{3.2}$$

where the supremum is taken over all finite partitions (i.e. disjoint families of sets) $\{A_j\} \subset \mathcal{B}(\mathbb{R}^d)$, all finite partitions $\{B_k\} \subset \mathcal{B}(\mathbb{R}^d)$, and all finite sets $\{\alpha_j\} \subset \mathbb{C}$, $\{\beta_k\} \subset \mathbb{C}$ (of cardinality equal to the cardinality of $\{A_j\}$ and $\{B_k\}$, respectively) such that $|\alpha_j| \leq 1$, $|\beta_k| \leq 1$ for all j and k. Since m is non-negative definite, a norm equivalent to (3.2) can be obtained with the partition $\{B_k\}$ replaced by $\{A_k\}$ and $\{\beta_k\}$ replaced by $\{\alpha_k\}$ [22]. If a set function m fulfills

$$\|m\|_V := \sup \sum_{j,k} |m(A_j, B_k)| < \infty, \tag{3.3}$$

supremum over finite partitions $\{A_j\} \subset \mathcal{B}(\mathbb{R}^d)$ and $\{B_k\} \subset \mathcal{B}(\mathbb{R}^d)$, it is said to be a measure of finite (Vitali) variation, denoted $m \in M(\mathbb{R}^{2d})$. We have the embeddings

$$M(\mathbb{R}^{2d}) \hookrightarrow BM(\mathbb{R}^d \times \mathbb{R}^d) \hookrightarrow PM(\mathbb{R}^{2d}) \tag{3.4}$$

where $PM(\mathbb{R}^{2d})$ denotes the set of pseudomeasures, i.e. the dual space of the Fourier algebra $\mathcal{F}L^1(\mathbb{R}^{2d})$ [16]. In fact, the left embedding follows from (3.2) and (3.3). The right embedding is proved in [10]. The subset of weakly harmonizable processes such that $m \in M(\mathbb{R}^{2d})$ in (3.1) are said to be *strongly harmonizable* [15, 22].

The integral (3.1) is a Morse-Transue (MT) integral, defined for a pair of functions by iterated Lebesgue integration [3, 20, 21, 22, 23, 25]. In general the iterated integral depends on the integration order. A pair of functions (f_1, f_2) is said to be *MT-integrable* if both iterated integrals exist and are equal. The MT integral differs in certain important respects from the Lebesgue (double) integral. It admits for example no Jordan decomposition. A sufficient condition for a pair of functions (f_1, f_2) to be MT-integrable [21] is that both iterated integrals exist and

$$\Lambda^*(|f_1|, |f_2|) < \infty, \tag{3.5}$$

where for non-negative functions f_1 and f_2 the superior integral Λ^* [20] is defined by

$$\Lambda^*(f_1,f_2) = \inf_{I^+ \ni p_j \geq f_j,\ j=1,2} \ \sup_{|u_j| \leq p_j, u_j \in C_c,\ j=1,2} \left| \iint u_1(\xi) u_2(\eta) m(d\xi, d\eta) \right|. \quad (3.6)$$

Here $I^+ = I^+(\mathbb{R}^d)$ denotes the set of non-negative lower semicontinuous functions, and $C_c = C_c(\mathbb{R}^d)$ the set of continuous functions of compact support. Since the function $f(\xi) = e^{it\xi}$ is bounded and measurable, the pair of functions $(e^{it\xi}, e^{-iu\eta})$ is MT-integrable for any bimeasure m [22, 23, 25], i.e. (3.1) is well defined for all $(t,u) \in \mathbb{R}^{2d}$.

In section 6 we will need the following lemma. We denote by $C(\mathbb{R}^d)$ the set of continuous and bounded functions.

Lemma 3.1. *Suppose $m \in BM(\mathbb{R}^d \times \mathbb{R}^d)$, $f, g \in C(\mathbb{R}^d)$, and $\kappa_2 \in \mathbb{R}^{2d \times 2d}$ is the matrix of a coordinate transformation which is block diagonal in the sense of*

$$\kappa_2 = \begin{pmatrix} \kappa_1 & 0 \\ 0 & \kappa_1 \end{pmatrix} \quad (3.7)$$

where $\kappa_1 \in \mathbb{R}^{d \times d}$ and $\det \kappa_1 = 1$. Then $m_{\kappa_1} := m \circ \kappa_2^{-1} \in BM(\mathbb{R}^d \times \mathbb{R}^d)$, and

$$\iint_{\mathbb{R}^{2d}} f(\kappa_1 \xi) g(\kappa_1 \eta) m(d\xi, d\eta) = \iint_{\mathbb{R}^{2d}} f(\xi) g(\eta) m_{\kappa_1}(d\xi, d\eta). \quad (3.8)$$

Proof. The pair $(f \circ \kappa_1, g \circ \kappa_1)$ is MT-integrable with respect to m, since $f \circ \kappa_1$ and $g \circ \kappa_1$ are bounded and measurable [25]. Since $\kappa_1^{-1} A \in \mathcal{B}(\mathbb{R}^d)$ if $A \in \mathcal{B}(\mathbb{R}^d)$, $m_{\kappa_1}(A,B) = m(\kappa_1^{-1}A, \kappa_1^{-1}B)$ defines a σ-additive measure of bounded variation in each variable separately. Obviously the Fréchet variation of m_{κ_1} equals the Fréchet variation of m. Hence $m_{\kappa_1} \in BM(\mathbb{R}^d \times \mathbb{R}^d)$ and (f,g) is MT-integrable with respect to m_{κ_1}. The set function $\mu(A) = \int_{\mathbb{R}^d} g(\eta) m(A, d\eta)$ is σ-additive and has bounded variation [3, 25]. Then by coordinate transformation of Lebesgue integrals on $\mathbb{R}^d$

$$\begin{aligned} \iint_{\mathbb{R}^{2d}} f(\kappa_1 \xi) g(\kappa_1 \eta) m(d\xi, d\eta) &= \int_{\mathbb{R}^d} f(\kappa_1 \xi) \left(\int_{\mathbb{R}^d} g(\kappa_1 \eta) m(d\xi, d\eta) \right) \\ &= \int_{\mathbb{R}^d} f(\kappa_1 \xi) \left(\int_{\mathbb{R}^d} g(\eta) \left(m \circ \begin{pmatrix} I & 0 \\ 0 & \kappa_1^{-1} \end{pmatrix} \right) (d\xi, d\eta) \right) \\ &= \int_{\mathbb{R}^d} f(\xi) \left(\int_{\mathbb{R}^d} g(\eta) m_{\kappa_1}(d\xi, d\eta) \right) \\ &= \iint_{\mathbb{R}^{2d}} f(\xi) g(\eta) m_{\kappa_1}(d\xi, d\eta). \end{aligned} \quad (3.9)$$

□

4. Spectral Representation of a Weakly Harmonizable Process

A process $X \in WH(\mathbb{R}^d, H)$ if and only if it can be written as an integral [15, 22]

$$X(t) = \int_{\mathbb{R}^d} e^{it\xi} Z(d\xi) \tag{4.1}$$

where $Z : \mathcal{B}(\mathbb{R}^d) \mapsto H$ is the *spectral vector measure* of the process. A vector measure [4, 6] is a σ-additive map from $\mathcal{B}(\mathbb{R}^d)$ to H (or more generally to a Banach space). A vector measure Z has automatically finite *semivariation* [6], which means that

$$\|Z\|_{sv} := \sup \left\| \sum_j \alpha_j Z(A_j) \right\|_H < \infty, \tag{4.2}$$

where the supremum is taken over all finite partitions $\{A_j\} \subset \mathcal{B}(\mathbb{R}^d)$ and finite sets $\{\alpha_j\} \subset \mathbb{C}$ such that $|\alpha_j| \leq 1$ for all j. We denote the set of vector measures $M_{sv}(\mathbb{R}^d, H)$. The relation between the bimeasure m and the vector measure Z is [15, 22]

$$E(Z(A)\overline{Z(B)}) = m(A, B), \; A, B \in \mathcal{B}(\mathbb{R}^d). \tag{4.3}$$

The integral (4.1) is a Dunford-Schwartz(-Bartle) integral [1, 6]. The formula (4.1) admits the interpretation that Z is the Fourier transform of the process X. Every $m \in BM(\mathbb{R}^d \times \mathbb{R}^d)$ defines $Z \in M_{sv}(\mathbb{R}^d, M(\mathbb{R}^d))$ by the isometric mapping $Z(A) = m(A, \cdot)$ [25]. $WH(\mathbb{R}^d, H)$ is a module over $\mathcal{L}(H)$, the set of bounded linear transformations on H, i.e. $T \in \mathcal{L}(H)$ and $X \in WH(\mathbb{R}^d, H)$ implies $T \circ X \in WH(\mathbb{R}^d, H)$ [22]. This is not true for the set of strongly harmonizable processes [22].

In Section 6 we will need the following lemma.

Lemma 4.1. *Suppose* $\Gamma \in M_{sv}(\mathbb{R}^{2d}, H)$ *and* $f, g \in C(\mathbb{R}^d)$. *Then* $f \otimes g$ *is integrable with respect to* Γ *and the integral can be written as an iterated integral*

$$\iint_{\mathbb{R}^{2d}} f(x)g(y)\Gamma(dx, dy) = \int_{y\in\mathbb{R}^d} g(y) \left(\int_{x\in\mathbb{R}^d} f(x)\Gamma(dx, dy) \right). \tag{4.4}$$

Proof. We use the theory of the Dunford-Schwartz integral as developed in [6], Chapter IV.10. The function $f \otimes g \in C(\mathbb{R}^{2d})$ is integrable with respect to Γ since it is bounded and measurable [1, 6]. We choose sequences $\{f_n\}_{n\geq1}$ and $\{g_n\}_{n\geq1}$ of *simple* functions, i.e. finite-range measurable functions, such that f_n converges uniformly to f and g_n converges uniformly to g. The simple functions have the form

$$\begin{aligned} f_n(x) &= \sum_j \alpha_{n,j} \chi_{A_{n,j}}(x), \\ g_n(y) &= \sum_j \beta_{n,j} \chi_{B_{n,j}}(y), \end{aligned} \tag{4.5}$$

where the sums are finite, $\{A_{n,j}\}_j \subset \mathcal{B}(\mathbb{R}^d)$ and $\{B_{n,j}\}_j \subset \mathcal{B}(\mathbb{R}^d)$ are partitions, and $\alpha_{n,j}, \beta_{n,j} \in \mathbb{C}$. The choice of f_n and g_n can be made such that $|f_n(x)| \le |f(x)|$ $\forall x$ and $|g_n(y)| \le |g(y)|$ $\forall y$. The integral of the simple function $f_n \otimes g_n$ is defined by

$$\iint_{\mathbb{R}^{2d}} f_n(x) g_n(y) \Gamma(dx, dy) = \sum_{i,j} \alpha_{n,j} \beta_{n,i} \Gamma(A_{n,j}, B_{n,i}). \tag{4.6}$$

From $|g_n(x)| \le |g(x)|$ we have $|f \otimes g_n(x,y)| \le |f \otimes g(x,y)|$ for all $(x,y) \in \mathbb{R}^{2d}$. Since $|f \otimes g|$ is a bounded measurable function it is integrable with respect to Γ [6]. Then, since $f \otimes g_n$ converges uniformly to $f \otimes g$, a dominated convergence theorem for vector measures (Theorem IV.10.10 in [6]) implies

$$\begin{aligned} \iint_{\mathbb{R}^{2d}} f(x) g(y) \Gamma(dx, dy) &= \lim_{n\to\infty} \iint_{\mathbb{R}^{2d}} f(x) g_n(y) \Gamma(dx, dy) \\ &= \lim_{n\to\infty} \sum_i \beta_{n,i} \iint_{\mathbb{R}^{2d}} f(x) \chi_{B_{n,i}}(y) \Gamma(dx, dy) \\ &= \lim_{n\to\infty} \sum_i \beta_{n,i} \iint_{\mathbb{R}^d \times B_{n,i}} f(x) \Gamma(dx, dy). \end{aligned} \tag{4.7}$$

Next we verify that the right hand side of (4.4) is well defined. For a fixed $B \in \mathcal{B}(\mathbb{R}^d)$ we define the set function $\Gamma_B(A) := \Gamma(A, B)$ which inherits σ-additivity and semivariation boundedness from Γ. Therefore we can integrate f with respect to Γ_B, and we obtain, again by the dominated convergence theorem,

$$\begin{aligned} \int_{\mathbb{R}^d} f(x) \Gamma_B(dx) &= \lim_n \int_{\mathbb{R}^d} f_n(x) \Gamma_B(dx) \\ &= \lim_n \sum_j \alpha_{n,j} \Gamma(A_{n,j}, B) \\ &= \lim_n \iint_{\mathbb{R}^d \times B} f_n(x) \Gamma(dx, dy) \\ &= \iint_{\mathbb{R}^d \times B} f(x) \Gamma(dx, dy) \\ &:= \Gamma_f(B), \end{aligned} \tag{4.8}$$

in the fourth equality using the dominated convergence theorem once more applied to the sequence $f_n \otimes 1$. The set function Γ_f is σ-additive [6] and has therefore bounded semivariation. Hence the right hand side integral of (4.4) is well defined.

Finally we obtain using (4.7), the dominated convergence theorem and (4.8)

$$\begin{aligned}\iint\limits_{\mathbb{R}^{2d}} f(x)g(y)\Gamma(dx,dy) &= \lim_{n\to\infty}\int_{\mathbb{R}^d} g_n(y)\Gamma_f(dy) \\ &= \int_{\mathbb{R}^d} g(y)\Gamma_f(dy) \\ &= \int_{y\in\mathbb{R}^d} g(y)\left(\int_{x\in\mathbb{R}^d} f(x)\Gamma(dx,dy)\right).\end{aligned} \tag{4.9}$$

□

5. The Wigner Distribution of a Deterministic Distribution

The Wigner distribution [9, 11] of a tempered distribution $f \in \mathcal{S}'(\mathbb{R}^d)$ is defined by

$$W(f) = \mathcal{F}_2(f \otimes \overline{f} \circ \kappa) \tag{5.1}$$

where κ denotes the $\mathbb{R}^{2d}$ coordinate transformation

$$\kappa(t,\tau) = (t+\tau/2, t-\tau/2),\ t \in \mathbb{R}^d,\ \tau \in \mathbb{R}^d, \tag{5.2}$$

and $\mathcal{F}_2$ denotes partial Fourier transformation in the second variable. We have $W(f) \in \mathcal{S}'(\mathbb{R}^{2d})$ since tensorization, coordinate transformation and partial Fourier transformation are bounded linear transformations on the Schwartz space $\mathcal{S}$. For $f \in L^2(\mathbb{R}^d)$ the Wigner distribution reduces to

$$W(f)(t,\xi) = \int_{\mathbb{R}^d} f(t+\tau/2)\overline{f(t-\tau/2)}e^{-i\tau\xi}d\tau \tag{5.3}$$

provided the Fourier transformation for $f \in L^1(\mathbb{R}^d)$ is defined by

$$\mathcal{F}f(\xi) = \widehat{f}(\xi) = \int_{\mathbb{R}^d} f(t)e^{-it\xi}dt, \tag{5.4}$$

which implies that the inverse Fourier transformation for $f \in L^1(\mathbb{R}^d)$ is defined by

$$\mathcal{F}^{-1}f(t) = (2\pi)^{-d}\int_{\mathbb{R}^d} f(\xi)e^{it\xi}dt. \tag{5.5}$$

6. The Wigner Distribution of a Weakly Harmonizable Gaussian Stochastic Process

From (5.1) we see that for $f \in \mathcal{S}'(\mathbb{R}^d)$ we can write

$$f \otimes \overline{f} \circ \kappa = \mathcal{F}_2^{-1}W(f). \tag{6.1}$$

We want to translate this formula to the case when f is replaced by a Gaussian weakly harmonizable process X. That is, we shall for a Gaussian $X \in WH(\mathbb{R}^d, H)$

prove the existence of a map $W : \mathbb{R}^d \times \mathcal{B}(\mathbb{R}^d) \mapsto H$ such that we have the partial inverse Fourier integral with respect to W

$$X^* \circ \kappa(t,\tau) = (2\pi)^{-d} \int_{\xi \in \mathbb{R}^d} e^{i\tau\xi} W(t, d\xi). \tag{6.2}$$

Then, according to (6.1) and definition (2.10), it is natural to interpret W as the Wigner distribution of X.

Remark 6.1. Strictly speaking W should be interpreted as the Wigner distribution of X if X^* were equal to $X \otimes \overline{X}$. However by (2.10) $X^* = X \otimes \overline{X} - E(X \otimes \overline{X})$. This modification can nevertheless be excused since we can add $\mathcal{F}_2(E(X \otimes \overline{X}) \circ \kappa)$, i.e. the Wigner distribution defined by (5.1) of the deterministic function $E(X \otimes \overline{X})$, to W to obtain a Wigner distribution of X with non-zero mean value. In the rest of the paper we concentrate on a W which fulfills (6.2) and call it the Wigner distribution of X.

Theorem 6.2. *Let X be a complex-valued symmetric Gaussian weakly harmonizable process $X : \mathbb{R}^d \mapsto H$. Then there exists a map $W : \mathbb{R}^d \times \mathcal{B}(\mathbb{R}^d) \mapsto H$ such that*

$$X^* \circ \kappa(t,\tau) = (2\pi)^{-d} \int_{\xi \in \mathbb{R}^d} e^{i\tau\xi} W(t, d\xi). \tag{6.3}$$

W is continuous in the first variable, and σ-additive with bounded semivariation in the second variable.

Proof. By (2.12) and (3.1) we can write r_{X^*} as the poly-MT-integral [3], defined by iterated integration,

$$\begin{aligned} &r_{X^*}(t_1, t_2; u_1, u_2) \\ &= \underset{\mathbb{R}^{4d}}{\int \cdots \int} e^{i\left((t_1,t_2)\cdot(\xi_1,\xi_2)-(u_1,u_2)\cdot(\eta_1,\eta_2)\right)} m(d\xi_1, d\eta_1)\overline{m(-d\xi_2, -d\eta_2)}. \end{aligned} \tag{6.4}$$

The iterated integral is independent of order since the integral (3.1) is independent of order. We define a set function on $\mathcal{B}(\mathbb{R}^d) \times \mathcal{B}(\mathbb{R}^d) \times \mathcal{B}(\mathbb{R}^d) \times \mathcal{B}(\mathbb{R}^d)$ by

$$m'(A_1, A_2; B_1, B_2) = m(A_1, B_1)\overline{m(-A_2, -B_2)}, \ A_1, A_2, B_1, B_2 \in \mathcal{B}(\mathbb{R}^d), \tag{6.5}$$

which inherits σ-additivity in each of the four variables separately from the corresponding property of m. Then, by a theorem of R. C. Blei [2], m' can be extended to a separately σ-additive map of bounded Fréchet variation on $\mathcal{B}(\mathbb{R}^{2d}) \times \mathcal{B}(\mathbb{R}^{2d})$. That is, there exists an extension of m', denoted m'', such that $m'' \in BM(\mathbb{R}^{2d} \times \mathbb{R}^{2d})$ and $m''(A_1 \times A_2; B_1 \times B_2) = m(A_1, B_1)\overline{m(-A_2, -B_2)}$ when $A_1, A_2, B_1, B_2 \in \mathcal{B}(\mathbb{R}^d)$.

Defining κ by (5.2), Lemma 3.1 gives

$$\begin{aligned} r_{X^*\circ\kappa}(t_1,\tau_1;t_2,\tau_2) &= E(X^*\circ\kappa(t_1,\tau_1)\overline{X^*\circ\kappa(t_2,\tau_2)}) \\ &= \int_{\mathbb{R}^{4d}}\cdots\int e^{i\left((t_1+\tau_1/2)\xi_1+(t_1-\tau_1/2)\xi_2-(t_2+\tau_2/2)\eta_1-(t_2-\tau_2/2)\eta_2\right)} m''(d\xi_1,d\xi_2;d\eta_1,d\eta_2) \\ &= \int_{\mathbb{R}^{4d}}\cdots\int e^{i(t_1,\tau_1)\cdot\kappa_1(\xi_1,\xi_2)-(t_2,\tau_2)\cdot\kappa_1(\eta_1,\eta_2)} m''(d\xi_1,d\xi_2;d\eta_1,d\eta_2) \\ &= \int_{\mathbb{R}^{4d}}\cdots\int e^{i\left((t_1,\tau_1)\cdot(\xi_1,\xi_2)-(t_2,\tau_2)\cdot(\eta_1,\eta_2)\right)} m''_{\kappa_1}(d\xi_1,d\xi_2;d\eta_1,d\eta_2), \end{aligned} \tag{6.6}$$

where

$$\kappa_1(\xi_1,\xi_2) = \left(\xi_1+\xi_2, \frac{\xi_1-\xi_2}{2}\right), \tag{6.7}$$

and $m''_{\kappa_1} = m''\circ\begin{pmatrix} \kappa_1^{-1} & 0 \\ 0 & \kappa_1^{-1} \end{pmatrix} \in BM(\mathbb{R}^{2d}\times\mathbb{R}^{2d})$ by Lemma 3.1. Hence according to the criterion (3.1) $X^*\circ\kappa(t,\tau)$ is weakly harmonizable and there exists therefore a spectral vector measure $\Gamma : \mathcal{B}(\mathbb{R}^{2d}) \mapsto H$ of bounded semivariation such that

$$X^*\circ\kappa(t,\tau) = \iint_{\mathbb{R}^{2d}} e^{i(\tau\xi+t\eta)}\Gamma(d\eta,d\xi). \tag{6.8}$$

By Lemma 4.1 we can write

$$X^*\circ\kappa(t,\tau) = (2\pi)^{-d}\int_{\xi\in\mathbb{R}^d} e^{i\tau\xi}(2\pi)^d\left(\int_{\eta\in\mathbb{R}^d} e^{it\eta}\Gamma(d\eta,d\xi)\right). \tag{6.9}$$

Thus we have a representation (6.3) with

$$W(t,B) = (2\pi)^d\int_{\eta\in\mathbb{R}^d} e^{it\eta}\Gamma(d\eta,B),\ t\in\mathbb{R}^d,\ B\in\mathcal{B}(\mathbb{R}^d). \tag{6.10}$$

$E(W)\equiv 0$ since $E(\Gamma)\equiv 0$. For fixed $B\in\mathcal{B}(\mathbb{R}^d)$, $\Gamma(\cdot,B)$ is a σ-additive vector measure of bounded semivariation. It follows from the dominated convergence theorem for vector measures [6] that $W(t,B)$ is continuous in the t variable for fixed $B\in\mathcal{B}(\mathbb{R}^d)$. For fixed $t\in\mathbb{R}^d$, $W(t,\cdot)$ is σ-additive [6], and has therefore bounded semivariation [6]. □

Next we treat the case of a real-valued Gaussian weakly harmonizable process.

Theorem 6.3. *Let X be a real-valued Gaussian weakly harmonizable process $X : \mathbb{R}^d \mapsto H$. Then there exists a map $W : \mathbb{R}^d\times\mathcal{B}(\mathbb{R}^d)\mapsto H$ such that*

$$X^*\circ\kappa(t,\tau) = (2\pi)^{-d}\int_{\xi\in\mathbb{R}^d} e^{i\tau\xi}W(t,d\xi). \tag{6.11}$$

W is continuous in the first variable, and σ-additive with bounded semivariation in the second variable.

Proof. By (2.13), (6.4) and (3.1) we can write r_{X^*} as the poly-MT-integral

$$r_{X^*}(t_1, t_2; u_1, u_2) = \int_{\mathbb{R}^{4d}} \cdots \int e^{i\left((t_1,t_2)\cdot(\xi_1,\xi_2)-(u_1,u_2)\cdot(\eta_1,\eta_2)\right)} m_1(d\xi_1, d\xi_2; d\eta_1, d\eta_2) \tag{6.12}$$

where

$$\begin{aligned} m_1(A_1, A_2; B_1, B_2) &= m(A_1, B_1)\overline{m(-A_2, -B_2)} + m(A_1, B_2)\overline{m(-A_2, -B_1)} \\ &= m'(A_1, A_2; B_1, B_2) + m_2'(A_1, A_2; B_1, B_2) \end{aligned} \tag{6.13}$$

where $m_2'(A_1, A_2; B_1, B_2) = m'(A_1, A_2; B_2, B_1)$. Then m_1 inherits σ-additivity in each of the four variables separately from the corresponding property of m. As in the proof of Theorem 6.2 the first term m' can be extended to a separately σ-additive map m'' of bounded Fréchet variation on $\mathcal{B}(\mathbb{R}^{2d}) \times \mathcal{B}(\mathbb{R}^{2d})$. Also the second term can be extended to a separately σ-additive map m_2'' of bounded Fréchet variation on $\mathcal{B}(\mathbb{R}^{2d}) \times \mathcal{B}(\mathbb{R}^{2d})$ by defining $m_2''(A; B) := m''(A; B^t)$ for all $A, B \in \mathcal{B}(\mathbb{R}^{2d})$ where $B^t = \{(x, y) \in \mathbb{R}^{2d} : (y, x) \in B\}$. The rest of the proof is identical to the proof of Theorem 6.2. □

7. The Wigner Distribution of Stationary Gaussian Processes

An important subset of the set of strongly harmonizable processes is the set of stationary processes. The defining property of a weakly stationary process X is that $E(X(t)) \equiv$ constant and there exists a function $r^s : \mathbb{R}^d \mapsto \mathbb{C}$ (index s for stationary) such that $r(t, u) = r^s(t - u)$ [15, 19, 22]. For Gaussian processes weak stationarity implies strict stationarity, i.e. translation invariance of the process probability law [5, 19]. Since r^s is a continuous non-negative definite function it can by Bochner's theorem [16] be written

$$r^s(t) = \int_{\mathbb{R}^d} e^{i\xi t} m^s(d\xi) \tag{7.1}$$

where m^s is a non-negative bounded measure. Comparison with (3.1) implies

$$m(\xi, \eta) = m^s\left(\frac{\xi + \eta}{2}\right) \delta_0(\xi - \eta) \tag{7.2}$$

where δ_0 denotes the Dirac measure at the origin, i.e. the spectral measure has support on the diagonal. The diagonality of m and (4.3) results in $E(Z(A)\overline{Z(B)}) = m^s(A \cap B)$, $A, B \in \mathcal{B}(\mathbb{R}^d)$.

In order to simplify the notation for computations involving δ_0 we will in the following sometimes denote $W(t, \xi)$ although $W(t, \cdot)$ is a vector measure and in general not well defined for $\xi \in \mathbb{R}^d$.

Theorem 7.1. *Let X be a complex-valued symmetric Gaussian stationary process $X : \mathbb{R}^d \mapsto H$. Then the covariance function of the Wigner distribution $W(t,\xi)$ has support on the diagonal of the (ξ_1, ξ_2) plane,*

$$E(W(t_1, A_1)\overline{W(t_2, A_2)}) = 0, \ A_1 \cap A_2 = \emptyset, \ A_1, A_2 \in \mathcal{B}(\mathbb{R}^d), \ t_1 \in \mathbb{R}^d, \ t_2 \in \mathbb{R}^d. \tag{7.3}$$

Furthermore $W(t,A)$ is weakly stationary in the t variable and fulfills $\forall t_0 \in \mathbb{R}^d$

$$r_W^s(t, A) := E(W(t_0 + t, A)\overline{W(t_0, A)}) = (2\pi)^d W(r^s)(t, A), \ t \in \mathbb{R}^d, \ A \in \mathcal{B}(\mathbb{R}^d), \tag{7.4}$$

where $W(r^s)$ is the deterministic Wigner distribution of r^s defined by (5.1).

Proof. We denote $t = (t_1, t_2) \in \mathbb{R}^{2d}$, $t_1 \in \mathbb{R}^d$, $t_2 \in \mathbb{R}^d$, and insert $r(t_1, u_1) = r^s(t_1 - u_1)$ into (2.12), which gives

$$r_{X^*}(t, u) = r_{X^*}(t_1, t_2; u_1, u_2) = r^s(t_1 - u_1)\overline{r^s(t_2 - u_2)} = r^s \otimes \overline{r^s}(t - u). \tag{7.5}$$

Thus X^* is weakly stationary and $r^s_{X^*} = r^s \otimes \overline{r^s}$. Hence, with κ defined by (5.2),

$$r_{X^* \circ \kappa}(t, u) = r_{X^*}(\kappa t, \kappa u) = r^s_{X^*} \circ \kappa(t - u) = r^s \otimes \overline{r^s} \circ \kappa(t - u), \tag{7.6}$$

which shows that $X^* \circ \kappa$ is also weakly stationary. Using (7.1) and $\overline{m^s} = m^s$ (since m^s is non-negative), we can write

$$\begin{aligned} r^s_{X^* \circ \kappa}(t) &= r^s \otimes \overline{r^s} \circ \kappa t \\ &= \iint_{\mathbb{R}^{2d}} e^{i(\eta_1(t_1 + t_2/2) - \eta_2(t_1 - t_2/2))} m^s(d\eta_1) m^s(d\eta_2) \\ &= \iint_{\mathbb{R}^{2d}} e^{i(t_1\xi_1 + t_2\xi_2)} m^s \otimes m^s \circ \kappa_2(d\xi_1, d\xi_2) \\ &= \int_{\mathbb{R}^{2d}} e^{it\xi} m^s \otimes m^s \circ \kappa_2(d\xi) \end{aligned} \tag{7.7}$$

where

$$\kappa_2(\xi_1, \xi_2) = (\xi_1/2 + \xi_2, -\xi_1/2 + \xi_2). \tag{7.8}$$

Thus the spectral measure of $X^* \circ \kappa$ is

$$m_{X^* \circ \kappa}(\xi, \eta) = m^s \otimes m^s \circ \kappa_2 \left(\frac{\xi + \eta}{2} \right) \delta_0(\xi - \eta), \ \xi, \eta \in \mathbb{R}^{2d}. \tag{7.9}$$

Since the assumptions of the theorem are stronger than the assumptions of Theorem 6.2, there is a spectral vector measure Γ as in (6.8), and we have by (4.3)

$$E(\Gamma(A)\overline{\Gamma(B)}) = m_{X^* \circ \kappa}(A, B), \ A \in \mathcal{B}(\mathbb{R}^{2d}), \ B \in \mathcal{B}(\mathbb{R}^{2d}). \tag{7.10}$$

Using (6.10), (7.9) and (7.10) we compute

$$\begin{aligned}
r_W(t_1,\xi_1;t_2,\xi_2) &= E(W(t_1,\xi_1)\overline{W(t_2,\xi_2)})\\
&= (2\pi)^{2d} \iint_{(\eta_1,\eta_2)\in\mathbb{R}^{2d}} e^{i(t_1\eta_1-t_2\eta_2)} E(\Gamma(d\eta_1,\xi_1)\overline{\Gamma(d\eta_2,\xi_2)})\\
&= (2\pi)^{2d} \int_{\eta\in\mathbb{R}^d} e^{i(t_1-t_2)\eta} m^s\otimes m^s\circ\kappa_2\left(d\eta,\frac{\xi_1+\xi_2}{2}\right)\delta_0(\xi_1-\xi_2).
\end{aligned} \tag{7.11}$$

Hence (7.3) is proved, and since the right hand side depends on t_1-t_2 only and $E(W)\equiv 0$ (because $E(\Gamma)\equiv 0$), we can also conclude that $W(t,A)$ is stationary in the t variable for fixed $A\in\mathcal{B}(\mathbb{R}^d)$. We obtain furthermore

$$\begin{aligned}
r_W^s(t,A) &:= E(W(t_0+t,A)\overline{W(t_0,A)})\\
&= (2\pi)^{2d}\int_{\eta\in\mathbb{R}^d} e^{it\eta} m^s\otimes m^s\circ\kappa_2\,(d\eta,A)\\
&= (2\pi)^{3d}\mathcal{F}_1^{-1}\left(m^s\otimes m^s\circ\kappa_2\right)(t,A).
\end{aligned} \tag{7.12}$$

From (7.1) we have in a distribution sense $m^s=\mathcal{F}^{-1}\widetilde{r^s}=(2\pi)^{-d}\mathcal{F}r^s$, where $\widetilde{r^s}(x):=r^s(-x)$ denotes coordinate reflection. Therefore we can compute with Fourier transforms in a distribution sense, using $\kappa_2^{-t}(x,y)=(x+y/2,-x+y/2)$, (5.2) and $\widetilde{r^s}=\overline{r^s}$,

$$\begin{aligned}
\mathcal{F}_1^{-1}(m^s\otimes m^s\circ\kappa_2) &= (2\pi)^{-2d}\mathcal{F}_1^{-1}(\mathcal{F}(r^s\otimes r^s)\circ\kappa_2)\\
&= (2\pi)^{-2d}\mathcal{F}_1^{-1}\mathcal{F}(r^s\otimes r^s\circ\kappa_2^{-t})\\
&= (2\pi)^{-2d}\mathcal{F}_2(r^s\otimes\overline{r^s}\circ\kappa)\\
&= (2\pi)^{-2d}W(r^s)
\end{aligned} \tag{7.13}$$

in the last equality using the definition (5.1). Thus insertion of (7.13) into (7.12) gives the desired result (7.4). □

The last theorem treats the case of a stationary real-valued process X, which gives a slight modification of Theorem 7.1.

Theorem 7.2. *Let X be a real-valued Gaussian stationary process $X:\mathbb{R}^d\mapsto H$. Then the covariance function of the Wigner distribution $W(t,\xi)$ has support on the union of the diagonal and the antidiagonal of the (ξ_1,ξ_2) plane. Furthermore $W(t,A)$ is weakly stationary in the t variable and fulfills $\forall t_0\in\mathbb{R}^d$, if $A\in\mathcal{B}(\mathbb{R}^d)$ is such that $A\cap(-A)=\emptyset$,*

$$\begin{aligned}
E(W(t_0+t,A)\overline{W(t_0,A)}) &= (2\pi)^d W(r^s)(t,A),\ \ t\in\mathbb{R}^d,\\
E(W(t_0+t,A)\overline{W(t_0,-A)}) &= (2\pi)^d W(r^s)(t,A),\ \ t\in\mathbb{R}^d,
\end{aligned} \tag{7.14}$$

where $W(r^s)$ is the deterministic Wigner distribution of r^s defined by (5.1).

Proof. Insertion of $r(t_1,u_1) = r^s(t_1-u_1)$ into (2.13) gives

$$r_{X^*}(t_1,t_2;u_1,u_2) = r^s \otimes r^s(t-u) + r^s \otimes r^s(t_1-u_2, t_2-u_1). \tag{7.15}$$

Thus X^* is not weakly stationary in this case. We have

$$\begin{aligned} r_{X^*\circ\kappa}(t,u) &= r_{X^*}(\kappa t, \kappa u) \\ &= r^s \otimes r^s \circ \kappa(t-u) + r^s \otimes r^s \left(t_1 - u_1 + \frac{t_2+u_2}{2}, t_1 - u_1 - \frac{t_2+u_2}{2} \right). \end{aligned} \tag{7.16}$$

The first term can be treated as in the proof of Theorem 7.1. The second term can be written with use of (7.1)

$$\begin{aligned} & r^s \otimes r^s \left(t_1 - u_1 + \frac{t_2+u_2}{2}, t_1 - u_1 - \frac{t_2+u_2}{2} \right) \\ &= \iint_{\mathbb{R}^{2d}} e^{i(\eta_1(t_1-u_1+(t_2+u_2)/2)-\eta_2(t_1-u_1-(t_2+u_2)/2))} m^s(d\eta_1) m^s(d\eta_2) \\ &= \iint_{\mathbb{R}^{2d}} e^{i((t_1-u_1)(\eta_1-\eta_2)+(t_2+u_2)(\eta_1+\eta_2)/2)} m^s(d\eta_1) m^s(d\eta_2) \\ &= \iint_{\mathbb{R}^{2d}} e^{i((t_1-u_1)\xi_1+(t_2+u_2)\xi_2)} m^s \otimes m^s \circ \kappa_2(d\xi_1, d\xi_2) \\ &= \iint_{\mathbb{R}^{2d}} e^{i((t_1,t_2)\cdot(\xi_1,\xi_2)+(u_1,u_2)\cdot(-\xi_1,\xi_2))} m^s \otimes m^s \circ \kappa_2(d\xi_1, d\xi_2) \\ &= \int \cdots \int_{\mathbb{R}^{4d}} e^{i(\xi t - \eta u)} m^s \otimes m^s \circ \kappa_2 \left(d\left(\frac{\xi_1+\eta_1}{2}\right), d\left(\frac{\xi_2-\eta_2}{2}\right) \right) \\ &\qquad \times \delta_0(\xi_1-\eta_1, \xi_2+\eta_2). \end{aligned} \tag{7.17}$$

Thus, using (7.9), the spectral measure of $X^* \circ \kappa$ is, with $\xi, \eta \in \mathbb{R}^{2d}$,

$$\begin{aligned} m_{X^*\circ\kappa}(\xi,\eta) &= m^s \otimes m^s \circ \kappa_2 \left(\frac{\xi_1+\eta_1}{2}, \frac{\xi_2+\eta_2}{2} \right) \delta_0(\xi_1-\eta_1, \xi_2-\eta_2) \\ &\quad + m^s \otimes m^s \circ \kappa_2 \left(\frac{\xi_1+\eta_1}{2}, \frac{\xi_2-\eta_2}{2} \right) \delta_0(\xi_1-\eta_1, \xi_2+\eta_2). \end{aligned} \tag{7.18}$$

Proceeding as in (7.10) and (7.11) gives

$$\begin{aligned} r_W(t_1,\xi_1;t_2,\xi_2) &= E(W(t_1,\xi_1)\overline{W(t_2,\xi_2)}) \\ &= (2\pi)^{2d} \iint_{(\eta_1,\eta_2)\in\mathbb{R}^{2d}} e^{i(t_1\eta_1 - t_2\eta_2)} E(\Gamma(d\eta_1,\xi_1)\overline{\Gamma(d\eta_2,\xi_2)}) \\ &= (2\pi)^{2d} \int_{\eta\in\mathbb{R}^d} e^{i(t_1-t_2)\eta} m^s \otimes m^s \circ \kappa_2 \left(d\eta, \frac{\xi_1+\xi_2}{2} \right) \delta_0(\xi_1 - \xi_2) \\ &+ (2\pi)^{2d} \int_{\eta\in\mathbb{R}^d} e^{i(t_1-t_2)\eta} m^s \otimes m^s \circ \kappa_2 \left(d\eta, \frac{\xi_1-\xi_2}{2} \right) \delta_0(\xi_1 + \xi_2). \end{aligned} \tag{7.19}$$

Thus the claimed support properties and the weak stationarity in the t variable is proved, and we can use (7.12) and (7.13) to prove (7.14). □

Acknowledgment. The author thanks the organizers of the Conference on Pseudo-Differential Operators and Related Topics held at Växjö University in 2004, Professors Paolo Boggiatto, Luigi Rodino and Joachim Toft, for the invitation to give a talk and for a very nice conference. The author also wishes to thank NuHAG at the Faculty of Mathematics of the University of Vienna for hospitality and for the kind offer to use their facilities during the preparation of this paper.

References

[1] R. G. Bartle, A general bilinear vector integral, *Studia Math.* **15** (1956), 337–52.

[2] R. C. Blei, Projectively bounded Fréchet measures, *Trans. Amer. Math. Soc.* **348** (11) (1996), 4409–32.

[3] R. C. Blei, *Fractional Dimensions and Bounded Fractional Forms*, Mem. Amer. Math. Soc. **57** (331) AMS, 1985.

[4] J. Diestel and J. J. Uhl, *Vector Measures*, Math. Surveys **15**, AMS, 1977.

[5] J. L. Doob, *Stochastic Processes*, Wiley, 1953.

[6] N. Dunford and J. T. Schwartz, *Linear Operators, Part I*, Interscience, New York and London, 1958.

[7] H. G. Feichtinger and W. Hörmann, Harmonic analysis of generalized stochastic processes on locally compact Abelian groups, Preprint, 1989.

[8] P. Flandrin, *Time-Frequency/Time-Scale Analysis*, Academic Press, 1999.

[9] G. B. Folland, *Harmonic Analysis in Phase Space*, Princeton University Press, 1989.

[10] C. C. Graham and B. M. Schreiber, Bimeasures algebra on LCA groups, *Pacific Journal of Math.* **115** (1) (1984), 91–127.

[11] K. Gröchenig, *Foundations of Time-Frequency Analysis*, Birkhäuser, 2001.

[12] W. Hörmann, *Generalized Stochastic Processes and Wigner Distribution*, Ph.D. Thesis, Universität Wien, 1989.

[13] S. Janson, *Gaussian Hilbert Spaces*, Cambridge University Press, 1997.

[14] A. J. E. M. Janssen, *Application of the Wigner Distribution to Harmonic Analysis of Generalized Stochastic Processes*, Mathematical Centre Tracts **114**, Mathematisch Centrum, Amsterdam, 1979.

[15] Y. Kakihara, *Multidimensional Second Order Stochastic Processes*, World Scientific, 1997.

[16] Y. Katznelson, *An Introduction to Harmonic Analysis*, Dover, 1976.

[17] B. Keville, *Multidimensional Second Order Generalised Stochastic Processes on Locally Compact Abelian Groups*, Ph.D. Thesis, University of Dublin, 2003.

[18] W. Martin, Time-frequency analysis of random signals, *Proc. ICASSP*, (1982), 1325–1328.

[19] K. S. Miller, *Complex Stochastic Processes*, Addison-Wesley, 1974.

[20] M. Morse and W. Transue, C-bimeasures and their superior integrals Λ^*, *Rend. Circolo Mat. Palermo* **2** (4) (1955), 270–300.

[21] M. Morse and W. Transue, C-bimeasures and their integral extensions, *Ann. Math.* **64**, (1956), 480–504.

[22] M. M. Rao, Harmonizable processes: structure theory, *L'Enseign. Math.* **28** (1982), 295–351.

[23] E. Thomas, L'integration par rapport à une mesure de Radon vectorielle, *Ann. Inst. Fourier (Grenoble)* **20** (1970), 55–191.

[24] P. Wahlberg, The random Wigner distribution of Gaussian stochastic processes with covariance in $S_0(\mathbb{R}^{2d})$, *J. Funct. Spaces Appl.* **3** (2005), 163–181.

[25] K. Ylinen, On vector bimeasures, *Ann. Mat. Pura Appl.* **117** (1978), 119–138.

Patrik Wahlberg
Department of Electroscience
Lund University
Box 118
SE-221 00 Lund
Sweden

and

NuHAG
Faculty of Mathematics
University of Vienna
Nordbergstrasse 15
A-1090 Vienna
Austria
e-mail: `patrik.wahlberg@ts.mah.se`

Operator Theory:
Advances and Applications, Vol. 164, 227–244

Reproducing Groups for the Metaplectic Representation

E. Cordero, F. De Mari, K. Nowak and A. Tabacco

Abstract. We consider the (extended) metaplectic representation of the semidirect product G of the symplectic group and the Heisenberg group. By looking at the standard resolution of the identity formula and inspired by previous work [5], [13], [4], we introduce the notion of admissible (reproducing) subgroup of G via the Wigner distribution. We prove some features of admissible groups and then exhibit an explicit example ($d = 2$) of such a group, in connection with wavelet theory.

Mathematics Subject Classification (2000). Primary 42C15; Secondary 17B45.

Keywords. Metaplectic representation, symplectic group, wavelets, Wigner distribution.

1. Introduction

Reproducing formulae based on, or inspired by, various versions of the resolution of the identity appear pervasively in the literature, from coherent states in physics[1] to group representations [6] and to wavelet and Gabor analysis [9]. In a very general and abstract sense, they can all be recast in a formula of the type

$$f = \int_H \langle f, \phi_h \rangle \, \phi_h \, dh, \qquad f \in \mathcal{H}, \tag{1.1}$$

where $\mathcal{H}$ is a Hilbert space and $h \mapsto \phi_h$ is an $\mathcal{H}$-valued measurable function on some measure space (H, dh). Of course, the cases of greatest interest concern Hilbert spaces of functions and measure spaces with additional structure such as Lie groups. A formula like (1.1) is known as *reproducing formula*.

This paper is part of an ongoing project that addresses several questions related to (1.1) in the case in which the ingredients are as follows. First, the Hilbert space is $L^2(\mathbb{R}^d)$. Secondly, H is a subgroup of the semidirect product G of the symplectic group and the Heisenberg group (of the appropriate dimensions). Thirdly,

This work was partially supported by the Progetto MIUR Cofinanziato 2002 "Analisi Armonica".

the map $h \mapsto \phi_h$ arises from the restriction to H of the (extended) metaplectic representation μ_e of G as applied to a fixed and suitable "window" $\phi \in L^2(\mathbb{R}^d)$. Thus, at least formally, (1.1) can be written

$$f = \int_H \langle f, \mu_e(h)\phi\rangle \mu_e(h)\phi \, dh, \qquad f \in L^2(\mathbb{R}^d).$$

The main question is: for which subgroups H of G does there exist a window $\phi \in L^2(\mathbb{R}^d)$ such that the above reproducing formula holds *for all* $f \in L^2(\mathbb{R}^d)$? Clearly, one looks for invariants or other general properties that will decide whether a group H enjoys the property or not. Further, in the affirmative case, one seeks conditions that single out the "good" windows, namely those for which the formula holds. These questions are far from being fully answered. A complete classification of reproducing subgroups in the case $d = 1$ is given in [5] and many interesting facts have been proved in [13] in a somewhat different setting. Some new results in higher dimensions are in [4]. For the relevance of the extended metaplectic representation in the context of harmonic analysis in phase space or time-frequency analysis, the reader is referred to [8] and [9], and the references therein.

2. Admissible Subgroups

2.1. Symplectic Group and Metaplectic Representation

The symplectic group is as usual

$$Sp(d, \mathbb{R}) = \left\{ g \in GL(2d, \mathbb{R}) : \ {}^t g J g = J \right\},$$

where $J = \left[\begin{smallmatrix} 0 & I_d \\ -I_d & 0 \end{smallmatrix}\right]$ defines the standard symplectic form

$$\omega(x, y) = {}^t x J y, \qquad x, y \in \mathbb{R}^{2d}. \tag{2.1}$$

The Lie algebra $\mathfrak{sp}(d, \mathbb{R})$ of $Sp(d, \mathbb{R})$ is therefore

$$\begin{aligned} \mathfrak{sp}(d, \mathbb{R}) &= \left\{ X \in \mathfrak{gl}(2d, \mathbb{R}) : \ {}^t X J + J X = 0 \right\} \\ &= \left\{ \begin{bmatrix} A & B \\ C & -\,{}^t A \end{bmatrix} : A \in \mathfrak{gl}(d, \mathbb{R}), B, C \in \mathrm{Sym}(d, \mathbb{R}) \right\}. \end{aligned}$$

The bracket in $\mathfrak{sp}(d, \mathbb{R})$ is the commutator of matrices. The Cartan involution θ on $\mathfrak{sp}(d, \mathbb{R})$ is defined by $\theta X = -\,{}^t X$, and it decomposes $\mathfrak{sp}(d, \mathbb{R})$ into its $+1$ and -1 eigenspaces, namely

$$\begin{aligned} \mathfrak{k} &= \{ X \in \mathfrak{sp}(d, \mathbb{R}) : \theta X = X \} \\ &= \left\{ X \in \mathfrak{sp}(d, \mathbb{R}) : \begin{bmatrix} A & B \\ -B & -\,{}^t A \end{bmatrix} : A \in \mathfrak{so}(d, \mathbb{R}), B \in \mathrm{Sym}(d, \mathbb{R}) \right\} \end{aligned}$$

and

$$\begin{aligned} \mathfrak{p} &= \{ X \in \mathfrak{sp}(d, \mathbb{R}) : \theta X = -X \} \\ &= \left\{ X \in \mathfrak{sp}(d, \mathbb{R}) : \begin{bmatrix} A & B \\ B & -\,{}^t A \end{bmatrix} : A, \, B \in \mathrm{Sym}(d, \mathbb{R}) \right\}, \end{aligned}$$

respectively. Here $\mathfrak{so}(d,\mathbb{R})$ denotes the Lie algebra of $d\times d$ skew-symmetric matrices. Thus $\mathfrak{sp}(d,\mathbb{R})=\mathfrak{k}\oplus\mathfrak{p}$, a direct sum of vector spaces. An immediate count gives $\dim\mathfrak{sp}(d,\mathbb{R})=d(2d+1)$, $\dim\mathfrak{k}=d^2$ and $\dim\mathfrak{p}=d^2+d$. The bracket relations $[\mathfrak{k},\mathfrak{k}]\subset\mathfrak{k}$, $[\mathfrak{k},\mathfrak{p}]\subset\mathfrak{p}$, $[\mathfrak{p},\mathfrak{p}]\subset\mathfrak{k}$ show that $\mathfrak{k}$ is a subalgebra, while $\mathfrak{p}$ is not. Moreover, $\mathfrak{k}$ is the Lie algebra of the connected Lie subgroup K defined as the fix-point set of the Cartan involution $\Theta g={}^tg^{-1}$ of $Sp(d,\mathbb{R})$ (observe that $d\Theta=\theta$). In practice,

$$K=Sp(d,\mathbb{R})\cap SO(2d)\simeq U(d)$$

is the unique maximal compact subgroup of $Sp(d,\mathbb{R})$, up to conjugation. If $d=2$ the full Lie algebra has dimension 10, whereas $\mathfrak{k}$ is given by

$$\mathfrak{k}=\left\{X\in\mathfrak{sp}(2,\mathbb{R}):\begin{bmatrix}bJ&\Sigma\\-\Sigma&bJ\end{bmatrix}:b\in\mathbb{R},\,\Sigma\in\mathrm{Sym}(2,\mathbb{R})\right\}\tag{2.2}$$

and has dimension 4.

The metaplectic representation μ of (the two-sheeted cover of) the symplectic group arises as intertwining operator between the standard Schrödinger representation ρ of the Heisenberg group $\mathbb{H}^d$ and the representation that is obtained from it by composing ρ with the action of $Sp(d,\mathbb{R})$ by automorphisms on $\mathbb{H}^d$ (see e.g.[8]). We briefly review its construction.

The Heisenberg group $\mathbb{H}^d$ is obtained by defining on $\mathbb{R}^{2d+1}$ the product

$$(z,t)\cdot(z',t')=(z+z',t+t'+\frac{1}{2}\omega(z,z')),$$

where ω is given in (2.1). We denote the translation and modulation operators on $L^2(\mathbb{R}^d)$ by

$$T_xf(t)=f(t-x)\quad\text{and}\quad M_\xi f(t)=e^{2\pi i\langle\xi,t\rangle}f(t).$$

The Schrödinger representation of the group $\mathbb{H}^d$ on $L^2(\mathbb{R}^d)$ is then defined by

$$\rho(x,\xi,t)f(y)=e^{2\pi it}e^{\pi i\langle x,\xi\rangle}e^{2\pi i\langle\xi,y\rangle}f(y-x)=e^{2\pi it}e^{\pi i\langle x,\xi\rangle}T_xM_\xi f(t),$$

where we write $z=(x,\xi)$ when we separate the space components x from the frequency components ξ of a point z in the phase space $\mathbb{R}^{2d}$. The symplectic group acts on $\mathbb{H}^d$ via automorphisms that leave the center $\mathbb{R}=\{(0,t):t\in\mathbb{R}\}$ of $\mathbb{H}^d$ pointwise fixed. The action $\varphi: Sp(d,\mathbb{R})\times\mathbb{H}^d\to\mathbb{H}^d$ is given by $\varphi(A,(z,t))=A\cdot(z,t)$, where

$$A\cdot(z,t)=(Az,t)\,.$$

Therefore, for any fixed $A\in Sp(d,\mathbb{R})$ there is a representation

$$\rho_A:\mathbb{H}^d\to\mathcal{U}(L^2(\mathbb{R}^d)),\qquad(z,t)\mapsto\rho\left(A\cdot(z,t)\right)$$

whose restriction to the center is a multiple of the identity. By the Stone-von Neumann theorem, $\rho_A\simeq\rho$. Hence there exists an intertwining unitary operator $\mu(A)$ for the two representations, namely

$$\rho_A=\mu(A)\circ\rho\circ\mu(A)^{-1}.$$

By Schur's lemma, μ is determined up to a phase factor. It turns out that the phase ambiguity is really a sign, so that μ lifts to a representation of the (double cover of the) symplectic group. It is the famous metaplectic or Shale-Weil representation. The representations ρ and μ can be combined and give rise to the extended metaplectic representation of the group

$$G = \mathbb{H}^d \times_\varphi Sp(d,\mathbb{R}),$$

the semidirect product of $\mathbb{H}^d$ and $Sp(d,\mathbb{R})$. The group law on G is

$$((z,t),A)\cdot((z',t'),A') = ((z,t)\cdot(Az',t'),AA')$$

and the extended metaplectic representation μ_e of G is

$$\mu_e\left((z,t),A\right) = \rho(z,t)\circ\mu(A).$$

For elements $Sp(d,\mathbb{R})$ in special form, the metaplectic representation can be computed explicitly in a simple way. For $f\in L^2(\mathbb{R}^d)$ we have

$$\mu\left(\begin{bmatrix} A & 0 \\ 0 & {}^tA^{-1}\end{bmatrix}\right) f(x) = (\det A)^{-1/2} f(A^{-1}x) \tag{2.3}$$

$$\mu\left(\begin{bmatrix} I & 0 \\ C & I\end{bmatrix}\right) f(x) = \pm e^{-i\pi\langle Cx,x\rangle} f(x) \tag{2.4}$$

$$\mu\,(J) = i^{d/2}\mathcal{F}^{-1}, \tag{2.5}$$

where $\mathcal{F}$ denotes the Fourier transform

$$\mathcal{F}f(\xi) = \int_{\mathbb{R}^d} f(x)e^{-2\pi i\langle x,\xi\rangle}\,dx, \qquad f\in L^1(\mathbb{R}^d)\cap L^2(\mathbb{R}^d).$$

2.2. Reproducing and Admissible Groups

The point of this paper is to look at the reproducing formulae that arise by restricting μ_e to subgroups H of G. A slight simplification in our formalism comes from the observation that the reproducing formula (2.6) is insensitive to phase factors: if we replace $\mu_e(h)\phi$ with $e^{is}\mu_e(h)\phi$ the formula is unchanged, for any $s\in\mathbb{R}$. The role of the center of the Heisenberg group is thus irrelevant, so that the "true" group under consideration is $\mathbb{R}^{2d}\times_\varphi Sp(d,\mathbb{R})$, which we denote again by G.

We write dh for the left Haar measure of the group H. Also, we shall always assume that the Haar measure of a compact group is normalized so that the total mass of the group is one.

Definition 2.1. We say that a connected Lie subgroup H of $G=\mathbb{R}^{2d}\times_\varphi Sp(d,\mathbb{R})$ is a *reproducing group* for μ_e if there exists a function $\phi\in L^2(\mathbb{R}^d)$ such that

$$f = \int_H \langle f,\mu_e(h)\phi\rangle\mu_e(h)\phi\,dh, \quad \text{for all } f\in L^2(\mathbb{R}^d). \tag{2.6}$$

Notice that we do require formula (2.6) to hold for all functions in $L^2(\mathbb{R}^d)$ for the same "window" ϕ, but we do not require the restriction of μ_e to H to be irreducible.

In [13], the authors consider subgroups D of $GL(d,\mathbb{R})$ and their actions on $\mathbb{R}^d$. Motivated by the analysis of the case in which D is the "$ax+b$" group, they say that a subgroup D is *admissible* if there exists a Borel measurable $h \in L^1(\mathbb{R}^d)$ such that $h \geq 0$ and

$$\int_D h(xa)\, d\mu(a) = 1 \qquad \text{for a.e. } x \in \mathbb{R}^d.$$

If D is the "$ax+b$" group, then any admissible wavelet ψ (in the usual Calderón sense) gives a function $h = |\hat{\psi}|^2$ for which the above formula holds, showing that "$ax+b$" is admissible. In [4], in the same context that we are considering here, we introduce a similar notion of admissibility via the Wigner distribution and prove a sufficient condition for a subgroup to be reproducing. We briefly explain this issue. The cross-Wigner distribution $W_{f,g}$ of $f, g \in L^2(\mathbb{R}^d)$ is defined by

$$W_{f,g}(x,\xi) = \int e^{-2\pi i \langle \xi, y\rangle} f(x + \frac{y}{2}) \overline{g(x - \frac{y}{2})}\, dy. \tag{2.7}$$

The quadratic expression $W_f := W_{f,f}$ is usually called the Wigner distribution of f. A crucial property of W is that it intertwines the (extended) metaplectic representation and the affine action on $\mathbb{R}^{2d}$ [8]. In other words:

$$W_{\mu_e(g)\phi}(x,\xi) = W_\phi\left(g^{-1} \cdot (x,\xi)\right), \qquad g \in G,$$

where the affine action $g \cdot (x,\xi)$ is defined by

$$g \cdot (x,\xi) = ((q,p),A) \cdot (x,\xi) = A^t(x,\xi) + {}^t(q,p). \tag{2.8}$$

The following result is proved in [4]:

Theorem 2.2. *If there exists a function ϕ such that the mapping*

$$h \to W_{\mu_e(h)\phi}(x,\xi) = W_\phi(h^{-1} \cdot (x,\xi)) \tag{2.9}$$

is in $L^1(H)$ for a.e. $(x,\xi) \in \mathbb{R}^{2d}$, and such that

$$\int_H |W_{\mu_e(h)\phi}(x,\xi)|\, dh \leq M \qquad \text{for a.e. } (x,\xi) \in \mathbb{R}^{2d}, \tag{2.10}$$

then (2.6) holds for all $f \in L^2(\mathbb{R}^d)$ if and only if the following admissibility condition is satisfied:

$$\int_H W_\phi(h^{-1} \cdot (x,\xi))\, dh = 1 \quad \text{for a.e. } (x,\xi) \in \mathbb{R}^{2d}. \tag{2.11}$$

The above discussion and Theorem 2.2 justify the following definition.

Definition 2.3. We say that a connected Lie subgroup H of $G = \mathbb{R}^{2d} \times_\varphi Sp(d,\mathbb{R})$ is an *admissible group* for μ_e if there exists a function $\phi \in L^2(\mathbb{R}^d)$ such that

$$\int_H W_\phi(h^{-1} \cdot (x,\xi))\, dh = 1 \quad \text{for a.e. } (x,\xi) \in \mathbb{R}^{2d}. \tag{2.12}$$

3. Geometric Features

The main geometric properties related to admissibility are described in [13, Prop. 2.3]. The arguments can be easily adapted to our setting and yield the following

Theorem 3.1. *Let $H \subset G$ be admissible. Then*

(i) *The stabilizer* $\mathrm{Stab}_z(H)$ *of the H-action on phase space is compact for a.e.* $z \in \mathbb{R}^{2d}$*;*

(ii) *If H satisfies* (2.10) *and* $H \subset Sp(d, \mathbb{R})$*, then H is not unimodular.*

Here, as is customary, $\mathrm{Stab}_z(H) = \{h \in H : h \cdot z = z\}$ and $g \cdot z$ is defined in (2.8). Likewise, we adopt the usual notation for orbits: $\mathcal{O}_z = \{h \cdot z : h \in H\}$, whenever there is no ambiguity about the group H acting on z.

Remark 3.2. (i) If the subgroup H has finite measure, then (2.6) holds if and only if condition (2.11) is satisfied. This is because for all $\phi \in L^2(\mathbb{R}^d)$ the Wigner distribution $W_\phi \in L^\infty$, and there exists $M > 0$ such that $|W_\phi(x, \xi)| \leq M$. Therefore

$$\int_{\mathbb{R}^{2d}} \int_H |W_{\mu_e(h)\phi}(x,\xi)\, W_f(x,\xi)|\, dh dx d\xi \leq M \left(\int_H dh \right) \int_{\mathbb{R}^{2d}} |W_f(x,\xi)| dx d\xi.$$

The only compact subgroups of G, however, are of the form $\{0\} \times K$ with K compact in $Sp(d, \mathbb{R})$. This is because the projection of G onto $\mathbb{R}^{2d}$ is a continuous group homomorphism and $\{0\}$ is the only compact additive subgroup of $\mathbb{R}^{2d}$. Hence, item (ii) of Theorem 3.1 applies and any such group cannot be admissible, hence reproducing.

(ii) The requirement $H \subset Sp(d, \mathbb{R})$ is essential for the non-unimodularity of the group H. In fact, one can find reproducing subgroups H of $\mathbb{R}^{2d} \times_\varphi Sp(d, \mathbb{R})$ that are unimodular. This case, for instance, arises in Gabor's analysis, where the subgroup is $H = \mathbb{R}^{2d}$, $\mu_e(q, p)\phi = T_q M_p \phi$ and $\phi(t) = 2^{d/4} e^{-\pi t^2}$.

The main contribution of this paper concerning the geometry of admissible groups is Corollary 3.3 below, where we give a dimension bound in the case $d = 2$. Its proof is postponed to Section 3.2 because it needs the Lie-algebraic considerations of the next section.

Corollary 3.3. *If $H \subset Sp(2, \mathbb{R})$ is admissible, then* $\dim(H) \leq 5$.

Some comments are in order. The natural question arises as to what happens in dimension one through five. First of all, it is clear from Theorem 3.1 that no one-dimensional subgroup of $Sp(2, \mathbb{R})$ can be admissible because any such group is abelian, hence unimodular. In dimension three and four there are indeed known examples of admissible groups. The TDS group discussed in Section 4 is a four dimensional admissible group. Consider next the three dimensional subgroup H of $Sp(2, \mathbb{R})$ consisting of the 4×4 matrices

$$\begin{bmatrix} a^{-1} I_2 & 0 \\ S_y & a I_2 \end{bmatrix},$$

where $a > 0$ and, $S_y = \left[\begin{smallmatrix} y_1 & y_2 \\ y_2 & -y_1 \end{smallmatrix}\right]$. It is isomorphic to $H' = \mathbb{R}^2 \rtimes \{\mathbb{R}^+ \cdot I_2\}$, the subgroup of the affine group of $\mathbb{R}^2$ with two dimensional translations and one dimensional dilations (here $\mathbb{R}^+ \cdot I_2$ are the positive scalar matrices). Furthermore, the restriction of the metaplectic representation to H is equivalent to the wavelet representation of H'. It follows from the results in [13] that H is admissible. We are presently unaware of admissible subgroups of $Sp(2,\mathbb{R})$ of dimension either two or five, and in fact we conjecture that no such examples exist. Clarifying this issue, as well as producing a full list of admissible subgroups (up to conjugation), is one of the main objectives of the research project this article refers to.

3.1. Compact Lie Subalgebras

We will prove below that there are only finitely many conjugation classes of (Lie algebras of) compact subgroups of $Sp(2,\mathbb{R})$ and that each of them has an explicit representative. Since the maximal compact subgroup K of $Sp(d,\mathbb{R})$ is the fixed point set of the involutive analytic automorphism Θ and it is compact and connected, the pair (G,K) is a Riemannian symmetric pair (see [10], p.209). Therefore, any compact connected Lie subgroup of G is conjugate to a subgroup of K (see [10], Theorem 2.1, Ch. VI). Consequently, if $\mathfrak{t} = \mathrm{Lie}(T)$ is the Lie algebra of the compact connected Lie subgroup T of G, then we may assume $T \subset K$ and $\mathfrak{t} \subset \mathfrak{k}$. In Proposition 3.4 below we classify the subalgebras of $\mathfrak{k}$ for $d = 2$, up to conjugation. In the sequel, we shall then refer to any such algebra as a canonical compact algebra.

First of all, we work out the bracket in $\mathfrak{k}$. For $T \in \mathfrak{k}$ we write $T = T_{b,\Sigma}$ using the parametrization introduced in (2.2). An immediate computation gives

$$[T_{b,\Sigma}, T_{b',\Sigma'}] = \begin{bmatrix} [\Sigma',\Sigma] & b[J,\Sigma'] + b'[\Sigma,J] \\ -b[J,\Sigma'] - b'[\Sigma,J] & [\Sigma',\Sigma] \end{bmatrix}. \tag{3.1}$$

We also write

$$\Sigma = \begin{bmatrix} m & n \\ n & p \end{bmatrix} = mE + nL + pF, \tag{3.2}$$

where evidently

$$E = \begin{bmatrix} 1 & 0 \\ 0 & 0 \end{bmatrix}, \qquad F = \begin{bmatrix} 0 & 0 \\ 0 & 1 \end{bmatrix} \qquad L = \begin{bmatrix} 0 & 1 \\ 1 & 0 \end{bmatrix}.$$

Since

$$\begin{aligned} [\Sigma',\Sigma] &= \{(p-m)n' - (p'-m')n\}\, J \\ [J,\Sigma] &= \begin{bmatrix} 2n & p-m \\ p-m & -2n \end{bmatrix} = 2n(E-F) + (p-m)L, \end{aligned} \tag{3.3}$$

it follows that if $T = T_{b,\Sigma}$, $T' = T_{b',\Sigma'}$ and if $[T,T'] = T_{b(T,T'),\Sigma(T,T')}$, then

$$b(T,T') = (p-m)n' - (p'-m')n \tag{3.4}$$

$$\Sigma(T,T') = 2(bn' - b'n)(E-F) + \left(b(p'-m') - b'(p-m)\right) L. \tag{3.5}$$

Next we consider the action of $SO(2)$ on $\mathfrak{k}$. Indeed, if

$$R_\theta = \begin{bmatrix} \cos\theta & \sin\theta \\ -\sin\theta & \cos\theta \end{bmatrix}, \tag{3.6}$$

denotes the standard rotation by the (real) angle θ, then it is easily checked that

$$SO(2) \simeq \left\{ k_\theta = \begin{bmatrix} R_\theta & 0 \\ 0 & R_\theta \end{bmatrix} : \theta \in [0, 2\pi] \right\} \subset Sp(2,\mathbb{R}) \cap SO(4) = K. \tag{3.7}$$

The adjoint action of K on $\mathfrak{k}$ restricts to the following action of $SO(2)$:

$$\operatorname{Ad} k_\theta \left(\begin{bmatrix} bJ & \Sigma \\ -\Sigma & bJ \end{bmatrix} \right) = k_\theta \begin{bmatrix} bJ & \Sigma \\ -\Sigma & bJ \end{bmatrix} k_{-\theta} = \begin{bmatrix} bJ & \operatorname{Ad} R_\theta(\Sigma) \\ -\operatorname{Ad} R_\theta(\Sigma) & bJ \end{bmatrix},$$

because $\operatorname{Ad} R_\theta(J) = J$. Parametrizing 2×2 symmetric matrices as in (3.2), we have

$$\begin{bmatrix} m \\ n \\ p \end{bmatrix} \overset{\operatorname{Ad} R_\theta}{\mapsto} \begin{bmatrix} m_\theta \\ n_\theta \\ p_\theta \end{bmatrix} = \begin{bmatrix} m(\cos\theta)^2 + p(\sin\theta)^2 + n\sin 2\theta \\ \frac{1}{2}(p-m)\sin 2\theta + n\cos 2\theta \\ m(\sin\theta)^2 + p(\cos\theta)^2 - n\sin 2\theta \end{bmatrix}.$$

Observe that under this action

$$\begin{bmatrix} p_\theta - m_\theta \\ 2n_\theta \end{bmatrix} = R_{-2\theta} \begin{bmatrix} p - m \\ 2n \end{bmatrix}.$$

Proposition 3.4. *Up to conjugation, the following is a complete list of the Lie algebras of the compact subgroups of $Sp(2,\mathbb{R})$.*

(i) *The 1-dimensional toral non-conjugate algebras*

$$\mathfrak{k}_{1,\lambda} = \left\{ t \begin{bmatrix} 0 & C_\lambda \\ -C_\lambda & 0 \end{bmatrix} : t \in \mathbb{R},\ C_\lambda = \left[\begin{smallmatrix} 1 & 0 \\ 0 & \lambda \end{smallmatrix}\right] \right\} \simeq \mathfrak{so}(2), \quad \lambda \in \mathbb{R};$$

(ii) *the (maximally compact) 2-dimensional Cartan subalgebra*

$$\mathfrak{k}_2 = \left\{ \begin{bmatrix} 0 & D_y \\ -D_y & 0 \end{bmatrix} : D_y = \begin{bmatrix} y_1 & 0 \\ 0 & y_2 \end{bmatrix} \right\} \simeq \mathfrak{so}(2) \times \mathfrak{so}(2);$$

(iii) *the 3-dimensional Lie algebra*

$$\mathfrak{k}_3 = \left\{ \begin{bmatrix} aJ & \Sigma_y \\ -\Sigma_y & aJ \end{bmatrix} : a \in \mathbb{R},\ \Sigma_y = \begin{bmatrix} y_1 & y_2 \\ y_2 & -y_1 \end{bmatrix} \right\} \simeq \mathfrak{so}(3)$$

(iv) *the maximal compact 4-dimensional algebra*

$$\mathfrak{k}_{\max} = \left\{ \begin{bmatrix} aJ & \Sigma \\ -\Sigma & aJ \end{bmatrix} : a \in \mathbb{R},\ \Sigma \in \operatorname{Sym}_2(\mathbb{R}) \right\} = \mathfrak{sp}(2,\mathbb{R}) \cap \mathfrak{so}(4) \simeq \mathfrak{u}(2);$$

Proof. (i) Clearly, the matter reduces to finding normal forms for vectors in $\mathfrak{k}$. These are well-known and follow from Williamson's classification of Hamiltonians (see [14] for the original paper and [2] for a modern account). The eigenvalues of $H \in \mathfrak{sp}(2,\mathbb{R})$ are of four possible kinds: zero eigenvalues, real pairs $(a,-a)$, quadruples $\pm a \pm ib$ and pairs of purely imaginary eigenvalues $(ib,-ib)$. By skew-symmetry, if $H \neq 0$ is conjugate to some element in $\mathfrak{k}$, then it has two pairs of imaginary eigenvalues, say $(ib_1,-ib_1)$ and $(ib_2,-ib_2)$, and one of b_1,b_2 must be non-zero. Different sets of eigenvalues correspond to non-conjugate matrices, and conversely, equal sets of eigenvalues correspond to conjugate matrices. Thus $H \in \mathfrak{k}$ is conjugate to a matrix of the form

$$H(b_1, b_2) = \begin{bmatrix} 0 & b_1E + b_2F \\ -(b_1E + b_2F) & 0 \end{bmatrix}$$

because $H(b_1, b_2) \in \mathfrak{k}$ and has eigenvalues $\{\pm ib_1, \pm ib_2\}$. It is straightforward to check that $wH(b_1, b_2)w^{-1} = H(b_2, b_1)$, where

$$w = \begin{bmatrix} 0 & L \\ -L & 0 \end{bmatrix} \in K.$$

We may thus assume $b_1 \neq 0$, so that $H(b_1, b_2)$ spans $\mathfrak{k}_{1,b_2/b_1}$.

(ii) First of all, any 2-dimensional compact Lie group is abelian. Secondly, it is well-known [12] that all the maximal abelian subalgebras of $\mathfrak{k}$ have dimension 2. (This actually follows also from the next part of this proof, where it is shown that no subalgebra of dimension 3 is abelian.) We may thus appeal to the conjugacy of maximal tori within compact connected Lie groups: two maximal abelian subalgebras of the Lie algebra of a compact connected Lie group K are conjugate via $\mathrm{Ad}(K)$ (see Theorem 4.34 in [12]). Hence there is a unique conjugacy class, that of

$$\mathfrak{t} = \left\{ \begin{bmatrix} 0 & 0 & a & 0 \\ 0 & 0 & 0 & b \\ -a & 0 & 0 & 0 \\ 0 & -b & 0 & 0 \end{bmatrix} : a, b \in \mathbb{R} \right\} = \mathfrak{k}_2. \tag{3.8}$$

This is the standard maximally compact Cartan subalgebra of $Sp(2, \mathbb{R})$.

(iii) Suppose now that $\mathfrak{t}$ is a 3-dimensional subalgebra of $\mathfrak{k}$. We shall denote by X, Y, Z a basis of $\mathfrak{t}$. First of all, observe that it is possible to assume $X = T_{1,\Sigma}$ and hence, by linear independence, $Y = T_{0,\Sigma_Y}$ and $Z = T_{0,\Sigma_Z}$. Otherwise, $X = T_{0,\Sigma_X}$, $Y = T_{0,\Sigma_Y}$ and $Z = T_{0,\Sigma_Z}$. Since X, Y, Z span a Lie algebra we could infer from (3.1) that $[\Sigma_X, \Sigma_Y] = [\Sigma_X, \Sigma_Z] = [\Sigma_Y, \Sigma_Z] = 0$. We would thus be given three mutually commuting diagonalizable matrices. and there would exist a basis in which they are all diagonal. Since they are 2×2, they could not be linearly independent, a contradiction. This proves our claim, so that until the end of the proof

$$X = T_{1,\Sigma}, \qquad Y = T_{0,\Sigma_Y}, \qquad Z = T_{0,\Sigma_Z}.$$

Next we look at $[\Sigma_Y, \Sigma_Z]$.

• Assume first that $[\Sigma_Y, \Sigma_Z] = 0$. By means of the $SO(2)$-action we can simultaneously diagonalize Σ_Y, Σ_Z by sending $X = T_{1,\Sigma}$ to another matrix of the same kind $T_{1,\Sigma'}$, that we rename $X = T_{1,\Sigma}$. Also, by linear independence, we assume that $Y = T_{0,E}$ and $Z = T_{0,F}$. Consequently, if we write Σ as in (3.2), subtracting if necessary $mY + pZ$ from X, we have $X = T_{1,nL}$ for some $n \in \mathbb{R}$. But then

$$[X, Y] = [\begin{bmatrix} J & nL \\ -nL & J \end{bmatrix}, \begin{bmatrix} 0 & E \\ -E & 0 \end{bmatrix}] = \begin{bmatrix} nJ & -L \\ L & nJ \end{bmatrix} = T_{n,-L}.$$

Now, E, F and L are linearly independent. Hence the only way in which $[X,Y]$ can be a linear combination of X,Y,Z is if $[X,Y]$ is actually equal to $nX = T_{n,n^2L}$, which in turn implies $n^2=-1$. This is impossible and so this case does not arise. Thus:

- $[\Sigma_Y,\Sigma_Z]=\alpha J$ for some $\alpha\neq 0$. By (3.1) we have $[Y,Z]=T_{\alpha,0}$. Suppose first that $X=T_{1,\Sigma}$ with $\Sigma\neq 0$. Since $[Y,Z]$ must lie in $\mathfrak{t}$ and since

$$[Y,Z]=T_{\alpha,0}=\alpha T_{1,\Sigma}-\alpha T_{0,\Sigma}=\alpha X-\alpha T_{0,\Sigma},$$

$T_{0,\Sigma}$ must be a linear combination of Y and Z. But then we may again change basis and take $X'=X-T_{0,\Sigma}=T_{1,0}$. In other words we assume $X=T_{1,0}$. In this case $[X,Y]=T_{0,[J,\Sigma_Y]}$ and $[X,Z]=T_{0,[J,\Sigma_Z]}$. Therefore, the subspace $\operatorname{span}\{\Sigma_Y,\Sigma_Z\}$ of $\operatorname{Sym}(2,\mathbb{R})$ must be invariant under $\operatorname{ad} J$, namely under the linear map

$$\begin{bmatrix} m\\ n\\ p\end{bmatrix}\mapsto\begin{bmatrix}0&2&0\\-1&0&1\\0&-2&0\end{bmatrix}\begin{bmatrix} m\\ n\\ p\end{bmatrix}.$$

Clearly, $\operatorname{ad} J$ has 0 as the only real eigenvalue with eigenvector $[1,0,1]$. Its orthogonal complement $\mathcal{S}=\{\,{}^t[y,z,-y]: z,y\in\mathbb{R}\}$ coincides with the image of $\operatorname{ad} J$ and has no proper invariant subspaces. Hence $\mathcal{S}$ is the only invariant plane. Therefore $\operatorname{span}\{\Sigma_Y,\Sigma_Z\}=\mathcal{S}$ and we conclude that

$$\mathfrak{t}=\left\{\begin{bmatrix}0&x&y&z\\-x&0&z&-y\\-y&-z&0&x\\-z&y&-x&0\end{bmatrix}: x,y,z\in\mathbb{R}\right\}=\mathfrak{k}_3$$

is the only 3 dimensional subalgebra of $\mathfrak{k}$, up to conjugation. This concludes the proof because the case of 4-dimensional algebras has already been discussed. □

3.2. Proof of Corollary 3.3

If $z=0$, $\operatorname{Stab}_0(H)=H$. Let $z\neq 0$ and assume that $T=\operatorname{Stab}_z(H)\subset Sp(2,\mathbb{R})$ is compact. First, we show that $\dim(T)\le 1$. We may assume that $\mathfrak{t}=\operatorname{Lie}(T)$ is one of the canonical compact algebras listed in Proposition 3.4. Indeed, given $\mathfrak{t}$, there exists a $g\in Sp(2,\mathbb{R})$ such that $\operatorname{Ad} g(\mathfrak{t})$ is canonical. Thus for all $X\in\mathfrak{t}$, we have

$$\exp(\operatorname{Ad} g(tX))gz=g\exp(tX)z=gz,\qquad \forall t\in\mathbb{R}$$

because $\exp(tX)\in T$. Hence $\operatorname{Ad} g(\mathfrak{t})\subset\operatorname{Lie}(\operatorname{Stab}_{gz}(H))$. Let $\mathfrak{t}=\operatorname{Lie}(T)$ denote a canonical algebra. Now $X\in\mathfrak{t}$ if and only if $\exp(tX)z=z$ for all $t\in\mathbb{R}$. This implies

$$0=\frac{d}{dt}\exp(tX)z\Big|_{t=0}=Xz.$$

It is an easy computation to check that, if $\dim(\mathfrak{t})=2,3,4$, then there exists a nonzero $X\in\mathfrak{t}$ with $\ker(X)=0$, so that $z\notin\ker(X)$. Therefore $\dim(\operatorname{Stab}_z(H))\le 1$ for every z for which $\operatorname{Stab}_z(H)$ is compact, that is for a.e. $z\in\mathbb{R}^4$ by virtue of Theorem 3.1. Finally, since $\mathcal{O}_z\subset\mathbb{R}^4$, $\dim(\mathcal{O}_z)\le 4$ and

$$4\ge\dim(\mathcal{O}_z)=\dim(H)-\dim(\operatorname{Stab}_z(H))\ge\dim(H)-1,$$

that is, $\dim(H) \leq 5$. □

4. The Translation-Dilation-Sheering Group (TDS)

We prove that the following 3-dimensional triangular group

$$H = \left\{ A_{t,\ell,y} := \begin{bmatrix} t^{-1/2} S_{\ell/2} & 0 \\ t^{-1/2} B_y & t^{1/2}\, {}^t S_{-\ell/2} \end{bmatrix} : t > 0, \ell \in \mathbb{R},\ y \in \mathbb{R}^2 \right\}, \tag{4.1}$$

is a reproducing subgroup of $Sp(2,\mathbb{R})$, where

$$B_y = \begin{bmatrix} 0 & y_1 \\ y_1 & y_2 \end{bmatrix}, \quad y = (y_1, y_2) \in \mathbb{R}^2; \quad S_\ell = \begin{bmatrix} 1 & \ell \\ 0 & 1 \end{bmatrix}, \quad \ell \in \mathbb{R}. \tag{4.2}$$

The matrix S_ℓ is called *sheering* matrix. We call H the TDS group, because we prove in Theorem 4.2 that the restriction of μ to it is equivalent to the wavelet representation (4.3) considered in [7] consisting of translation, dilation and sheering operators. This construction gives rise to the so-called "contourlet frames".

In [4] we show that the TDS group is admissible via Theorem 2.2. Here we give a second, direct proof. First, though, we review the connection between this subgroup of $Sp(2,\mathbb{R})$ and the 2-dimensional wavelet theory in [7] alluded to above. The sheering operator on functions is given by

$$\left(S_\ell f\right)(x) = f\left({}^t S_\ell x\right), \qquad f \in L^2(\mathbb{R}^2),$$

where S_ℓ is as in (4.2). These are the ingredients of the contourlet frames. As for curvelets, one allows dilation and translation operations, but the angular selectivity is achieved by a sheering operation rather than a rotation [7].

Let L denote the 2-dimensional subgroup of $Sp(2,\mathbb{R})$ given by

$$L = \left\{ \begin{bmatrix} t & 0 \\ -\ell t & t \end{bmatrix} : t > 0,\ \ell \in \mathbb{R} \right\} \subset Sp(2,\mathbb{R}).$$

We consider its natural action on $\mathbb{R}^2$, that is the semidirect product $H = \mathbb{R}^2 \times_\varphi L$. This action has two open orbits $\mathcal{O}_+$ and $\mathcal{O}_-$ in $\mathbb{R}^2$, where $\mathcal{O}_+ = \{(x_1, x_2)\,;\, x_2 > 0\}$ and $\mathcal{O}_- = \{(x_1, x_2)\,;\, x_2 < 0\}$. The wavelet representation ν is given by

$$\nu(t, y, \ell) f = \left(T_y D_t S_\ell\right) f, \qquad f \in L^2(\mathbb{R}^2), \tag{4.3}$$

but it is more convenient to view ν in the frequency domain, namely

$$\pi(t, y, \ell) f(u) = (\mathcal{F} \circ \nu(t, y, \ell) f)(u) = e^{-2\pi i \langle y, u \rangle} D_{-t}\, {}^t S_{-\ell} f\,(u)\,. \tag{4.4}$$

We have $\pi = \pi_{\mathcal{O}_+} \oplus \pi_{\mathcal{O}_-}$, where $\pi_{\mathcal{O}_+}$ and $\pi_{\mathcal{O}_-}$ are the subrepresentations of π obtained by restriction to $L^2(\mathcal{O}_+)$ and $L^2(\mathcal{O}_-)$, respectively. For $\pi_{\mathcal{O}_+}$, the admissibility condition for a wavelet ϕ such that $\hat{\phi} \in L^2(\mathcal{O}_+)$ is

$$\int_0^\infty \int_{\mathbb{R}} \left| \frac{\hat{\phi}(\xi_1, \xi_2)}{\xi_2} \right|^2 d\xi_1 d\xi_2 < \infty$$

and similarly for $\pi_{\mathcal{O}_-}$ (see [3] for more details).

We check that $H = G_0G_1$, where

$$G_0 = \left\{ g_0(t,y) = \begin{bmatrix} t^{-1/2} & 0 \\ t^{-1/2}B_y & t^{1/2} \end{bmatrix} : t > 0, y \in \mathbb{R}^2 \right\}$$

$$G_1 = \left\{ g_1(\ell) = \begin{bmatrix} S_\ell & 0 \\ 0 & S_{-\ell} \end{bmatrix} : \ell \in \mathbb{R} \right\}.$$

Indeed, for $y = (y_1, y_2) \in \mathbb{R}^2$, it is straightforward to see that:

$$g_1(\ell)g_0(t,(y_1,y_2))g_1(\ell)^{-1} = g_0(t,(y_1, y_2 - 2\ell y_1)). \tag{4.5}$$

This means that G_1 normalizes G_0 and hence that $H = G_0G_1$ is a semidirect product, with product law given by

$$g(t,(y_1,y_2),\ell)g(r,(z_1,z_2),s) = g(tr,(y_1 + tz_1, y_2 + tz_2 - 2\ell tz_1), s + \ell).$$

Since G_0 is normal in H, one has the obvious isomorphism $H/G_0 \simeq G_1$.

The computation of the left Haar measure on H is a straightforward exercise:

$$dh(t,(y_1,y_2),\ell) = \frac{dt}{t^3}\, dy_1\, dy_2\, d\ell. \tag{4.6}$$

In order to compute the metaplectic representation on H, we observe first that the matrix $A_{t,y,\ell}$ in (4.1) can be written as the product of a diagonal matrix $D_{t,\ell}$ and a lower triangular matrix $L_{t,y,\ell}$ as follows

$$A_{t,y,\ell} = D_{t,\ell}L_{t,y,\ell} = \begin{bmatrix} t^{-1/2}S_{\ell/2} & 0 \\ 0 & t^{1/2}\,{}^tS_{-\ell/2} \end{bmatrix} \begin{bmatrix} I & 0 \\ t^{-1}\,{}^tS_{\ell/2}B_yS_{\ell/2} & I \end{bmatrix}.$$

We then use the fact that μ is a representation and formulae (2.3) and (2.4) to obtain that for $f \in L^2(\mathbb{R}^2)$

$$\begin{aligned} \mu(A_{t,y,\ell})f(x) &= \mu(D_{t,\ell}L_{t,y,\ell})f(x) = t^{1/2}(L_{t,y,\ell}f)(t^{1/2}S_{-\ell/2}x) \\ &= t^{1/2}e^{-i\pi\langle {}^tS_{\ell/2}B_yx, S_{-\ell/2}x\rangle}\, f(t^{1/2}S_{-\ell/2}x) \\ &= t^{1/2}e^{-i\pi\langle B_yx,x\rangle}\, f(t^{1/2}S_{-\ell/2}x). \end{aligned} \tag{4.7}$$

In the following we denote

$$\mathbb{R}^2_+ = \{(x_1,x_2) : x_2 > 0\}, \qquad \dot{\mathbb{R}}^2_+ = \{(x_1,x_2) : x_1 \neq 0,\, x_2 > 0\}.$$

Similarly we define $\mathbb{R}^2_-$ and $\dot{\mathbb{R}}^2_-$. We shall be concerned with the mapping

$$\Psi : \mathbb{R}^2 \to \mathbb{R}^2 \qquad x \mapsto \left(x_1x_2, \frac{x_2^2}{2} \right), \tag{4.8}$$

whose properties are summarized in the following elementary proposition.

Proposition 4.1. *The mapping (4.8) defines diffeomorphisms $\Psi : \dot{\mathbb{R}}^2_\pm \to \dot{\mathbb{R}}^2_+$ and is such that $\Psi(-x) = \Psi(x)$. Further, it satisfies:*

(a) *the Jacobian of Ψ at $x = (x_1,x_2) \in \dot{\mathbb{R}}^2_\pm$ is $J_\Psi(x) = x_2^2$;*
(b) *the Jacobian of Ψ^{-1} at $u = (u_1,u_2) = \Psi(x_1,x_2)$ is $J_{\Psi^{-1}}(u) = (2u_2)^{-1} = x_2^{-2}$;*
(c) *$\Psi^{-1}(t^2S_{2\ell}u) = tS_\ell\Psi^{-1}(u)$ for every $t > 0$ and every $u \in \dot{\mathbb{R}}^2_\pm$;*

(d) $\langle B_y x, x\rangle = 2\langle y, \Psi(x)\rangle$ *for every* $x \in \dot{\mathbb{R}}^2_\pm$ *and every* $y \in \mathbb{R}^2$.

The following theorem shows the equivalence between the wavelet representation π and the metaplectic representation μ.

Theorem 4.2. *Let* $u \in \dot{\mathbb{R}}^2_+$, *and extend the map*

$$\mathcal{Q}f(u) = |2u_2|^{-1/2} f(\Psi^{-1}(u_1, u_2))$$

as an even function to $\mathbb{R}^2 \setminus \{x_1x_2 = 0\}$. *This is an isometry of* $L^2_{\mathrm{even}}(\mathbb{R}^2)$ *onto itself that intertwines the representations* π *and* μ, *that is* $\pi(g) \circ \mathcal{Q} = \mathcal{Q} \circ \mu(g)$ *for every* $g \in H$.

Proof. Let $f \in L^2_{\mathrm{even}}(\mathbb{R}^2)$. Then, by item (b) in Proposition 4.1,

$$\begin{aligned}
\|\mathcal{Q}f\|_2^2 &= \int_{\mathbb{R}^2} |\mathcal{Q}f(u)|^2 \, du \\
&= 2\int_{\mathbb{R}^2_+} \frac{1}{2u_2} \left|f(\Psi^{-1}(u_1, u_2))\right|^2 \, du_1 du_2 \\
&= 2\int_{\mathbb{R}^2_+} |f(x_1, x_2)|^2 \, dx_1 dx_2 \\
&= \|f\|_2^2.
\end{aligned}$$

Thus $\mathcal{Q}$ is an isometry. By (4.4) and item (c) in Proposition 4.1,

$$\begin{aligned}
\pi(t, y, \ell)\,(\mathcal{Q}f)\,(u) &= te^{-2\pi i\langle y,u\rangle}\; \mathcal{Q}f\,(tS_{-\ell}u) \\
&= \frac{t}{|2tu_2|^{1/2}} e^{-2\pi i\langle y,u\rangle}\; f\left(\Psi^{-1}\,(tS_{-\ell}u)\right) \\
&= \frac{t^{1/2}}{|2u_2|^{1/2}} e^{-2\pi i\langle y,u\rangle}\; f\left(t^{1/2}S_{-\ell/2}\Psi^{-1}(u)\right).
\end{aligned}$$

Finally, by (4.8) and item (d) in Proposition 4.1,

$$\begin{aligned}
\mathcal{Q}\left(\mu(t, y, \ell)f\right)(u) &= \frac{1}{|2u_2|^{1/2}} \left(\mu(t, y, \ell)f\right)\left(\Psi^{-1}(u)\right) \\
&= \frac{t^{1/2}}{|2u_2|^{1/2}} e^{-i\pi\langle B_y\Psi^{-1}(u),\Psi^{-1}(u)\rangle} f\left(t^{1/2}S_{-\ell/2}\Psi^{-1}(u)\right) \\
&= \frac{t^{1/2}}{|2u_2|^{1/2}} e^{-2i\pi\langle y,u\rangle} f\left(t^{1/2}S_{-\ell/2}\Psi^{-1}(u)\right),
\end{aligned}$$

as desired. ∎

Theorem 4.3. *The identity*

$$\int_H |\langle f, \mu(h)\phi\rangle|^2 \, dh = c_\phi \, \|f\|_2^2 \tag{4.9}$$

holds for every $f \in L^2(\mathbb{R}^2)$ if and only if the function ϕ satisfies the following two admissibility conditions:

$$c_\phi = 4 \int_{\mathbb{R}^2_+} |\phi(x)|^2 \frac{dx}{x_2^4} = 4 \int_{\mathbb{R}^2_+} |\phi(-x)|^2 \frac{dx}{x_2^4} \tag{4.10}$$

and

$$\int_{\mathbb{R}^2_+} \phi(x)\overline{\phi(-x)}\, \frac{dx}{x_2^4} = 0. \tag{4.11}$$

We need the following introductory equality of Plancherel type.

Lemma 4.4. *Let $h \in L^2(\mathbb{R}^2)$ be a function which vanishes outside some annulus $c < \|x\| < C$, with $0 < c < C < \infty$. Then*

$$\int_{\mathbb{R}^2} \left| \int_{\mathbb{R}^2} h(x) e^{2\pi i \langle y, \Psi(x)\rangle}\, dx \right|^2 dy = \int_{\mathbb{R}^2_+} |h(x) + h(-x)|^2 \frac{dx}{x_2^2}.$$

Proof. By Proposition 4.8, denoting Ψ the mapping $(x_1, x_2) \mapsto (x_1 x_2, \frac{1}{2}x_2^2)$ regardless of the domain on which we look at it, we obtain

$$\begin{aligned}
\int_{\mathbb{R}^2} h(x) e^{2\pi i \langle y, \Psi(x)\rangle}\, dx &= \left(\int_{\dot{\mathbb{R}}^2_+} + \int_{\dot{\mathbb{R}}^2_-} \right) h(x) e^{2\pi i \langle y, \Phi(x)\rangle}\, dx \\
&= \int_{\dot{\mathbb{R}}^2_+} h(x) e^{2\pi i \langle y, \Psi(x)\rangle}\, dx + \int_{\dot{\mathbb{R}}^2_+} h(-x) e^{2\pi i \langle y, \Psi(-x)\rangle}\, dx \\
&= \int_{\dot{\mathbb{R}}^2_+} [h(x) + h(-x)]\, e^{2\pi i \langle y, \Psi(x)\rangle}\, dx \\
&= \int_{\dot{\mathbb{R}}^2_+} \left[h(\Psi^{-1}(u)) + h(-\Psi^{-1}(u))\right] e^{2\pi i \langle y, u\rangle} \frac{du}{2u_2} \\
&= \int_{\mathbb{R}^2} \frac{\chi_+(u)}{2u_2} \left[h(\Psi^{-1}(u)) + h(-\Psi^{-1}(u))\right] e^{2\pi i \langle y, u\rangle}\, du,
\end{aligned}$$

where χ_+ is the characteristic function of $\dot{\mathbb{R}}^2_+$. By the Plancherel formula we obtain

$$\begin{aligned}
&\int_{\mathbb{R}^2} \left| \int_{\mathbb{R}^2} \frac{\chi_+(u)}{2u_2} \left[h(\Psi^{-1}(u)) + h(-\Psi^{-1}(u))\right] e^{2\pi i \langle y, u\rangle}\, du \right|^2 dy \\
&= \int_{\mathbb{R}^2} \left| \frac{\chi_+(u)}{2u_2} \left[h(\Psi^{-1}(u)) + h(-\Psi^{-1}(u))\right] \right|^2 du \\
&= \int_{\dot{\mathbb{R}}^2_+} \left| h(\Psi^{-1}(u)) + h(-\Psi^{-1}(u)) \right|^2 \frac{du}{4u_2^2} \\
&= \int_{\dot{\mathbb{R}}^2_+} |h(x) + h(-x)|^2 \frac{dx}{x_2^2},
\end{aligned}$$

as desired. □

Proof of Theorem 4.3. Using (4.6) for the left Haar measure and (4.7) for the metaplectic representation, the left-hand side of (4.9) becomes

$$\int_H |\langle f, \mu(h)\phi\rangle|^2\, dh = \int_{\mathbb{R}^2_+}\int_{\mathbb{R}^2}\left|\int_{\mathbb{R}^2} f(x)t^{1/2}e^{\pi i\langle B_y x, x\rangle}\overline{\phi(t^{1/2}S_{-\ell/2}x)}\, dx\right|^2 t^{-3}dydtd\ell$$
$$= \int_{\mathbb{R}^2_+}\int_{\mathbb{R}^2}\left|\int_{\mathbb{R}^2} f(x)t^{1/2}e^{2\pi i\langle y, \Psi(x)\rangle}\overline{\phi(t^{1/2}S_{-\ell/2}x)}\, dx\right|^2 t^{-3}dydtd\ell, \tag{4.12}$$

where we have used property (d) of Ψ stated in Proposition 4.1.

We take $f \in L^2(\mathbb{R}^2)$ vanishing outside a ring and apply Lemma 4.4 to the right-hand side of (4.12). The computation of $|h(x) + h(-x)|^2$, for the function $h(x) = f(x)t^{1/2}\overline{\phi(t^{1/2}S_{-\ell/2}x)}$ is immediate and we thus obtain

$$\int_H |\langle f, \mu(h)\phi\rangle|^2\, dh = \int_{\mathbb{R}^2_+}\int_{\mathbb{R}^2}\Big\{|f(x)|^2 t|\phi(t^{1/2}S_{-\ell/2}x)|^2$$
$$+ |f(-x)|^2 t|\phi(-t^{1/2}S_{-\ell/2}x)|^2$$
$$+2\mathcal{R}e\, f(x)\overline{f(-x)}t\phi(-t^{1/2}S_{-\ell/2}x)\overline{\phi(t^{1/2}S_{-\ell/2}x)}\Big\}\frac{dx}{x_2^2}t^{-3}dtd\ell.$$

First, we consider f with $f(x_1, x_2) = 0$ for $x_2 < 0$. We perform the change of variables $(t, \ell) \mapsto y = (y_1, y_2)$, given by $t^{1/2}S_{-\ell/2}x = y$. Hence $dtd\ell = 4x_2^{-2}dy$ and

$$\int_H |\langle f, \mu(h)\phi\rangle|^2\, dh = \int_{\mathbb{R}^2_+} |f(x)|^2\left(\int_{\mathbb{R}^2_+} |\phi(y)|^2\frac{4}{y_2^4}\, dy\right) dx$$
$$= \|f\|_2^2\left(\int_{\mathbb{R}^2_+} |\phi(y)|^2\frac{4}{y_2^4}\, dy\right).$$

If $f(x_1, x_2) = 0$ for $x_2 > 0$, then arguing in a similar way we obtain

$$\int_H |\langle f, \mu(h)\phi\rangle|^2\, dh = \int_{\mathbb{R}^2_+} |f(x)|^2\left(\int_{\mathbb{R}^2_+} |\phi(-y)|^2\frac{4}{y_2^4}\, dy\right) dx$$
$$= \|f\|_2^2\left(\int_{\mathbb{R}^2_+} |\phi(-y)|^2\frac{4}{y_2^4}\, dy\right).$$

Therefore, if the window function ϕ fulfills (4.10), then (4.9) holds for every f which vanishes either for $x_2 < 0$ or for $x_2 > 0$. Take now a bounded function f with support in some annulus $c < \|x\| < C$. Then

$$2\mathcal{R}e\, f(x)\overline{f(-x)}t\phi(-t^{1/2}S_{-\ell/2}x)\overline{\phi(t^{1/2}S_{-\ell/2}x)}\frac{1}{x_2^2}$$

is integrable with respect to the measure $dx_1dx_2t^{-3}dtd\ell$ and its integral is easily seen to vanish. Finally,

$$\begin{aligned}\int_{\mathbb{R}^2_+}\int_{\mathbb{R}^2_+} & 2\mathcal{R}e\, f(x)\overline{f(-x)}t\phi(-t^{1/2}S_{-\ell/2}x)\overline{\phi(t^{1/2}S_{-\ell/2}x)}\frac{1}{x_2^2}dx_1dx_2t^{-3}dtd\ell \\ &= 4\int_{\mathbb{R}^2_+} f(x)\overline{f(-x)}dx\int_{\mathbb{R}^2_+}\overline{\phi(y)}\phi(-y)\frac{dy}{y_2^4},\end{aligned}$$

and (4.11) follows.

Conversely, if conditions (4.10) and (4.11) are satisfied and if f is a function as in the assumptions of Lemma 4.4, then all the terms of (4.12) are integrable and the reproducing formula (4.9) is true for all $f \in L^2(\mathbb{R}^2)$. To see this, take $f \in L^2(\mathbb{R}^2)$ and let f_n be a sequence of functions as in Lemma 4.4 which tends to f in the L^2-norm. The sequcnce $F(f_n) = \langle f_n, \mu(h)\phi\rangle$ is a Cauchy sequence in $L^2(H, dh)$ which tends pointwise to $F(f) = \langle f, \mu(h)\phi\rangle$. Since (4.9) holds for all f_n, it follows that it also holds for f. □

Example. We finish by giving an example of admissible wavelets. Take a Meyer wavelet ψ (see, e.g., [11]). It is the Schwartz function defined by

$$\hat{\psi}(x) = e^{-ix/2}\sin(\omega(x))$$

where $\omega \in C_c^\infty(\mathbb{R})$ is an even function with support in $\{x : \frac{2\pi}{3} < |x| < \frac{8\pi}{3}\}$. The function $g(x) = \hat{\psi}(x)$ satisfies:

$$c_g := \int_0^{+\infty}\frac{|g(x)|^2}{x^4}\,dx = \int_0^{+\infty}\frac{|g(-x)|^2}{x^4}\,dx < +\infty.$$

Next we consider an asymmetric step function f and observe that

$$\int_{-\infty}^{+\infty}|f(x)|^2\,dx = \int_{-\infty}^{+\infty}|f(-x)|^2 = 2.$$

The admissible wavelet ϕ is defined by:

$$\phi(x_1, x_2) = f(x_1)g(x_2)$$

It follows that:

$$\begin{aligned}\int_{-\infty}^{+\infty}\int_0^{+\infty}|\phi(x)|^2\,\frac{dx}{x_2^4} &= \int_{-\infty}^{+\infty}\int_0^{+\infty}|f(x_1)g(x_2)|^2\,\frac{dx_2}{x_2^4}dx_1 \\ &= \left(\int_0^{+\infty}\frac{|g(x_2)|^2}{x_2^4}\,dx_2\right)\left(\int_{-\infty}^{+\infty}|f(x_1)|^2\,dx_1\right) \\ &= c_g\cdot 2\end{aligned}$$

and similarly for the integral in $-x$. Finally

$$\begin{aligned}\int_{-\infty}^{+\infty}\int_{0}^{+\infty} \phi(x)\overline{\phi(-x)}\,\frac{dx}{x_2^4} &= \int_{-\infty}^{+\infty}\int_{0}^{+\infty} f(x_1)g(x_2)\overline{f(-x_1)g(-x_2)}\,\frac{dx_2}{x_2^4}dx_1 \\ &= \left(\int_{0}^{+\infty} \frac{g(x_2)\overline{g(-x_2)}}{x_2^4}\,dx_2\right)\left(\int_{-\infty}^{+\infty} f(x_1)f(-x_1)\,dx_1\right)\end{aligned}$$

and the latter integral vanishes due to the asymmetry of f.

References

[1] S. T. Ali, J. P. Antoine, J. P. Gazeau, *Coherent States, Wavelets and Their Generalizations*, Springer-Verlag, New York, 2000.

[2] V. I. Arnold, *Mathematical Methods of Classical Mechanics*, Springer-Verlag, New York, 1978.

[3] D. Bernier and K. F. Taylor, Wavelets from square-integrable representations. *SIAM J. Math. Anal.* **27** (2) (1996), 594–608.

[4] E. Cordero, F. De Mari, K. Nowak and A. Tabacco, Analytic features of reproducing groups for the metaplectic representation, to appear.

[5] F. De Mari and K. Nowak, Analysis of the affine transformations of the time-frequency plane, *Bull. Austral. Math. Soc.* **63** (2) (2001), 195–218.

[6] J. Dixmier, *Les C^*-Algèbres et leurs représentations*, Gauthier-Villars Éditeur, Paris, 1969.

[7] M. N. Do and M. Vetterli, New tight frames of curvelets and optimal representations of objects with piecewise C span 2 singularities, *IEEE Trans. Signal Proc.*, to appear.

[8] G. B. Folland, *Harmonic Analysis in Phase Space*, Princeton University Press, 1989.

[9] K. Gröchenig, *Foundations of Time-Frequency Analysis*, Birkhäuser, Boston, 2001.

[10] S. Helgason, *Differential Geometry, Lie Groups and Symmetric Spaces*, Academic Press, New York, 1978.

[11] E. Hernández and G. L. Weiss, *A First Course on Wavelets*, CRC Press, Boca Raton, 1996.

[12] A. W. Knapp, *Lie Groups Beyond an Introduction*, Second Edition, Birkhäuser, Boston, 2002.

[13] R. S. Laugesen, N. Weaver, G. L. Weiss, and E. N. Wilson, A characterization of the higher dimensional groups associated with continuous wavelets, *J. Geom. Anal.* **12** (1) (2002), 89–102.

[14] J. Williamson, On an algebraic problem, concerning the normal forms of linear dynamical systems, *Amer. J. Math.* **58** (1) (1936), 141–163.

E. Cordero
Dipartimento di Matematica
Politecnico di Torino
Corso Duca degli Abruzzi, 24
10129 Torino
Italy
e-mail: cordero@dm.unito.it

F. De Mari
DIPTEM
Piazzale J. F. Kennedy, Pad. D.
16129 Genova
Italy
e-mail: demari@dima.unige.it

K. Nowak
Department of Mathematics and Computer Science
Drexel University
3141 Chestnut Street
Philadelphia, PA 19104-2875
USA
e-mail: knowak@cs.drexel.edu

A. Tabacco
Dipartimento di Matematica
Politecnico di Torino
Corso Duca degli Abruzzi, 24
10129 Torino
Italy
e-mail: anita.tabacco@polito.it